KING, Elbert A. Space geology: an introduction. Wiley, 1976. 349p ill map index 75-45357. 16.95. ISBN 0-471-47810-5. C.I.P.

King is a geologist and teacher who spent six years at the NASA Manned Spacecraft Center in Houston. He begins with a thorough discussion of meteorites and tektites and then proceeds to the effects of meteorites on planetary surfaces, in chapters about craters, terrestrial impact craters, and impact metamorphism. Other chapters summarize and interpret recent findings about the Moon, Mars, other planets and moons, asteroids, and comets. Mineralogical and chemical analyses of samples and morphologies of planetary surfaces are discussed throughout the text. There are numerous high-quality black-and-white photographs, several charts and diagrams, a glossary, an author index, a subject index, and an extensive list of references at the chapter ends. Well written and highly recommended for all college libraries.

SPACE
GEOLOGY
AN INTRODUCTION

About the Author

ELBERT A. KING received his Ph.D. in Geology from Harvard University in 1965. From 1963-1966, Dr. King served with NASA as Instructor in Astronaut Training Courses in Geology, and was affiliated with the Space Agency from 1967-1969 as Curator of the Lunar Receiving Laboratory, Johnson Space Center, and as a member of the Preliminary Sample Examination Team for the Apollo 11 Mission. From 1969-1974, he was Chairman and Associate Professor of Geology at the University of Houston. A recipient of a Special Commendation from the Geological Society of America (1973) and a Fellow of the Meteoritical Society, Dr. King presented an invited paper about the Moon before The Royal Society, and has been widely published in professional journals. He is currently Professor of Geology at the University of Houston.

SPACE
GEOLOGY
AN INTRODUCTION

Elbert A. King

University of Houston

JOHN WILEY & SONS, Inc.,   New York • London • Sydney • Toronto

Library of Congress Cataloging in Publication Data:

King, Elbert A
  Space geology: an introduction.

  Includes bibliographical references and index.
   1. Solar system.   2. Lunar geology.   3. Planets —
Surfaces.   4. Meteorite craters.   I. Title.
QB501.K56   550′.999   75-45357
ISBN 0-471-47810-5

Printed in the United States of America

10 9 8 7 6 5 4 3 2 1

Preface

This book is intended for use as the main textbook in a senior or first-year graduate level course in space geology (astrogeology, planetary geology or the like). Although many excellent articles have been written on various aspects of this broad topic, the literature is badly scattered and is, for the most part, rather recent. Thus, numerous references are included for the use of the student and instructor, as well as suggested readings and sources at the end of each chapter. Some of the most important and interesting articles have appeared in relatively obscure and unobtainable journals and proceedings. Where there is a choice, the more obtainable work has been listed for ease in locating supplemental material. This volume is *not* a treatise and therefore no attempt has been made to make the references complete. They are intended rather to give the reader some useful points of entry into a *vast* literature.

The approach of this book generally is qualitative but "hard" data are emphasized, such as samples, images and physical measurements. The depth of coverage, in part, reflects my own interests, but is mostly controlled by the amount of geologic data available that is related to the individual topic. For course presentations, quantitative treatments of virtually each topic can be incorporated from the suggested readings and from specific references.

With the tremendous momentum of the space program, the volume of data concerning the space related aspects of geology has increased at an enormous rate in recent years. I have retained some historical perspective, but this material necessarily is limited. The portrayal of the various topics as they are currently understood is much more important.

The entire volume has benefited substantially from the critical reviews and constructive comments of John Butler (University of Houston), Carleton

Moore (Arizona State University), Bob Phinney (Princeton University), J. V. Smith (University of Chicago), and George Wetherill (Carnegie Institution). To these colleagues, I am deeply indebted. My sincere thanks go to a larger number of my co-workers and friends who provided photographs, figures, and other illustrative material. These are individually credited where they are used. I especially appreciate the assistance of Mike Daily and Dave Pettus, who provided me with excellent photographic and darkroom support for many of my own illustrations and who printed negatives received from many others. Trude V. V. King provided determined and excellent assistance in proofreading and indexing.

It is my hope that this book will introduce many students to the fascinating aspects of space geology and the opportunities for new discoveries and interpretations in this exciting field — the geological exploration of space.

Elbert King

Houston, Texas

Elbert A. King is Professor of Geology at the University of Houston. Prior to joining its faculty in the fall of 1969, he worked for six years at the NASA, Manned Spacecraft Center, serving as Curator of the Lunar Receiving Laboratory during his last two years there. He was a member of the Lunar Sample Preliminary Examination Team for Apollo 11 and has been a principal investigator for lunar sample analysis for all of the Apollo missions. His other research and professional activities have included the photogeologic mapping of a large area of Mars, mineralogic and petrologic research with tektites, meteorites and shocked terrestrial rocks, and astronaut geology training. Dr. King received his Ph.D. degree in geology from Harvard University in 1965, after receiving his B.S. and M.A. degrees in geology from The University of Texas.

Contents

1. Meteorites

1. Meteorites   1

INTRODUCTION   1
STONY METEORITES   7
  Chondrites   7
    Carbonaceous Chondrites   14
    Olivine-hypersthene and Olivine-bronzite chondrites   21
    Enstatite Chondrites   28
    Origin of Chondrules   30
  Achondrites   35
IRON METEORITES   41
  Introduction   41
  Classification   43
  Cooling Rates   47
  Shock Effects   49
STONY-IRON METEORITES   50
  Pallasites   51
  Mesosiderites   54
MICROMETEORITES   55
ORIGINS OF METEORITES   58
  Ages   58
  Orbits   60
REFERENCES AND NOTES   61
SUGGESTED READING AND GENERAL REFERENCES   67

2. Tektites   69

INTRODUCTION   69
OCCURRENCES, CHEMISTRY AND PHYSICAL PROPERTIES   69
ORIGIN   71
REFERENCES AND NOTES   79
SUGGESTED READING AND GENERAL REFERENCES   80

3. Craters   81

INTRODUCTION   81
IMPACT CRATERS   82
   Cratering Mechanics   82
     Sequence of Events   82
     Rankine-Hugoniot Equations   88
     Energy Partitioning   89
REFERENCES AND NOTES   91
SUGGESTED READING AND GENERAL REFERENCES   93

4. Terrestrial Impact Craters   95

INTRODUCTION   95
METEOR CRATER, ARIZONA   98
ODESSA CRATERS, TEXAS   102
HAVILAND CRATER, KANSAS   103
THE CANADIAN CRATERS   104
RIES KESSEL, STEINHEIM BASIN AND STOPFENHEIM KUPPEL, GERMANY   112
THE AUSTRALIAN CRATERS   115
   Wolf Creek Crater   115
   Boxhole Crater   116
   Dalgaranga Crater   116
   Gosses Bluff   116
   Henbury Craters   119
ARABIAN CRATERS   120
CRITERIA FOR RECOGNITION OF IMPACT STRUCTURES   122
REFERENCES AND NOTES   124
SUGGESTED READING AND GENERAL REFERENCES   129

5. Impact Metamorphism

131

INTRODUCTION 131
MINERALOGIC AND PETROLOGIC CRITERIA 132
   Coesite 132
   Stishovite 134
   Baddeleyite 134
   Lechatelierite 136
   Other Minerals 138
   Crystal Structure Damage 138
   Diaplectic Glass 139
   Planar Features 139
   Kink Bands 141
   Undulatory Extinction 142
   Fused Rock Glass 143
   Breccia 143
STRUCTURAL CRITERIA 146
   Shatter Cones 146
INTERPRETATION OF SHOCK HISTORY 147
REFERENCES AND NOTES 150
SUGGESTED READING AND GENERAL REFERENCES 151

6. The Moon

153

INTRODUCTION 153
TELESCOPIC AND SPACECRAFT IMAGERY 155
   History 155
   Geologic Mapping 156
   Nature of the Lunar Surface 162
   Estimation of Surface Ages 167
   Surface Transient Phenomena 169
LUNAR SAMPLE ANALYSES 169
INTRODUCTION 169
   Crystalline Rocks with Igneous Textures 171
      Rock-Forming Minerals 180
      Chemical Compositions 184
      Ages 191
      Crystallization History 192
      Origin of Lunar Magmas 196
      Shock Metamorphism 196
   Uncohesive Particulate Material 198

Regularly Shaped Glass   198
Glassy Agglutinates   201
Modal Analyses   202
Grain Size Analyses   203
Physical Properties   206
Breccias and Microbreccias   207
Soil Breccias   208
Autometamorphosed Breccias   210
Lunar Chondrules   210
Relative Abundance   214
Geochemistry, Ages, Physical Properties   214
Organic and Biologic Analyses   219
LUNAR GEOPHYSICS   219
Structure of the Lunar Crust   219
State of the Lunar Interior   221
Moonquakes and Lunar Tectonism   222
Mass Concentrations   224
Laser Altimetry   226
Orbital Gamma-Ray and X-Ray Analyses   227
ORIGIN OF THE MOON   227
REFERENCES AND NOTES   229
SUGGESTED READING AND GENERAL REFERENCES   243

7. Mars

7. Mars   245
INTRODUCTION AND HISTORY   245
Surface Features   248
Volcanic Features   248
Aeolian Features   251
Channels   254
Canyons   256
Scarps and Cliffs   257
Polar Areas and Caps   258
Craters   261
GEOPHYSICAL OBSERVATIONS   265
SURFACE AND BULK COMPOSITION   265
ATMOSPHERE   265
MOONS   268
FUTURE EXPLORATION   272
REFERENCES AND NOTES   272
SUGGESTED READING AND GENERAL REFERENCES   275

8. Asteroids   277

INTRODUCTION AND HISTORY   277
ORBITS   278
  Belt Asteroids   278
  Trojan Asteroids   279
  Mars-Crossing Asteroids   280
  Apollo Asteroids   281
  Asteroids with Closely Similar Orbits   281
OPTICAL STUDIES   282
REFERENCES AND NOTES   285
SUGGESTED READING AND GENERAL REFERENCES   288

9. Comets   289

INTRODUCTION AND HISTORY   289
ORBITS   291
COMPOSITION   292
STRUCTURE, SIZE, AND ORIGINS   293
REFERENCES AND NOTES   293
SUGGESTED READING AND GENERAL REFERENCES   295

10. Other Planets and Moons   297

INTRODUCTION   297
MERCURY   297
VENUS   302
JUPITER   304
SATURN   311
URANUS, NEPTUNE AND PLUTO   311
REFERENCES AND NOTES   311
SUGGESTED READING AND GENERAL REFERENCES   312

11. Comparative Planetology   315

INTRODUCTION   315
BULK COMPOSITIONS   316
DEGREE OF DIFFERENTIATION   317
INTERNAL STATE   318
TECTONIC STYLES   318
REFERENCES AND NOTES   319

Glossary   321

Name Index   333

Subject Index   343

Naturally great excitement was produced in the village at the fall of such a strange visitor and many were the conjectures as to where it came from, nor was I surprised when I heard that several of the people made deep salaam to the stone when they first caught sight of it, and treated it as if it were a messenger from the gods.   *Babu Umesh Chandra Rai, 1906*

1. Meteorites

INTRODUCTION

Our only extraterrestrial samples, except for the rocks recently returned from the Moon by the Apollo Project, are meteorites. Accounts of stones falling from the sky are recorded in ancient Chinese, Greek, and Roman literature, but the origin of these stones commonly was ascribed to strong winds which supposedly blew the stones into the sky from whence they fell. One of the most holy Moslem relics is the black stone of the Kaaba, preserved on sacred ground at Mecca. This stone is widely reputed to be a meteorite and even thought to be the oldest preserved fragment from an observed fall.[1] As recently as 1808, Thomas Jefferson was quoted in reference to a report of a meteorite fall; ". . . It may be very difficult to explain how the stone you possess came into the position in which it was found. But is it easier to explain how it got into the clouds from whence it was supposed to have fallen? The actual fact however is the thing to be established, and this I hope will be done by those whose situations and qualifications enable them to do it."[2] Once meteorites were recognized to be a genuine natural phenomenon, substantial debates ensued as to the places and processes of their origins. As recently as 1879, Ball (1) seriously proposed that meteorites were terrestrial fragments that had been placed in near-Earth space by immense volcanic explosions early in Earth history.

[1]Dietz and McHone recently have cast doubt on this view, but the matter is still unresolved (Meteoritics, vol. 9, p. 173-179).
[2]Letter to Mr. Daniel Salmon, February 15, 1808.

Our understanding of meteorites has progressed slowly and is still in a rather primitive state compared with our current understanding of terrestrial rocks and lunar samples. Although a vast descriptive and analytical literature of meteorites has been accumulated, the critical lack of certainty as to the place or places of origin of the various types of meteorites (and other "field information" usually considered essential in the interpretation of terrestrial rocks and their petrogenetic relationships) has severely limited interpretations of modes of meteorite origin, and possible genetic relations between different meteorite types.

More than 1700 different meteorites are preserved in collections throughout the world. Large collections of meteorites are maintained and actively curated by the Smithsonian Institution, British Museum, Soviet Academy of Sciences, Paris Museum of Natural History, Center for Meteorite Studies of Arizona State University, Harvard and Yale Universities, Chicago Field Museum, and a number of other institutions. A meteorite fall may consist of thousands of stone fragments or individual pieces of iron, but all of these individual pieces are considered part of the same meteorite. In virtually all cases, multiple falls are known to result from the breakup of a larger individual piece into two or more fragments under aerodynamic pressure from the hypervelocity entry of the meteoroid into the dense portion of the Earth's atmosphere (Fig. 1). Reported falls of single meteorites are common. If a recovered meteorite was not observed during its luminous passage through the atmosphere as a meteor or fireball (bolide), but is recognized by its texture, mineralogy, chemistry, surface features or unusual occurrence, it is termed a "find" (as opposed to a "fall"). By international convention, meteorities are named for the closest post office, geographic feature, or community to the location of the fall or find that is easily located on a small scale map of the area if the longitude and latitude of the locality are given.

Meteorites traditionally have been categorized in the broadest sense by the relative amounts of metal and silicate that they contain. Hence, the terms iron, stony-iron, and stone are widely used. Irons are composed predominantly of metallic nickel-iron, but may contain silicate and other inclusions. Stones are composed chiefly of silicates, predominantly iron-magnesium silicates, but may contain more than 24 percent metallic nickel-iron. The stony-irons are a group of uncommon, ill-defined meteorites that have a relatively large amount of nickel-iron and contain a variety of silicate minerals. The relative abundances of these three groups of meteorites are best represented by the simple statistics of observed falls (Table I). Significant selection factors are involved in the recognition of finds by untrained persons, who initially recover the majority of finds. Irons are relatively easy to recognize and to distinguish from ordinary terrestrial rocks. They are ferromagnetic, commonly rusty in surface appearance, have a high specific gravity (ranging from approximately 7.8 to 7.98), and have bright,

Figure 1   The Lost City meteor, photographed by the Prairie Network of the Smithsonian Institution. The meteor occurred at 2014 CST on January 3, 1970, and was photographed from four separate observation points. The meteor was observed initially at a height of 86 km at a velocity of 14.2 km/sec. The apparently discontinuous nature of the meteor is due to the action of the "chopping shutter," which is used to determine the meteor's velocity. From all of the photographs similar to this one, the impact area was accurately predicted and several fragments of stony meteorite were recovered, including one individual that weighted 9.83 kg. (Courtesy of R. E. McCrosky, Smithsonian Astrophysical Observatory.)

Table I   Abundances of Meteorite Falls and Finds by Meteorite Type[a]

	Falls		Finds		Total	
Type	Number	Percent	Number	Percent	Number	Percent
Irons	43	6	551	54	594	33
Stony-irons	12	<2	58	6	70	4
Stones	723	93	404	40	1127	63
Totals	778		1013		1791	

[a]Data from Hey, M. H. (1966) Catalogue of Meteorites, third revised and enlarged edition, British Museum, Pub. No. 464, 637 p.

white to light gray metallic luster on a freshly filed or ground surface. Also, the irons seem to weather less rapidly than many of the stones and thus will persist longer at surface conditions, thereby allowing more opportunity for their recognition. Although most stones have many characteristics by which they can be readily identified (fusion crust, presence of chondrules, minor amount of metal, etc.), they are not so conspicuous among terrestrial rocks and require the attention of a more careful or trained observer for their recognition.[3] Note the relative abundance of irons and stones in the falls and finds columns of Table I.

The mineralogy of meteorites has been reviewed by Mason (2) who lists more than 80 minerals now known from the various types of meteorites (Table II).

[3]The total number of meteorites that have been recognized and studied is embarrassingly small. New and unstudied specimens are sorely needed. If you have a suspected meteorite or know of one, there are numerous museum curators and university researchers (including the author) who will be pleased to have the opportunity to examine the suspected meteorite and to establish its identity.

Table II   The Minerals of Meteorites (CcI, II, III = carbonaceous chondrites, Type I, II, III; Ce = enstatite chondrites; Ae = enstatite achondrites)[a]

Name	Formula	Occurrence
Alabandite	$(Mn,Fe)S$	Accessory in some Ce and Ae
Andradite	$Ca_3Fe_2Si_3O_{12}$	Accessory in Allende (CcIII)
Awaruite	Ni_3Fe	Accessory in Odessa iron and Allende (CcIII)
Barringerite	$(Fe,Ni)_2P$	Accessory in Ollague pallasite
Bloedite	$Na_2Mg(SO_4)_2 \cdot 4H_2O$	Accessory in Ivuna (CcI)
Brezinaite	Cr_3S_4	Accessory in Tucson iron
Brianite	$CaNa_2Mg(PO_4)_2$	Accessory in some irons
Calcite	$CaCO_3$	Accessory in CcI and CcII

Table II (Cont'd) The Minerals of Meteorites (CcI, II, III= carbonaceous chondrites, Types I, II, III; Ce= enstatite chondrites; Ae= enstatite achondrites)[a]

Name	Formula	Occurrence
Carlsbergite	CrN	Accessory in many irons
Chalcopyrite	$CuFeS_2$	Accessory in Karoonda (CcIII)
Chaoite	C	Rare, in ureilites
Chlorapatite	$Ca_5(PO_4)_3Cl$	Accessory in many meteorites
Chromite	$FeCr_2O_4$	Accessory in most meteorites
Clinopyroxene	$(Ca,Mg,Fe)SiO_3$	Common in stones and stony-irons
Cohenite	$(Fe,Ni)_3C$	Accessory in many irons and in Ce
Copper	Cu	Common as an accessory
Cordierite	$Mg_2Al_4Si_5O_{18}$	Accessory in Allende (CcIII)
Cristobalite	SiO_2	Accessory, mainly in Ce
Daubreelite	$FeCr_2S_4$	Accessory in Ce, Ae, and many irons
Diamond	C	Present in ureilites and irons
Djerfisherite	$K_3CuFe_{12}S_{14}$	Accessory in some Ce, Ae, and Toluca iron
Dolomite	$CaMg(CO_3)_2$	Accessory in CcI
Epsomite	$MgSO_4 \cdot 7H_2O$	Prominent in CcI
Farringtonite	$Mg_3(PO_4)_2$	Accessory in some pallasites
Gentnerite	$Cu_8Fe_3Cr_{11}S_{18}$	Accessory in Odessa iron
Graftonite	$(Fe,Mn)_3(PO_4)_2$	Rare accessory in some irons
Graphite	C	Common accessory in irons and some stones
Grossular	$Ca_3Al_2Si_3O_{12}$	Accessory in Allende (CcIII)
Gypsum	$CaSO_4 \cdot 2H_2O$	Accessory in CcI and CcII
Haxonite	$Fe_{23}C$	Accessory in many irons
Heazlewoodite	Ni_3S_2	Accessory in Odessa iron
Hercynite	$(Fe,Mg)Al_2O_4$	Accessory in some CcIII
Hibonite	$CaAl_{12}O_{19}$	Accessory in some CcII and CcIII
Ilmenite	$FeTiO_3$	Accessory in many stones and stony-irons
Kamacite	(Fe,Ni)	In irons, stony-irons and most chondrites
Krinovite	$NaMg_2CrSi_3O_{10}$	Rare accessory in a few irons
Lawrencite	$(Fe,Ni)Cl_2$	Accessory in some meteorites
Lonsdaleite	C	Rare, in ureilites and irons
Mackinawite	FeS_{1-x}	Common as an accessory
Magnesite	$(Mg,Fe)CO_3$	Accessory in CcI
Magnetite	Fe_3O_4	Accessory in Cc
Majorite	$Mg_3(MgSi)Si_3O_{12}$	In Coorara and Tenham chondrites
Melilite	$Ca_2(Mg,Al)(Si,Al)_2O_7$	In CcIII
Merrihueite	$(K,Na)_2Fe_5Si_{12}O_{30}$	Rare accessory in Mezö-Madaras chondrite
Monticellite	$Ca(Mg,Fe)SiO_4$	Accessory in Sharps chondrite
Nepheline	$NaAlSiO_4$	Accessory in a few chondrites
Niningerite	$(Mg,Fe)S$	Accessory in some Ce

Table II (Cont'd)   The Minerals of Meteorites (CcI, II, III= carbonaceous chondrites, Types I, II, III; Ce= enstatite chondrites; Ae= enstatite achondrites)[a]

Name	Formula	Occurrence
Oldhamite	CaS	Accessory in Ce and Ae
Olivine	$(Mg,Fe)_2SiO_4$	Common in stones and stony-irons
Orthopyroxene	$(Mg,Fe)SiO_3$	Common in stones and stony-irons
Osbornite	TiN	Accessory in Ae
Panethite	$(Ca,Na)_2(Mg,Fe)_2(PO_4)_2$	Accessory in Dayton iron
Pentlandite	$(Fe,Ni)_9S_8$	Accessory, mainly in CcII and CcIII
Perovskite	$CaTiO_3$	Accessory in CcIII
Perryite	$(Ni,Fe)_5(Si,P)_2$	Accessory in Ce and Ae
Plagioclase	$(Na,Ca)(AlSi)_4O_8$	Common in stones and stony-irons
Potash feldspar	$(K,Na)AlSi_3O_8$	Rare accessory in a few irons
Pyrite	FeS_2	Accessory in Karoonda (CcIII)
Pyrrhotite	$Fe_{1-x}S$	Accessory in CcI
Quartz	SiO_2	Accessory in some eucrites and Ce
Rhönite	$CaMg_2TiAl_2SiO_{10}$	Accessory in Allende (CcIII)
Richterite	$Na_2CaMg_5Si_8O_{22}F_2$	Rare accessory in a few irons, and in Abee (Ce)
Ringwoodite	$(Mg,Fe)_2SiO_4$	In Coorara and Tenham chondrites
Roedderite	$(K,Na)_2Mg_5Si_{12}O_{30}$	Rare accessory in Ce and irons
Rutile	TiO_2	Rare accessory
Sarcopside	$(Fe,Mn)_3(PO_4)_2$	Rare accessory in some irons
Schreibersite	$(Fe,Ni)_3P$	Accessory in irons, stony-irons, and some chondrites
Serpentine (or chlorite)	$(Mg,Fe)_6Si_4O_{10}(OH)_8$	Matrix of CcI and CcII
Sinoite	Si_2N_2O	Rare in some Ce
Sodalite	$Na_8Al_6Si_6O_{24}Cl_2$	Accessory in some CcIII
Sphalerite	$(Zn,Fe)S$	Accessory in Ce and some irons
Spinel	$MgAl_2O_4$	Accessory mainly in Cc
Stanfieldite	$Ca_4(Mg,Fe)_5(PO_4)_6$	Accessory in some stony-irons
Sulfur	S	Accessory in CcI
Taenite	(Fe,Ni)	As for kamacite
Tridymite	SiO_2	Accessory in some stones, stony-irons, and irons
Troilite	FeS	Present in most meteorites
Ureyite	$NaCrSi_2O_6$	Rare accessory in some irons
Whewellite	$CaC_2O_4 \cdot H_2O$	Accessory in Murchison (CcII)
Whitlockite	$Ca_9MgH(PO_4)_7$	Accessory in many meteorites
Wollastonite	$CaSiO_3$	Accessory in Allende (CcIII)
Yagiite	$(K,Na)_2(Mg,Al)_5(Si,Al)_{12}O_{30}$	Rare accessory in Colomera iron
Zircon	$ZrSiO_4$	Rare accessory

[a]From B. Mason (1972) The Mineralogy of Meteorites: Meteoritics, vol. 7, No. 3, p. 309-326. For further definition of the meteorite types and their symbols in the "Occurrence" column, see the applicable following sections on individual meteorite types.

STONY METEORITES

CHONDRITES

The chondrites, stony meteorites containing chondrules, numerically dominate the falls, and may be the most abundant class of meteorite in the Solar System at the present time (Fig. 2). Chondrules are mostly spherical to subspherical bodies, composed chiefly of silicates, that range in size from less than 0.1 to more than 20 mm. Mineralogically, chondrules commonly are composed of olivine, pyroxene, plagioclase, glass, troilite, nickel-iron, and combinations of these minerals. Texturally, chondrules may be aggregates of crystals, single crystals, wholly glass, or crystal and glass in a wide range of proportions and textures in seemingly endless variety (Fig. 3). Many chondrules appear to have been free fluid drops that assumed a spherical or nearly spherical shape because of surface tension and later solidified and crystallized. However, other chondrules clearly have not had this sort of history, but may be rounded clasts, or have had other more complex origins (see page 30 on origin of chondrules).

Figure 2   The Rosebud stony meteorite, an olivine-hypersthene chondrite. This stone apparently was oriented through much of its atmospheric passage and has a well formed ablation surface with a stagnation point at the top, a roughly conical shape, and well preserved fusion crust. Its weight is 56.8 kg, and its maximum dimension is approximately 45 cm. (University of Texas collection.)

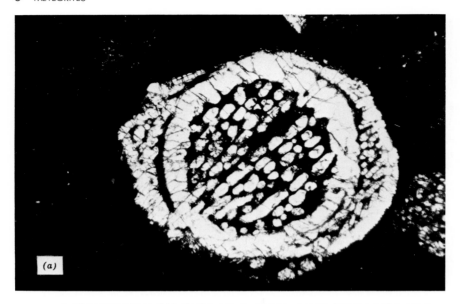

Figure 3   Some of the variety of textures and mineralogy in common types of meteoritic chondrules. *(a)* Barred olivine-glass chondrule with a thick inner rim and a thinner outer portion, from the Pueblito de Allende meteorite. All of the olivine is in optical continuity, and the glass is dark brown and turbid. *(b)* An olivine-glass chondrule with euhedral to subhedral olivine crystals in a matrix of transparent light brown glass that is devitrified in part. Notice the rimmed portion of the chondrule at the lower left with elongate olivine crystals, from the Pueblito de Allende meteorite.

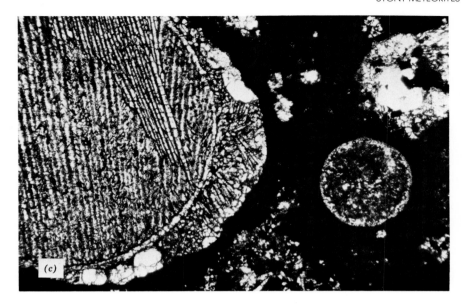

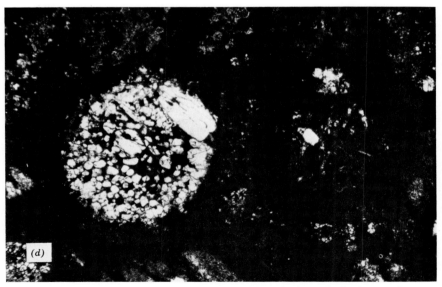

(c) Left: a large barred pyroxene-glass chondrule with a thin rim and adhering matrix;
right: a smaller fine grained olivine, pyroxene, glass chondrule with an indistinct rim, from
the Pueblito de Allende meteorite. *(d)* Left: chondrule composed of randomly oriented
olivine grains in a matrix of turbid brown glass; right: a dark chondrule containing olivine,
plagioclase, pyroxene, and very fine grained matrix, possibly a lithic fragment, from the
Pueblito de Allende meteorite.

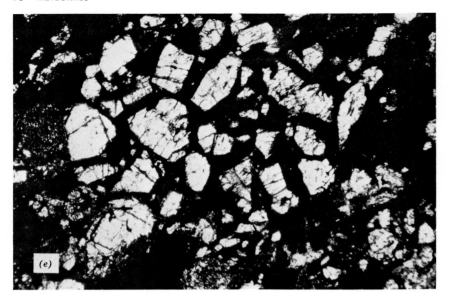

Figure 3 (Cont'd) *(e)* Chondrule composed of many individual crystals of olivine, randomly oriented, in a dark fine grained matrix, from the Weldona meteorite. *(f)* Radiating clinopyroxene and brown turbid glass chondrule, from the Weldona meteorite. The scale is the same in each frame, length of field of view is 3 mm. The large proportion of opaque matrix in frames *(a)* through *(d)* is typical of standard thickness petrographic sections of carbonaceous chondrites.

   The classification of chondrites is based on their chemistry and mineralogy. The most widely accepted classification was first proposed in 1920 by Prior (3) and was later modified by Mason (4) and includes the following classes: enstatite chondrites; bronzite (or olivine-bronzite) chondrites; hypersthene (or olivine-hypersthene) chondrites; and carbonaceous chondrites.[4] The chemistry of chondrites approximates that of mafic and ultramafic igneous rocks, at least for the more common varieties. Table III shows representative analyses of the two most abundant classes of chondrites.

   Chemical and mineralogical regularities in the chondrites were recognized early by Prior (3), and these observations led to the formulation of Prior's Rules, which can be stated as follows:

1. The smaller the volume of nickel-iron in a chondrite, the greater must be the nickel to iron ratio in the metal.
2. The smaller the volume of nickel-iron in a chondrite, the greater must be the iron to magnesium ratio in ferromagnesian silicates.

[4]Mason also proposed the class "olivine-pigeonite chondrites," but later recommended that this term be abandoned.

Table III   Average Major Element Chemical Compositions of the Ordinary Chondrite Meteorites[a]

	Component (Weight %)	Olivine-Bronzite	Olivine-Hypersthene
Silicates	SiO_2	36.41	39.70
	MgO	23.09	24.58
	FeO	8.87	14.33
	Al_2O_3	2.60	2.81
	CaO	1.87	1.92
	Na_2O	0.93	0.94
	K_2O[b]	0.10	0.11
	Cr_2O_3	0.33	0.41
	MnO	0.26	0.26
	TiO_2	0.11	0.11
	P_2O_5	0.18	0.22
	H_2O	0.30	0.24
Metals	Fe	17.45	7.13
	Ni	1.68	1.07
	Co	0.10	0.07
	P	0.05	0.04
	FeS	5.67	6.06

[a]Analytical data as selected by Urey and Craig (8).
[b]K_2O values after Geiss and Hess (1958) Ap. Jour., vol. 127, p. 224.

Earlier work with chondrites and their classification was done notably by Rose (5), who first recognized the chondrites as a distinct group, Tschermak (6), who was a superb petrographer, and Brezina (7), who eventually subdivided the chondrites into more than 30 classes. The Rose-Tschermak-Brezina classification was widely used by students of meteorites until about 1960. The fallacies and superficial basis of much of the Rose-Tschermak-Brezina classification had been pointed out by Prior (3), but the system was not abandoned finally until the comprehensive review of meteorites by Mason (4).

Prior pointed out that the class to which any individual chondritic meteorite belonged could be established by any one of three analyses:

1. The MgO to FeO ratio in the bulk analysis.
2. The volume of and the nickel to iron ratio of the metal.
3. The MgO to FeO ratio in the HCl insoluble silicates.

However, one of the difficulties in applying chemical criteria to the classification of meteorites is the difficulty in obtaining reliable and representative analyses. A critical review of analyses of stony meteorites by Urey and Craig (8) resulted in their disregard of approximately two thirds of the published analyses of chondritic meteorites as unreliable in some respect, an action that understandably offended a number of analysts. Although reliable chemical analyses are now more abundant, there still remains a tremendous problem in sampling chondritic meteorites. By any modern view of chondrites, they are clastic rocks that are accumulations of individual chondrules, rock fragments and grains of matrix, which may have a wide range of composition. Meteorite investigators and analysts often try to consume as little material as possible in analyses, because meteorites are rare and unique objects. Therefore, many analyses are performed on a total sample of only a few grams. Such analyses commonly are taken to be "representative" of the total meteorite. *If* it is desirable to know the approximate bulk composition of the meteorite, it is certainly preferable to pulverize and homogenize 50 to 100 grams of chondrite and take an aliquot of this larger sample for analysis, *if* there is sufficient material available. Even this procedure is not effective for some of the less homogeneous meteorites. These procedures usually are acceptable for falls, but finds, which may have had their composition modified by a small to very considerable amount of weathering, are a more difficult problem. Although analyses of weathered specimens may be acceptable for certain elements that are contained in the relatively stable minerals of the meteorite, the amount of metal, relative amounts of Fe, FeO, and Fe_2O_3, total nickel, water, organic, and carbon contents (as well as concentrations of some other elements) are particularly suspect and should be treated with great caution.

A more recent attempt to classify the petrologic types of chondrites by Van Schmus and Wood (9) is based on chemistry, mineralogy, and texture (Table IV). This classification is now widely used but has some ambiguities and

Table IV Petrologic Classification of Chondritic Meteorites[a]

	Petrologic Types					
	1	2	3	4	5	6
Homogeneity of olivine and pyroxene compositions	—	Greater than 5% mean deviations		Less than 5% mean deviations to uniform	Uniform	
Structural state of low-Ca pyroxene	—	Predominately monoclinic		Abundant monoclinic crystals	Orthorhombic	
Degree of development of secondary feldspar	—	Absent		Predominately as microcrystalline aggregates	Clear, interstitial grains	
Igneous glass	—	Clear and isotropic primary glass; variable abundance		Turbid if present	Absent	
Metallic minerals (maximum Ni content)	—	(<20%) Taenite absent or very minor		Kamacite and taenite present (>20%)		
Sulfide minerals (average Ni content)	—	>0.5%		<0.5%		
Overall texture	No chondrules	Very sharply defined chondrules		Well-defined chondrules	Chondrules readily delineated	Poorly defined chondrules
Texture of matrix	All fine-grained, opaque	Much opaque matrix	Opaque matrix	Transparent microcrystalline matrix	Recrystallized matrix	
Bulk carbon content	~2.8%	0.6–2.8%	0.2–1.0%		<0.2%	
Bulk water content	~20%	4–18%			<2%	

[a] After Van Schmus and Wood (9).

difficulties in application, especially in the carbonaceous chondrites. An underlying assumption of this classification, is a model for the progressive metamorphism of chondrites (10) to account for indistinctness of chondrules, relative homogeneity of olivine composition and other criteria shown in Table IV. The petrologic type is combined with the bulk chemistry group of the meteorite (E = enstatite chondrites, C = carbonaceous chondrites, H = olivine-bronzite chondrites, L = olivine-hypersthene chondrites, LL = ''amphoteric'' chondrites) to give the full designation, for example C3, LL4, H6, etc.

Carbonaceous Chondrites. There are few specimens of carbonaceous chondrites that have been recovered and studied. Almost all of these ~30 specimens are observed falls. However, with the recovery of the Pueblito de Allende meteorite in Mexico, in February, 1969 (11), the total weight in scientific hands of this meteorite class increased by several fold. More than 2000 kg of stones from this fall were recovered, and the mass of the incoming body has been estimated to have been as much as 20 metric tons.

The monitoring of atmospheric pressure waves, and noise in various other wavelengths for the primary purpose of the detection of atmospheric nuclear detonations, as well as for ballistic missile and satellite reentry work, has resulted in reports of large numbers of atmospheric events that apparently are due to the infall of fragile meteorites that are not recovered on the Earth's surface. The carbonaceous chondrites, particularly Types I and II, commonly are fragile, and falls of some stones may be easily broken up on impact with the ground or in handling. Some very high energy events, such as the Revelstoke fall in British Columbia in 1965 (12) that had a total energy of 10^{19}-10^{20} ergs, have resulted in the recovery of only very small amounts of meteorite. At Revelstoke only 1 gram of carbonaceous chondrite was recovered after an intensive search; however, ''black dust'' seen on the snow but not collected may have been from the meteorite. It seems reasonable to conclude that the carbonaceous chondrites and/or some other very fragile meteorite type(s) are much more abundant than is indicated by their representation in the statistics of recovered falls.

The first analysis of a carbonaceous chondrite was performed by Berzelius in 1834, on a sample from the fall at Alais, France in 1806. The results were so different from those obtained on other meteorites that Berzelius doubted the meteoritic origin of the specimen. Most of our modern knowledge of the major element composition of carbonaceous chondrites is due to the work of Wiik (13). Wiik reported 11 new analyses of carbonaceous chondrites (Table V) and showed that carbonaceous chondrites could be divided into three groups based on their major element compositions (Table VI). Wiik called these groups simply Types I, II, and III, a classification that is widely used today. He also pointed out that if the analyses were recalculated on a volatile-free basis, the compositions of carbonaceous chondrites are virtually

Table V Compositions of Carbonaceous Chondrites (All Values in Weight Percent)[a]

Component	Orgueil	Ivuna	Mighei	Nawapali	Haripura	Santa Cruz	Murray	Ornans	Cold Bokkeveld	Lance	Mokoia
Fe	0.00	0.00	0.00	0.00	0.00	0.00	0.00	0.72	0.00	2.19	0.00
Ni	0.00	0.00	0.00	0.00	0.00	0.00	0.00	—	0.00	1.50	0.00
Co	0.00	0.00	0.00	0.00	0.00	0.00	0.00	—	0.00	0.07	0.00
FeS	15.07	18.38	10.05	7.67	14.93	9.04	7.67	6.43	8.16	6.49	6.74
SiO_2	22.56	22.71	27.81	27.08	26.39	29.36	28.69	33.52	27.33	33.23	33.40
TiO_2	0.07	0.07	0.08	0.09	0.08	0.12	0.09	—	0.08	0.13	0.10
Al_2O_3	1.65	1.62	2.15	2.09	2.27	2.19	2.19	—	2.29	2.93	2.51
MnO	0.19	0.23	0.21	0.17	0.19	0.19	0.21	0.18	0.19	0.20	0.19
FeO	11.39	9.45	19.13	20.76	15.12	22.34	21.08	—	20.17	24.80	25.43
MgO	15.81	16.10	19.46	17.89	18.04	21.16	19.77	23.87	18.73	23.54	23.98
CaO	1.22	1.89	1.66	2.20	2.01	2.30	1.92	1.99	1.56	2.64	2.56
Na_2O	0.74	0.75	0.63	0.54	0.70	0.50	0.22	0.55	0.61	0.58	0.51
K_2O	0.07	0.07	0.05	0.16	0.07	0.12	0.04	0.17	0.05	0.14	0.04
P_2O_5	0.28	0.41	0.30	0.26	0.23	0.32	0.32	0.35	0.30	0.32	0.38
H_2O^+	19.89	18.68	12.86	16.41	13.70	9.23	9.98	0.25	15.17	1.40	2.07
H_2O^-						1.10	2.44	0.18			
Cr_2O_3	0.36	0.33	0.36	0.38	0.45	0.39	0.44	0.50	0.42	0.49	0.52
NiO	1.23	1.34	1.53	1.54	1.71	1.64	1.50	0.00	1.49	0.00	1.64
CoO	0.06	0.06	0.07	0.06	0.08	0.08	0.08	0.00	0.08	0.00	0.08
C	3.10	4.83	2.48	2.50	4.00	2.54	2.78	—	1.30	0.46	0.47
Loss on ignition	6.96	4.10	0.36	2.03	—	—	0.62	—	2.23	—	—
Sum	100.65	101.02	99.19	101.83	99.97	102.62	100.04	—	100.16	101.11	100.62

Meteorite Name

[a]Analyses by H. B. Wiik (13).

Table VI   Mean Values of Selected Constituents of the Different Types of Carbonaceous Chondrites[a]

Type	Meteorite Names	SiO$_2$	MgO	FeO	C	H$_2$O	FeS
I	Orgueil, Tonk, Ivuna	22.56	15.21	9.77	3.54	20.08	16.52
II	Cold Bokkeveld, Nogoya, Mighei, Nawapali, Haripura, Boriskino, Santa Cruz, Murray	27.57	19.18	20.28	2.46	13.35	8.66
III	Felix, Lance, Mokoia, Warrenton, Ornans	33.58	23.74	24.20	0.46	0.99	6.05

[a]From Wiik (13).

identical to each other. Carbonaceous chondrites, in general, are characterized by their high contents of volatile elements and compounds, which include water, sulfur, and rare gases. They mostly have large carbon contents as the name implies, except for many of the Type III specimens. The mineralogy and petrology of carbonaceous chondrites is characterized by a lack or an exceedingly small percentage of free metal, only minor amounts of troilite, and a wide range of chondrule size, texture, and mineralogy (Fig. 3a-3d). Type I carbonaceous chondrites do not contain chondrules, and as Mason (4) notes, this is "an awkward contradiction." However, these carbonaceous meteorites clearly should be grouped with the carbonaceous chondrites because of their very close chemical and mineralogical similarities with Type II specimens. The colors of carbonaceous chondrites are mostly dark gray to black, and some specimens (such as Mokoia) have a greenish tint. Some Type III specimens are lighter in color ranging from light to medium gray. The specific gravities of Type I samples are the lowest of all meteorites, approximately 2.2, but Type II and III samples range from around 2.5 to 3.7.

The groundmass of carbonaceous meteorites is dark and commonly almost opaque if viewed in standard thickness thin sections (0.03 mm). However, if special sections are prepared that are approximately one third as thick (Fig. 4), the microcrystalline groundmass becomes much more visible. The fine grain size of the groundmass commonly makes optical mineral identifications and petrography difficult. Also the range of minerals that occur in the groundmass is rather varied. Type I and II specimens commonly contain serpentine and other hydrous minerals. The groundmass of the Pueblito de Allende meteorite (Type III) is composed largely of olivine, more fayalitic than the olivine in the chondrules, and magnetite. In fact, the overall medium gray color of the meteorite probably is caused in part by the abundance of magnetite in the groundmass of Pueblito de Allende, and not the low carbon content which is only 0.27 weight percent.

Figure 4   Ultrathin section (~10 μm) of the Murchison, Australia carbonaceous chondrite (Type II). Much structure is visible in the matrix of the meteorite that is not visible in standard thickness (30 μm) petrographic thin sections. Notice the broken chondrules and angular fragments. Plane polarized light, length of field of view is 2.8 mm. (Section courtesy of Grover Moreland.)

A common feature of Type II and III carbonaceous chondrites is the presence of high calcium-aluminum inclusions, aggregates, and chondrulelike bodies. These inclusions were recognized in the Kaba meteorite by Sztrokay et al. (14) in 1961, and later by others. But their ubiquity and possible importance were not fully realized until the recovery of Pueblito de Allende, in which these inclusions are large, abundant, and obvious. Marvin et al. (15) and Clarke et al. (11), and others have described the mineralogy and chemistry of the Ca-Al rich inclusions in the Pueblito de Allende meteorite. The major phases present vary between different inclusions but include melilite or spinel, diopside and other clinopyroxenes, anorthite, perovskite, enstatite, grossularite, nepheline, sodalite, Ca-Al rich glass, and cordierite (Fig. 5). It seems likely that these Ca-Al rich inclusions represent very high temperature mineral assemblages, and are condensates from the solar nebula. Recent oxygen isotope analyses have been interpreted to indicate that some of the Ca-Al rich inclusions may contain pre-Solar System material. The lowest measured isotopic composition of Sr^{87}/Sr^{86} in the Solar System (0.69877 ± 0.00002) has come from high Ca-Al inclusions in the Allende

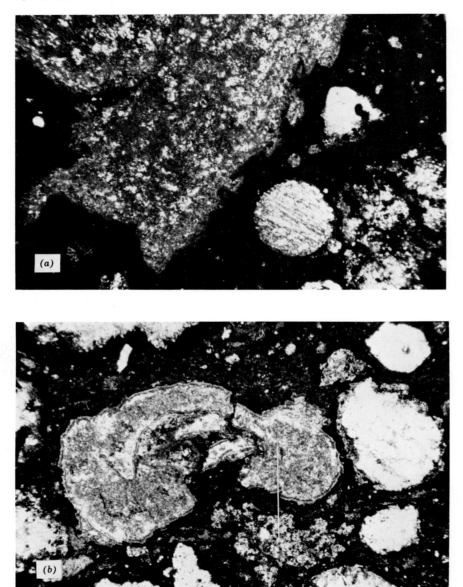

Figure 5   Calcium-aluminum rich inclusions in the Pueblito de Allende meteorite. *(a)* Top left: melilite, anorthite, spinel, and glass inclusion with irregular outline. *(b)* Bottom center: rounded inclusion composed of glass, diopside, anorthite, and minor perovskite. Lengths of fields of view are 3 mm.

meteorite (16). The study of these bodies currently is an active area of research.

Another discovery that resulted from investigations of the Pueblito de Allende meteorite is the presence of void spaces and bubble cavities in many of the chondrules (Fig. 6). Some of the chondrules contain spherical void spaces that are as much as one third the diameter of the total chondrule. Chondrules with voids subsequently have been recognized in several carbonaceous meteorites. This observation is important to some models of origin of chondritic meteorites. It certainly can be interpreted that the Pueblito de Allende meteorite, and probably most other carbonaceous chondrites, were never portions of the deep interior of a planetary body after chondrule formation; otherwise, the chondrules with the voids should have collapsed under high lithostatic pressure.

As in most other chondritic meteorites, the carbonaceous chondrites contain inclusions of fragments of other meteorite types, fragments of the same type, or both with different mineralogy and/or texture (Fig. 7). Little effort has gone into breaking up large fragments of carbonaceous chondrites to look for these inclusions because of the scarcity of material. However, in the Pueblito de Allende meteorite, inclusions of fragments of different carbonaceous meteorite types are common, and there is ample material available to break up a few stones to look for and remove these inclusions for analyses. Some of the inclusions are Type II carbonaceous chondrite fragments (an observation also made in the Leoville meteorite, which like Pueblito de Allende is a Type III), and it has been demonstrated that different dark inclusions in the meteorite have different contents of organic compounds, as well as major element and mineralogic compositions.

The analysis of carbonaceous chondrites for organic compounds has a long history, but definitive and repeatable work essentially has been lacking until the application of organic mass spectrometry and combined gas chromatography. Much of the recent mass spectrometric work with carbonaceous chondrites has been concerned with the identity, relative abundances, and origin of organic compounds that these meteorites contain. This information *may* have bearing on the origin(s) of life in the Solar System, but such analyses certainly yield information on abiogenic synthesis of complex carbon compounds early in Solar System history. Contamination of the meteorites by terrestrial organic compounds and organisms during and after fall has presented the organic analysts with a monumental problem. Much of the information in the literature is contradictory, both in terms of the data derived from the same specimen and in interpretation of these data. The amount of organic compounds in most carbonaceous chondrites is small and the analyses are difficult. However, a small number of carbonaceous chondrites have been analyzed shortly after fall, and even these few

Figure 6   Bubble cavities in a large chondrule (lower center) in the Pueblito de Allende carbonaceous chondrite seen on a sawed surface. The diameter of the large chondrule is approximately 5 mm.

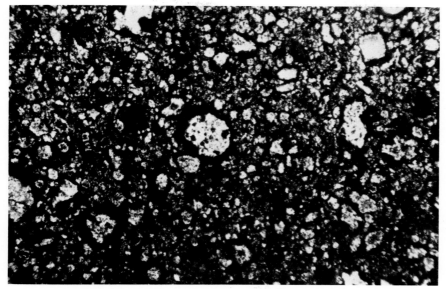

Figure 7 Photomicrograph of a lithic inclusion of a different carbonaceous chondrite (Type II) in the Pueblito de Allende meteorite (Type III). Notice that the sizes of the chondrules are much smaller than in the host meteorite, and that the mineralogy of most chondrules is much simpler, mostly single crystals or aggregates of a few crystals of olivine. Such inclusions also have been reported in the Leoville chondrite (Type III). Length of field of view is approximately 3 mm.

specimens contain a large and varied suite of carbon compounds (Fig. 8). Both aliphatic and aromatic hydrocarbons have been reported; also, a number of amino acids are present in some samples (Table VII). The general conclusion of most investigators at the present time is that these compounds are abiogenic in origin, but that the organic chemistry of the early Solar System is much more complex than was previously suspected.

There are neither precise observations nor photographs of carbonaceous chondrite falls that would allow limits to be placed on their place(s) of origin in the Solar System. Investigators speculate that they may be cometary and asteroidal debris.

Olivine-Hypersthene and Olivine-Bronzite Chondrites. These are the most abundantly recovered varieties of all the stony meteorites, and these two classes together are appropriately termed "ordinary chondrites." The major difference between the olivine-bronzite and olivine-hypersthene chondrites is the total iron content and the distribution of iron between metallic and silicate phases. The olivine-hypersthene chondrites have a lesser total iron content and a lesser amount of free metal than do the olivine-bronzite chondrites (see Tables III and VIII). The difference in the amount of free metal is sufficiently

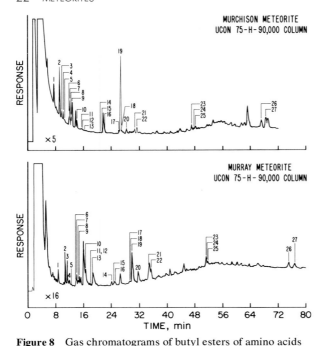

Figure 8 Gas chromatograms of butyl esters of amino acids
from the Murchison (upper) and Murray (lower)
carbonaceous chondrites. Identifications are: (1) isovaline,
(2) α-aminoisobutyric acid, (3) D-valine, (4) L-valine,
(5) N-methylalanine, (6) D- α-aminobutyric acid, (7) D-alanine,
(8) L- α-aminobutyric acid, (9) L-alanine,
(10) N-methylglycine, (11) N-ethylglycine, (12) D-norvaline,
(13) L-norvaline, (14) D-β-amino-isobutyric acid,
(15) L- β-aminoisobutyric acid, (16) β-aminobutyric acid,
(17) D-pipecolic acid, (18) L-pipecolic acid, (19) glycine,
(20) β-alanine, (21) D-proline, (22) L-proline, (23) γ-aminobutyric
acid, (24) D-aspartic acid, (25) L-aspartic acid,
(26) D-glutamic acid, (27) L-glutamic acid. From Kvenvolden
et al. (1971) Advances in Organic Geochemistry, p. 387-401.
(Reprinted by permission of Pergamon Press Ltd.)

large that the two types mostly can be distinguished by visual inspection of a
sawed surface. The forsterite-fayalite contents of the olivine in ordinary
chondrites have been studied extensively by Mason (18; Fig. 9). It is apparent
that these meteorites conform to Prior's Rules. The measurement of mean
olivine composition is very useful information for the classification of an
ordinary chondrite. Normally, this is established easily by X-ray powder
diffraction analysis of a bulk sample. Also, the homogeneity of olivine
compositions is used by Van Schmus and Wood (9) as a criterion for

Table VII Hydrocarbons and Amino Acids Identified in the Murchison Meteorite[a]

Hydrocarbon Types	Amino Acids
Normal alkanes	Glycine
Monomethyl alkanes	Alanine
Dimethyl alkanes	Valine
Alkyl cyclohexanes	"Leucine"
Other alkyl cyclohexanes	Proline
Polycyclic alkanes	Aspartic
Olefins	Glutamic
Benzenes (and alkyl-)	Sarcosine
Biphenyls naphthalenes	β-Alanine
(and alkyl-)	N-Methyl alanine
Antracenes (and alkyl-)	N-Ethyl alanine
Phenantrenes (and alkyl-)	α-Amino-N-butyric
Acenaphthenes	α-Amino-isobutyric
Fluoroanthenes	β-Amino-N-butyric
Pyrenes	β-Amino-isobutyric
	α-Amino-N-butyric
	Isovaline
	Norvaline
	Pipecolic acid

[a]Compiled from the analyses of several research groups. Not a complete list, but representative of the major compounds. After Oro, 1972 (17).

petrologic types in all of the chondritic meteorites; however, this is more difficult to observe by X-ray diffraction and must be established with care. Generally, a large number of electron microprobe analyses of both chondrule and matrix olivine is required.

Mineralogically, the olivine-bronzite and olivine-hypersthene chondrites are similar in their modal analyses, except for the amount of free nickel-iron. Mason (4) gives the mineralogic modal analyses for these two types as shown in Table VIII. In addition, specimens from both groups may contain minor amounts of whitlockite, chromite, apatite, clinopyroxene, and minerals of even lesser abundance.

The division of 93 "superior analyses" of stony meteorites into H and L (high and low iron) groups by Urey and Craig (8) on the basis of oxidized iron and iron contained in metal and sulfide (Fig. 10), illustrates the difference in total iron and oxidized iron between the two groups. All olivine-hypersthene chondrites are L type, and the bronzite chondrites are H type. The genetic significance of this systematic division is not known, but it is at least partially related to the way in which the two groups are defined.

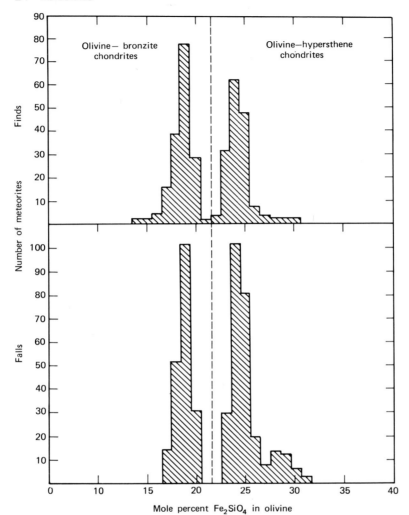

Figure 9   Histograms showing the distribution of olivine compositions in 791 falls and finds of ordinary chondrites. It is apparent that there are two distinct populations. (From Mason, 18; Courtesy The American Museum of Natural History.)

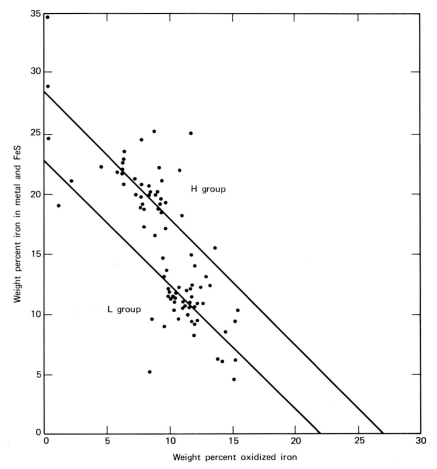

Figure 10 High (H) and low (L) iron groups of Urey and Craig (8) as determined by plotting 93 "superior" analyses of chondritic meteorites (modified from Mason, 4). The two diagonal lines indicate where the analyses should plot if both groups conformed exactly to Prior's Rules.

Table VIII   Range of Mineralogic Modal Analyses in Ordinary Chondrites

Mineral	Olivine-Hypersthene Chondrites	Olivine-Bronzite Chondrites
Olivine	35-60 vol. %	25-40 vol. %
Orthopyroxene[a]	25-35	20-35
Plagioclase	5-10	5-10
Nickel-iron	1-10	16-21
Troilite	~5	~5

[a]The orthopyroxene is dominantly either hypersthene (>20% $FeSiO_3$) or bronzite (10-20% $FeSiO_3$) as the names of the two groups indicate; however, these varieties are not defined with the same limits as in routine petrology (bronzite 12-30% $FeSiO_3$).

The amphoterites (LL chemical group of Van Schmus and Wood) are regarded as a separate type of ordinary chondrite by some authors but are lumped together with the olivine-hypersthene chondrites (L group) as a subvariety by others. They are the most oxidized of the ordinary chondrites,[5] and are considered here with the olivine-hypersthene chondrites.

The chondritic structure of most ordinary chondrites is apparent on a fracture surface. Metallic nickel-iron and troilite are visible as small grains with metallic luster. The meteorite color when fresh and unweathered ranges from light gray to black. However, as weathering proceeds the specimens generally become darker in color, and many weathered, dark brown specimens are common in most meteorite collections. The specific gravities of olivine-hypersthene chondrites range from about 3.4 to 3.6, and olivine-bronzite chondrites range from about 3.6 to 3.9, again mostly reflecting the differences in total iron and metallic iron contents.

As in all chondritic meteorites, the textural, mineralogical, and size range of individual chondrules is extremely large (Fig. 11). However, in the ordinary chondrites, olivine, orthopyroxene, and glass dominate the chondrule mineralogy. Many ordinary chondrites contain petrographic evidence of shock damage either throughout the total meteorite or in included individual chondrules and rock fragments (Fig. 12, also see page 49). Much of the dark glass and maskelynite, as well as troilite-rich veins, in the ordinary chondrites, is thought to have originated as shocked and impact injection material.

Great importance has been attached to the texture of the chondrules and matrix by Van Schmus and Wood (9) in assigning petrologic type (Table IV)

[5]Although first thought to be achondrites by Prior, the amphoterites were shown by Kvasha (1958, Chem. Erde, vol. 19, p. 249-274) to contain chondrules. Later work, most notably by Keil and Fredriksson (1964, Jour. Geophys. Res., vol. 69, p. 3487-3515) established that this is a chemically identifiable group.

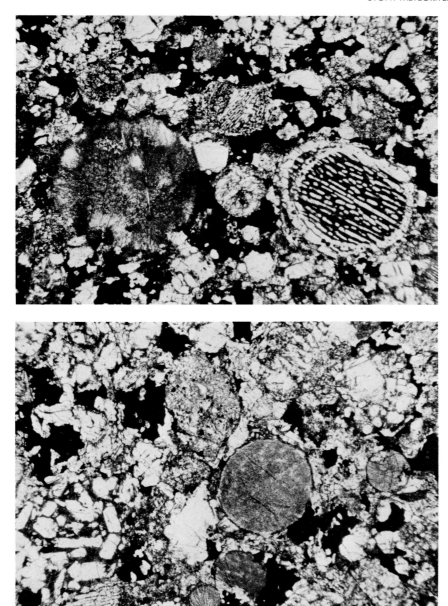

Figure 11   Two photomicrographs of thin sections of the Faucett meteorite, a typical olivine-bronzite chondrite, illustrating the variety, abundance, and size range of chondrules even in a very small area. The fine-grained gray chondrules are mostly orthopyroxene. Most of the other chondrules are various textures of olivine and dark turbid glass. The lengths of the fields of view are 3 mm, plane polarized light.

Figure 12   Badly shock damaged olivine in an ordinary chondrite showing extremely undulatory extinction, dislocations, and glide planes. Such fragments are common in most chondrites. Length of field of view is 1.4 mm, crossed polarizers.

to an individual meteorite. Certainly the distinctness of chondrules from matrix has a wide textural range in the ordinary chondrites (Fig. 13), and an impressive case has been made for this difference being due to some sort of progressive metamorphism or recrystallization or both that probably results from different thermal histories. Chondrites with very distinct chondrules and relatively heterogeneous olivine compositions are termed "unequilibrated," and only those with only moderately distinct to barely discernible chondrules and homogeneous mineral compositions are called "equilibrated."

Enstatite Chondrites. The Hvittis meteorite, which fell in southwestern Finland in 1901, was described by Borgström (19) as the first known meteorite composed mostly of enstatite, but the term "enstatite chondrite" was first used for this class of stony meteorite by Prior (3). The enstatite chondrites are characterized mineralogically and chemically by virtually no olivine, free silica as quartz, tridymite and cristobalite in small amounts, approximately 55 percent enstatite or protoenstatite, virtually all of the iron being present as metallic nickel-iron, and an abundance of sulfur (hence, sulfides) as compared with the ordinary chondrites. These meteorites, which number less than 20 specimens, are *highly* reduced. The iron is almost totally reduced to metal, and this metal also contains silicon. Calcium, chromium, and manganese may be present in the sulfide in significant amounts (20). The total iron contents of the enstatite chondrites range to greater values than the other chondrite

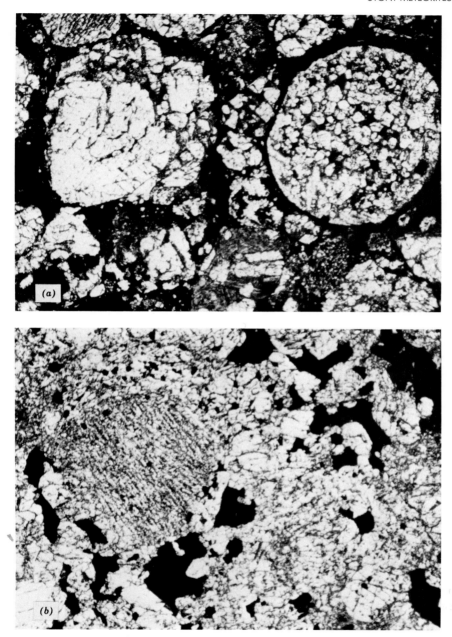

Figure 13   Examples of very distinct chondrules *(a)* in the Mezö Madaras unequilibrated chondrite and indistinct chondrules *(b)* in the Tourinnes la Grosse equilibrated and recrystallized chondrite. Plane polarized light, lengths of both fields of view are approximately 3 mm.

groups, commonly more than 30 percent. Many of the meteorites that are classed as enstatite chondrites (E group) on the basis of their chemistry have few or no chondrules. The color of fresh enstatite chondrites ranges from gray to almost black. Specific gravities range mostly from 3.51 to 3.57, and correlate well with total iron content.

A comprehensive review of the data on enstatite chondrites was published by Mason (21), together with his own investigations. He concluded that the enstatite chondrites form a well-defined group of chondrites distinct in chemistry and mineralogy. They do not have the uniformity of iron content shown in the analyses of other chondrites, ranging from 20.7 to 35.0 weight percent. Mason observes that as the number and distinctness of chondrules decreases, the iron content decreases, and the coarseness of crystallinity increases. Furthermore, clinoenstatite is the dominant mineral phase in the iron-rich enstatite chondrites, and enstatite is the dominant phase in iron-poor specimens. Plagioclase is absent or present only in extremely small amounts in the iron-rich enstatite chondrites. Grant (22) has made an intensive study of free silica in meteorites and finds that quartz, tridymite, and cristobalite are relatively abundant in the enstatite chondrites (Figs. 14, 15; Table IX), in contrast with the other chemical groups of chondrites.

The largest enstatite chondrite that has been recovered is Abee, which fell in Alberta, Canada, just north of Edmonton. This fall was a single stone weighing 107 kg, almost four times the weight of the largest specimen previously known, Indarch, from the USSR (Fig. 16).

Origin of Chondrules.   The key to understanding the origin of chondrites lies in identifying and understanding the process that forms chondrules. Many hypotheses have been presented for the origin of chondrules during the past 100 years, and some of these ideas clearly are incorrect in the light of present data. We shall mention here only some of the more interesting, plausible, and historic ideas.

Spherulitic crystallization of chondrules from homogeneous or nearly homogeneous silicate magmas and melts has been proposed and supported by a number of authors. Brezina (23) proposed, in 1885, that chondrules resulted from the arrested, rapid crystallization of molten silicate, and that the chondrules were formed as a result of a special type of magmatic segregation. This idea, with minor variations, was revived and supported by Roy (24) and Ringwood (25). Roy suggested that chondrules are crystallization products of a silicate melt, but that many of the chondritic rocks resulting from cooling of the melt have had later metamorphism and deformation histories. Ringwood proposed that spherulitically crystallized magma solidified at temperatures of less than 1000°C with high partial pressures of CO_2 and H_2O to yield chondritic rocks.

Mason (4, 26) suggested that chondrules are formed by the thermal metamorphism of the amorphous and/or serpentine-rich groundmass of Type I carbonaceous chondrites at temperatures near or above 400° C.

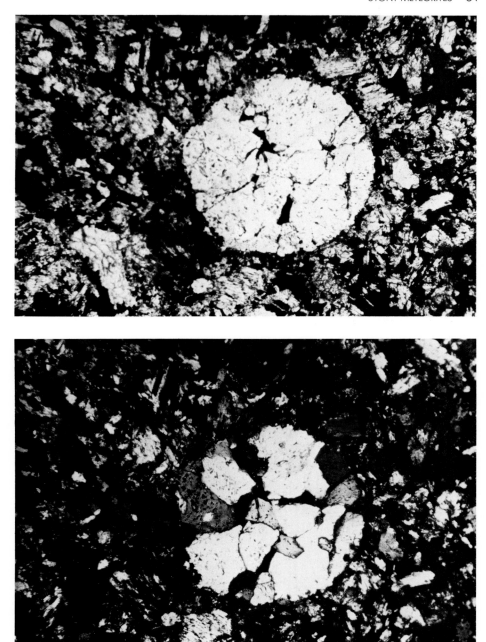

Figure 14   Quartz chondrule in the St. Mark's enstatite chondrite. Top: plane polarized light; bottom: crossed Nicols to show the polycrystalline structure of the chondrule. The chondrule is approximately 0.2 mm in diameter. (Courtesy of Ray W. Grant.)

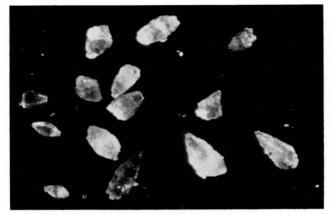

Figure 15 Quartz crystals separated from the St. Mark's enstatite chondrite. Notice that most of the crystals are euhedral and a few are doubly terminated. The largest crystal is approximately 0.06 mm long. (Courtesy of Ray W. Grant.)

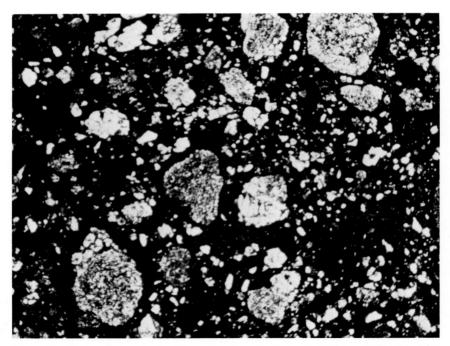

Figure 16 Photomicrograph of a thin section from the Indarch, USSR enstatite chondrite. The opaque areas are mostly metallic nickel-iron. Plane polarized light, length of field of view is approximately 3 mm.

Table IX   Meteorites That Contain Free Silica[a]

Meteorite Name	Classification	Silica Polymorph
Abee, Alberta, Canada	Enstatite chondrite	quartz, cristobalite
Atlanta, Louisiana	Enstatite chondrite	tridymite
Barratta, Australia	Hypersthene chondrite	quartz
Bath, South Dakota	Bronzite, chondrite	cristobalite
Bishopville, South Carolina	Aubrite	tridymite
Cachari, Argentina	Eucrite	tridymite
Carbo, Mexico	Octahedrite	cristobalite
Chaves, Portugal	Howardite	tridymite
Clover Springs, Arizona	Mesosiderite	tridymite
Crab Orchard Mtns, Tennessee	Mesosiderite	tridymite
Daniel's Kuil, South Africa	Enstatite chondrite	tridymite
Esterville, Iowa	Mesosiderite	tridymite
Frankfort, Alabama	Howardite	tridymite
Hainholz, Germany	Mesosiderite	tridymite
Homestead, Iowa	Bronzite chondrite	tridymite
Hvittis, Finland	Enstatite chondrite	cristobalite
Indarch, USSR	Enstatite chondrite	tridymite
Johnstown, Colorado	Hypersthene chondrite	tridymite
Juvinas, France	Eucrite	tridymite
Kendall County, Texas	Hexahedrite	tridymite
Khairpur, Pakistan	Enstatite chondrite	tridymite
Mincy, Missouri	Mesosiderite	tridymite
Moore County, North Carolina	Eucrite	tridymite
Morristown, Tennessee	Mesosiderite	tridymite
Pasamonte, New Mexico	Eucrite	tridymite, cristobalite
Pillistfer, Estonia	Enstatite chondrite	tridymite
Sioux Country, Nebraska	Howardite	tridymite, quartz
Stannern, Czechoslovakia	Eucrite	quartz
Steinback, Germany	Siderophyre	tridymite
St. Mark's, South Africa	Enstatite chondrite	quartz
Toluca, Mexico	Octahedrite	quartz
Vaca Muerta, Chile	Mesosiderite	tridymite

[a]From Grant (22). A number of other meteorites have been reported to contain free silica, but many of the older reports are single grains in mineral separates and may be contamination. Other enstatite chondrites *reported* to contain free silica are: Adhi Kot, Pakistan; Bethune, Colorado; Jajh deh Kot Lalu, Pakistan; Kota-Kota, Malawi; Pesyanoe, Siberia; Saint-Sauveur, France.

Several scientists have suggested that chondrules result from simple condensation in the primitive solar nebula, but there are many difficulties with physical-chemical models for this origin.

That chondrules may be cooled fluid silicate drops that were produced by lightning discharges in the dusty, primitive solar nebula has been theorized by

Whipple (27). This idea has been accepted as a reasonable theory by many meteorite researchers.

Tschermak (6) concluded that some chondrules were rounded mineral grains and rock clasts, but thought that other textural types of chondrules were solidified silicate droplets and spherules. Urey and Craig (8) proposed that chondrules were produced from the impact of two asteroids, one considerably larger than the other, by the cooling and crystallization of silicate droplets that were melted in the impact. These solidified and partially solidified droplets then fell back to the surface of the larger asteroid. Recent experimental work on the crystallization of silicate droplets, that have been fused and then allowed to supercool, has reproduced the textures of common varieties of meteoritic chondrules (28).

King et al. (29) have reported the occurrence of chondrules and chondrulelike bodies in the samples returned from the Moon by Apollo 14 (see page 210), and have noted the possible implications for the origin of chondritic meteorites. They conclude that chondrules have formed in the lunar samples by processes that accompany the impact of a large body on the lunar surface. The processes suggested include the crystallization of silicate droplets and spherules formed by fusion of lunar rocks in the impact, a variation of the mechanism proposed by Urey and Craig, and the rounding of rock clasts, an extension of the suggestion by Tschermak for the origin of chondrules in the Soko-Banja chondrite.

Observations of the petrography, especially textures, in Apollo 14 samples have led to the interpretation that breccias from the Fra Mauro Formation could be derived from a single cooling unit more than 100 m thick. A convincing recrystallization sequence of lithologic types is shown in the Apollo 14 samples (see page 210). The difference between these samples and the rocks from the earlier Apollo 11 and Apollo 12 missions is that the Apollo 14 landing site is on the ejecta blanket from the Imbriun Basin. Thus, surrounding large impact craters, there probably are thick impact-emplaced base surge and fallback deposits. These deposits may be emplaced at elevated temperatures (as much as 800°C or more), and the differences in cooling rates in the various portions of the cooling unit could account for different textures, just as in terrestrial ash flow tuffs. Such a setting on a planetary surface would provide the thermal gradient required to produce the Van Schmus and Wood petrologic types of chondrites (Fig. 17). Fluid drops that were thrown above the surface could supercool and crystallize, glass spherules incorporated into the hot deposit could anneal and devitrify, and movement of particles with the base surge would provide ample opportunity for the rounding of clasts.

If the temperatures were sufficiently high for an extended period, there would be opportunity for recrystallization surrounding the hottest fragments and some diffusion. Much of the recrystallization in chondrites is concentric to chondrules. Also, many chondrites contain abundant petrographic evidence of shock history. The shocked minerals may be common throughout the

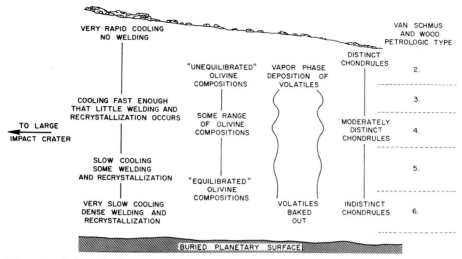

Figure 17   Schematic diagram of the model of King et al. (29) for the origin and interrelation of the Van Schmus and Wood petrologic types of chondrites. (Reprinted from *Chondrules in Apollo 14 samples and size analyses of Apollo 14 and 15 fines* by E. A. King, Jr., J. C. Butler and M. F. Carman by permission of the MIT Press, Cambridge, Mass.)

meteorite, or may occur only in certain chondrules and clasts. Most chondrites are known to contain clasts, broken chondrules, and chondrule fragments (Fig. 18). This indicates that many chondrites may be the result of multiple impact and possibly other events.

ACHONDRITES

Stony meteorites without chondrules are appropriately termed "achondrites," except for the Type I carbonaceous chondrites and some of the enstatite chondrites as previously discussed. Only about 70 achondrites have been recovered and studied, and most of them are falls. These meteorites understandably are difficult for untrained observers to identify, except for the distinctive vitreous black fusion crusts on fresh falls. They generally are classified on the basis of texture, mineralogy, and chemistry. The achondrites do not contain abundant nickel-iron (it is totally absent in some specimens) and tend to be more coarsely crystalline than other stony meteorites. Some achondrites have the textures of ordinary igneous rocks, and many others are breccias containing angular clasts of different mineralogies and textures. Some brecciated achondrites contain angular fragments of chondrites.

The achondrites were divided by Prior (3) into two groups based on calcium content: calcium-poor achondrites (0-3 wt % CaO); and calcium-rich

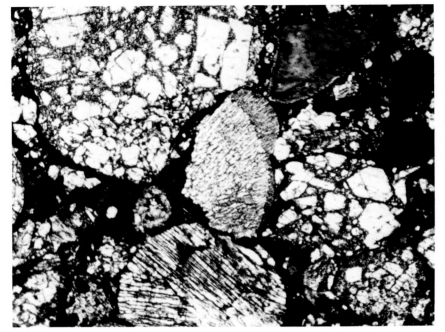

Figure 18 Photomicrograph of a thin section of the Mezö Madaras chondrite showing an obvious chondrule fragment in the center of the field of view. Such fragments are common in virtually all chondrites. Plane polarized light, length of field of view is approximately 3 mm.

achondrites (5-25 wt % CaO). Within each of these major groups, Prior named a number of subdivisions. These subdivisions subsequently were modified somewhat by others, but the system remains essentially intact. The classification of achondrites is listed by Mason (4) as shown in Table X.

Wahl (30) and Michel (31) were early contributors to the petrography of achondrites, and Wahl (32) later discussed some of the general petrography

Table X Classification of Achondritic Stony Meteorites

    A. Calcium-poor achondrites
        1. Enstatite achondrites (aubrites)
        2. Hypersthene achondrites (diogenites)
        3. Olivine achondrites (chassignites)
        4. Olivine-pigeonite achondrites (ureilites)
    B. Calcium-rich achondrites
        1. Pyroxene-plagioclase achondrites (eucrites and howardites)
        2. Augite achondrites (angrites)
        3. Diopside-olivine achondrites (nakhlites)

and properties of the brecciated meteorites. Many individual descriptions and contributions treat only a single specimen or a few samples of achondrites. Mason (4) includes a general summary of the achondrites. A comprehensive review and investigation of the eucrites and howardites by Duke and Silver (34) emphasizes the brecciated structure of many specimens, and they propose a classification of eucrites and howardites (together called "basaltic achondrites") based on mineralogy and texture. This view, which emphasizes the difference between monomict and polymict breccias, has been widely recognized.

In numbers, the calcium-rich achondrites dominate the achondrites, especially the eucrites and howardites of which more than 40 specimens are known (Fig. 19). Some of the other classes are represented only by a single stone (angrites and chassignites), and only two nakhlites (Fig. 20) have been recognized. The aubrites and diogenites total about 10 specimens each. Some meteorites in these groups are interesting for the very large orthopyroxene crystals that they contain, as much as several centimeters long. One of the aubrites, Peña Blanca Springs, is of special interest because it was a well observed fall (35). The meteorite fell, in 1946, into a swimming pool that had been constructed by damming up a small creek near the headquarters of a large West Texas ranch. More than 70 kg of the stone were recovered, and

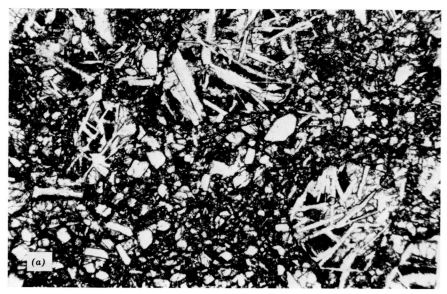

Figure 19   Two fields of view of thin sections of the Pasamonte eucrite: *(a)* showing the general clastic texture of the meteorite with obvious clasts of igneous texture. Also note the crude layering and sorting. Plane polarized light, length of field of view is approximately 3 mm.

Figure 19 (Cont'd)   *(b)* Igneous texture of a clast in Pasamonte with abundant plagioclase, augite, and glassy groundmass. Plane polarized light, length of field of view is approximately 1.4 mm.

Figure 20   Photomicrograph of the Nakhla meteorite, a rare variety of calcium-rich achondrite, showing the granular igneous texture and numerous crystals of olivine and diopside. USNM 426, plane polarized light, length of field of view is approximately 2.0 mm.

the phenomena accompanying the daylight fall are well documented. The Cumberland Falls meteorite is also of special interest because it is a brecciated enstatite achondrite that contains breccia fragments of a dark chondrite (Fig. 21).

The major element chemical composition of the basaltic achondrites is not too revealing of their relations to other meteorite groups, except that some of the mesosiderites contain stony portions of similar composition. Some examples of eucrite and howardite analyses are presented in Table XI. It should be emphasized that sampling of the brecciated eucrites and howardites for chemical analyses is a very serious problem, and that the analyses reported may only approximate the bulk composition of the total meteorite.

Because of the absolute scarcity of other types of achondrites, most of the work that bears on the origin of achondrites has been done with the eucrites and howardites. The textures of many of the individual breccia fragments in the basaltic achondrites, especially the eucrites, are those of ordinary terrestrial igneous rocks such as gabbros, diabases, and basalts (Fig. 19). Many of the eucrites contain fragments with ophitic to subophitic texture,

Figure 21   A fragment of the Cumberland Falls brecciated enstatite achondrite with a sawed surface showing fragments of large enstatite crystals and dark chondritic clasts. Long dimension of the specimen is approximately 20 cm. (USNM 604, courtesy of the Smithsonian Institution.)

Table XI   Analyses of some Eucrites and Howardites[a] All Values in Weight Percent

	Howardites[b]		Eucrites[c]		Range in Eucrites
	Yurtuk	Petersburg	Pasamonte	Juvinas	
SiO_2	49.45	49.21	48.59	49.32	48.6–49.6
TiO_2	—	—	0.65	0.68	0.4– 1.0
Al_2O_3	9.66	11.05	12.70	12.64	11.7–13.9
Fe_2O_3	—	—	—	—	—
FeO	15.61	20.41	19.58	18.49	15.3–20.1
MgO	17.40	8.13	6.77	6.83	5.4– 7.4
CaO	6.39	9.01	10.25	10.32	8.6–11.5
Na_2O	0.31	0.82	0.45	0.42	0.4– 0.9
K_2O	—	—	0.05	0.05	0.04–0.22
Cr_2O_3	0.04	0.42	0.33	0.30	0.06–0.9
MnO	0.72	—	0.56	0.53	0.3– 0.8
P_2O_5	0.01	—	0.10	0.09	0.09–0.16
Fe (metal)	—	—	—	0.04	0.00–0.05
FeS	—	0.16	0.06	0.53	0.03–0.57
$H_2O(+)$	—	—	0.27	0.02	
$H_2O(-)$	—	—	0.01	0.03	
Total	99.59	99.21	100.37	100.29	

[a]Modified from Duke and Silver (34).
[b]Analyses from Urey and Craig (8).
[c]A. D. Maynes, Analyst.

and zoning of plagioclase and pyroxene occurs in some specimens. One mineralogical difference between the basaltic achondrites and the chondrites is that the plagioclase in the basaltic achondrites is mostly anorthite, instead of the dominant oligoclase of the chondrites. This is easily a result of the high calcium content of the calcium-rich achondrites. The igneous textures of many individual fragments in the eucrites together with the very old ages (4.5 $\times$ 10^9 yr), which are based on the time required to evolve the observed Pb^{207}-Pb^{206} composition from that of Canyon Diablo troilite as measured by Patterson (36), and other isotopic data have led Duke and Silver (34) to conclude that magmatic processes were initiated very early in Solar System history. However, the possibility that some of these specimens are crystallized impact melts should not be overlooked. The process that caused the brecciation in both the monomict and polymict breccias is speculative, but it probably is the shock effects accompanying the impacts of meteorites into the regolith of some unknown planetary or asteroidal surface. Petrographic evidence of shock effects in most of the eucrites is not obvious, but a few of the eucrites contain diaplectic maskelynite (see Impact Metamorphism, page 139), which is a reliable criterion for shock damage. In the Shergotty eucrite, virtually all of the plagioclase has been converted to maskelynite (37), indicating extreme shock (see Impact Metamorphism, Fig. 6).

The very old crystallization ages of the basaltic achondrites, the short cosmic-ray exposure ages of 2.8 to 6.0 × 10⁶ yr (34), unique chemistry, and other considerations led Duke and Silver (34) to suggest that the basaltic achondrites may have originated from the Moon as secondary ejecta from meteoroid impacts. They speculated that the howardites might be fragments of the lunar highlands, and that the eucrites might have originated from the maria. This suggestion was widely accepted by meteorite workers, but the lunar samples returned from the Apollo landings have not confirmed this idea.

Whatever the origin(s) of the achondrites, they constitute a group that, for the most part, are chemically and texturally distinct from the more common classes of stony meteorites. Further work on the achondrites may help to provide information on early Solar System conditions and processes.

IRON METEORITES

INTRODUCTION

Masses of meteoritic iron have been found far back in antiquity. These objects were sometimes kept as matters of curiosity, but often were worked or used in some way as a tool. A number of museums contain early iron implements that apparently have been fashioned from meteoritic iron. Meteoritic iron, of course, is easily distinguished from ordinary terrestrial rocks (Fig. 22); hence, many irons are finds. In fact, more than 90 percent of the approximately 500 irons known are finds. A recent complication in the recognition of irons by untrained observers is the widespread iron and other metallic slag that results from numerous industrial processes, especially from the steel industry.

The largest single meteoritic masses are among the irons, which are also the second most abundant meteorite type. A single mass of weathered meteoritic nickel-iron in Southwest Africa, the Hoba iron, is estimated to weigh approximately 60 tons, and may have weighed 100 tons (39) at the time it fell. There are many irons weighing more than one ton that are preserved in museums and other collections. Several large irons are in the collection of the American Museum of Natural History. These include three of the Cape York, Greenland irons, the largest of which weighs 31 metric tons. The School of Mines in Mexico City has a number of large iron meteorites that have been found in Mexico. Some of the large irons are lying at or near their original discovery sites because they are in remote locations and are very difficult to move. It also appears that some of the very large terrestrial meteorite impact structures have resulted from the capture of large iron bodies, but no large impact crater is *known* to have been caused by the impact of a large stony mass.

Figure 22   Exterior surfaces of two weathered irons. *(a)* A piece of the Henbury, Australia meteorite showing the most common type of surface seen on iron meteorites. Scale is in inches. *(b)* A fragment of the Carthage octahedrite in which the weathering has brought out the octahedral structure. Centimeter scale. (Both photographs courtesy of Arizona State University.)

Table XII   Major Element Chemical Composition of some Iron Meteorites[a]

	Hexahedrite	Octahedrite	Ataxite
	Coya Norte	Canyon Diablo	Tlacotepec
Ni	5.65	8.19	16.56
Co	0.40	0.44	0.69
P	0.29	0.34	0.04
C	0.004	0.026	0.004
S	0.032	0.009	0.004
Cu	0.014	0.028	0.001

[a]Analyses from Moore et al. (40). Virtually all of the remainder is iron. All values in weight percent.

Iron meteorites are composed chiefly of two minerals, kamacite (body centered cubic alpha iron, also called ferrite) and taenite (face centered cubic gamma iron, also called austenite). However, minor amounts of other minerals such as graphite, schreibersite, daubreelite, cohenite, chromite, and troilite generally are present. Thus, besides Fe and Ni, the elements Co, S, P, Cu, Cr, and C usually are reported in analyses of meteoritic iron (40), but will rarely total more than 2 weight percent (Table XII). Troilite, graphite, and schreibersite commonly form large distinct inclusions (Fig. 23) in the nickel-iron, and the true abundance of S, P, and C in some meteorites probably has been reported as a lower value than the true value because some workers have avoided inclusions in sampling irons for analyses. A few iron meteorites contain large silicate inclusions. These inclusions may contain unusual minerals such as potassium-rich feldspar (41), and a few inclusions have been recognized that contain chondrules (42).

CLASSIFICATION

The systematic study of iron meteorites did not evolve as rapidly as the work with stony meteorites. Much of the earlier work is entirely descriptive and is summarized, along with excellent photographs and systematic metallography[6] of iron meteorites, in Perry (43). A brief summary of iron meteorites and their properties is included in Mason (4).

The general classification of iron meteorites is mostly a function of their nickel contents. Although most iron meteorites have a fairly limited range of nickel content, the total range is rather large (Fig. 24). Irons with nickel contents in the range from about 6 to 14 percent display a prominent pattern of kamacite and taenite lamellae called Widmanstätten structure (Fig. 25). The

[6]Both the terms metallography and petrography are used for the study of the mineralogy and petrography of iron meteorites in reflected light.

Figure 23 Polished and etched slab of the Odessa medium octahedrite showing the Widmanstätten structure, Neumann bands, and inclusions. The inclusions are mostly graphite (dark gray to black) and troilite (medium gray). Schreibersite occurs adjacent to the inclusions, and is especially abundant around the inclusion at lower right. Length of field of view is approximately 7 cm.

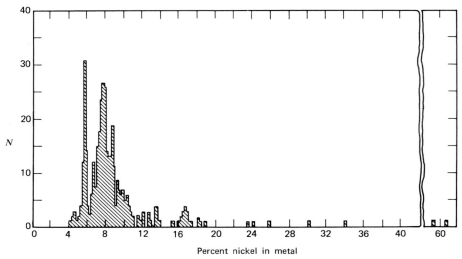

Figure 24 Histogram illustrating the number of iron meteorites *(N)* with various weight percents of nickel. (After Yavnel, 1958; Mason 4.)

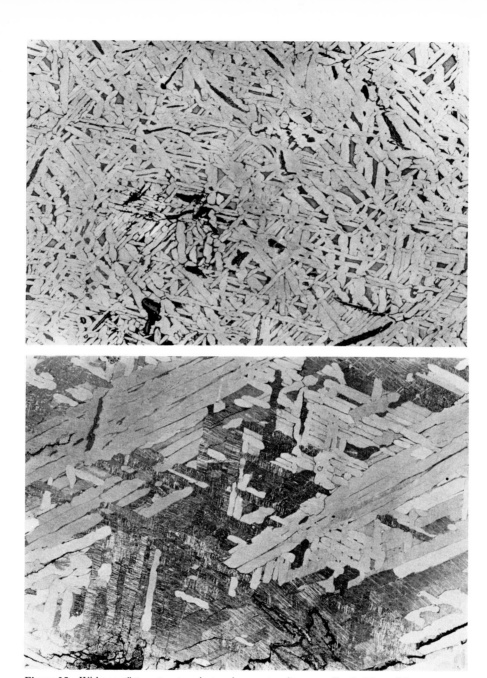

Figure 25 Widmanstätten structure in two iron meteorites: *top,* Turtle River, Minnesota; *bottom,* Xiquipilco, Mexico, the widths of the individual kamacite lamellae correlate with the nickel content of the meteorite: wide lamellae for lesser nickel contents; narrow lamellae for greater nickel contents. Both of these irons are classified as medium octahedrites. Widths of fields of view are approximately 7 cm.

45

lamellae of nickel iron phases are parallel to the faces of an octahedron {*111*}, and the pattern that is obtained is a function of the orientation of the cut surface (44). The Widmanstätten structure cannot be observed unless the surface has been polished and lightly etched with acid, commonly a very dilute solution of nitric acid. This structure was first observed in an iron meteorite by Count Alois de Widmanstätten, of Vienna, in 1808. The presence of a Widmanstätten figure is a good criterion for the recognition of most iron meteorites. Although this type of structure is only rarely achieved in terrestrial alloys, more than 80 percent of the iron meteorites display it. Iron meteorites with this octahedral arrangement of kamacite and taenite lamellae appropriately are termed "octahedrites." There is a good correlation of nickel content with width of kamacite lamellae, wide lamellae for low nickel content and fine, narrow lamellae for high nickel content. Iron meteorites with less than about 6 percent nickel do not have Widmanstätten structure, and are mainly large single crystals of kamacite, although some consist of more than one crystal. These meteorites tend to have the cubic (hexahedral) cleavage of the body centered kamacite and are thus called "hexahedrites." There is a gradational transition from the hexahedrites to the octahedrites, and some specimens are on the border and could be classified as either coarsest octahedrites or granular hexahedrites. Even within the octahedrites the width of kamacite lamellae can be correlated with bulk nickel content. If the nickel content is more than about 14 percent, the fine octahedrite Widmanstätten structure disappears, and only a fine ataxitic intergrowth of taenite and kamacite is visible.[7] However, if the nickel content is very great, approximately 25 to 65 percent, the ataxite is almost certain to be mostly taenite with only small inclusions of kamacite and a few other minerals. The experimental phase diagram for iron-nickel alloys (Fig. 26) readily explains the major relations in the irons, at least, in a qualitative sense (45).

Another means of classifying iron meteorites and of attempting to understand their genetic relations is based on the contents of two very minor elements in irons — gallium, and germanium (46). By using gallium and germanium concentrations, Wasson has attempted to identify genetic relations between different irons and to establish the number of parent bodies (Fig. 27).

COOLING RATES

A considerable amount of research on octahedrites has been directed toward deducing the cooling rates of these bodies from the small-scale distribution of nickel in the mineral phases. With the development of the electron microprobe, it was readily determined that the distribution of nickel between

---

[7]The fine intergrowth of taenite and kamacite was called "plessite" in the older literature when it was thought to be a distinct mineral phase. However, this nomenclature is still in use as a convenient term for this fine intergrowth.

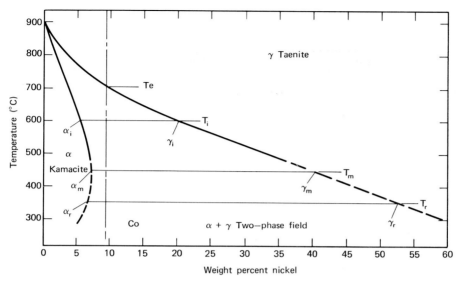

Figure 26 The iron-nickel phase diagram showing the stability fields of kamacite (α-iron), taenite (γ-iron), and the two-phase (α + γ) region. The vertical broken line represents a cooling path for a bulk composition of 9 weight percent nickel. The various isothermal lines indicate the compositions of the coexisting phases for equilibrium conditions. Undercooling will affect the proportions and compositions of phases such that they will not conform to the equilibrium "lever rule." (Courtesy of J. I. Goldstein.)

kamacite and taenite, and even within individual crystals of each of these phases, is inhomogeneous. The taenite and plessite have greater contents of Ni near the margins of the crystals, and kamacite is depleted in Ni at the margins of crystals that adjoin taenite or plessite (Fig. 28). If the following data are known; (a) bulk nickel content of the meteorite, (b) nickel diffusion coefficients in kamacite and taenite as a function of temperature, (c) the size of the kamacite and taenite crystals in question, and (d) distribution of nickel across the crystals, then it is possible to calculate the nucleation temperature of the kamacite and the cooling history of the meteorite. The slower the meteorite cools, the greater the opportunity for Ni diffusion to the central portion of taenite crystals; hence, the greater should be the final nickel content at the centers of taenite crystals (47). The cooling rates for iron meteorites that are obtained by this method indicate that the octahedrites cooled 1°C to 100°C per million years (between about 600°C and 400°C), and that the parent bodies of the octahedrites must have been rather small, with radii of about 50 to 200 km. If these calculations, and necessary assumptions, are correct, it makes some asteroids likely candidates for the sources of certain iron meteorites. However, there are a few uncertainties in this

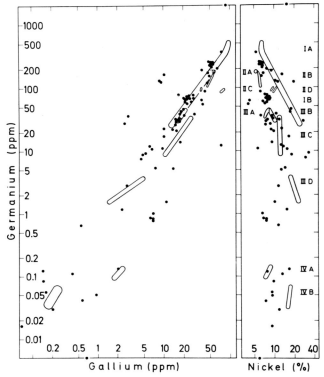

Figure 27 Groupings of gallium-germanium and nickel-germanium concentrations in iron meteorites. The enclosed areas represent the major groups (approximately 80 percent of the analyzed irons) and the dots are individual meteorites that have "anomalous" concentrations. Analyses of more than 420 iron meteorites are represented on the plot. (Courtesy of John Wasson.)

method. One of them involves the diffusion of nickel in kamacite and taenite with minor amounts of nitrogen, phosphorus, and carbon present. The presence of minor amounts of these elements in metallic systems, as are present in the iron meteorites, may significantly affect the diffusion coefficients of nickel, and more experimental work is needed. Also, we assume that the total body from which the octahedrites were derived was nickel-iron. Actually, we do not know this, and the body could have had an outer layer or crust of silicate-rich material or some entirely different structure. An observed fall of an octahedrite with sufficient photographic coverage such that the pre-infall orbit could be calculated would be very useful in determining whether or not some of the iron meteorites originate from asteroids.

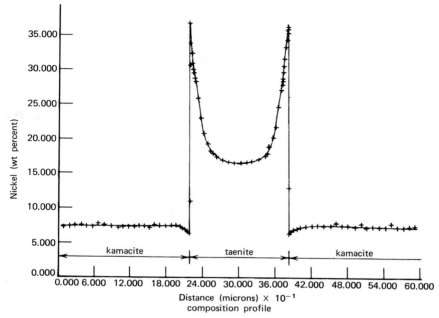

Figure 28   Nickel concentrations across kamacite-taenite-kamacite in the Grant meteorite. Notice the depletion of nickel at the boundary between the kamacite and taenite (Agrell effect) and the "M-shaped" distribution of nickel concentrations across the taenite crystal. (Courtesy of J. I. Goldstein.)

SHOCK EFFECTS

Neumann bands are thin lamellae of mechanically twinned nickel-iron, and they are present in all hexahedrites and in the kamacite lamellae of most octahedrites. The twin plane is {112} in the body centered kamacite, and Neumann bands do not commonly occur in face centered taenite. These twin lamellae probably are caused by shock related mechanical deformation that took place at temperatures less than 600°C, as has been pointed out by Uhlig (48).

Textural and mineralogical evidence for shock in the Canyon Diablo meteorite and other irons has been provided by several workers (49). The occurrence of diamond in the Canyon Diablo iron led to a controversy as to whether the diamond indicated that the iron had come from the very high pressure interior of a large planetary body (50), or whether the diamonds were produced by the shock wave resulting from the impact of the meteorite (51). Correlation of the occurrence of diamond and other textural criteria for shock in individual specimens and crystallographic investigation of the diamond,

Figure 29   Mechanically deformed, shrapnel-like fragment of the Henbury, Australia meteorite. Such highly deformed fragments are found associated with most impact craters caused by iron meteorites. The internal crystal structure of these fragments is greatly distorted.

which is a hexagonal polymorph that easily is made experimentally by shock, has led to general acceptance of the latter view.

Line broadening of X-ray diffraction film patterns of kamacite has been used as an indicator of impact-induced shock metamorphism, but line broadening may also result from simple mechanical deformation and elastic strain (52). Some iron meteorite fragments definitely have suffered strong mechanical deformation (Fig. 29), and it may not be possible to ascertain whether or not they have been subjected to explosive shock by this technique alone.

STONY-IRON METEORITES

INTRODUCTION

In 1863, Maskelyne designated the stony-irons as a separate class of meteorites, and his designation has been widely accepted. This is a rather ill-defined group, and the different subdivisions have little in common except that they are composed of mixtures of metal and silicates. The stony-irons comprise only 4 percent of the known meteorites (4). There are two groups that contain most of the stony-irons: pallasites (olivine-metal) and mesosiderites (olivine-bronzite-metal). Two other special classifications are recognized by Mason to accommodate unique specimens: Siderophyre, for the Steinback meteorite which is composed of orthopyroxene and minor amounts of tridymite enclosed in a groundmass of nickel-iron; and Lodranite, for the Lodran meteorite which is a friable mass of olivine, ortho-pyroxene, and nickel-iron in approximately equal amounts.

PALLASITES

The first pallasite was recognized as an unusual object by P. S. Pallas in 1772, who had the specimen transported to the Academy of Sciences in St. Petersburg. It later was examined by Chladni (53), who concluded that it must be an extraterrestrial object. At least 43 specimens are now known. Mineralogically, the pallasites are rather simple, composed of olivine crystals and/or fragments in a matrix of nickel-iron. Brezina (7) divided the pallasites into four classes based on the shape of the olivine crystals (Figs. 30 and 31);

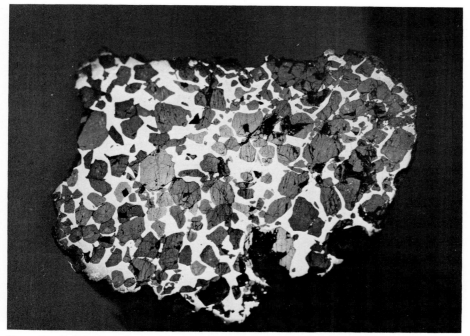

Figure 30   Polished specimen of the Ollague, Bolivia pallasite (USNM 2190). Notice that many of the olivine grains have crystal faces and some are almost euhedral. In other pallasites the olivine may be totally anhedral or fragmental. Specimen length is approximately 12 cm.

Figure 31   Weathered surface of the Salta, Argentina pallasite (USNM 1333), showing prominent olivine grains in raised relief on a background of rusty nickel-iron. Length of field of view is approximately 5 cm.

however, this classification generally has been discarded. In 1916, Prior (54) showed that the pallasites obeyed his rules that he had established in the chondrites for iron and nickel contents of mineral phases. Lovering et al. (55) reported on the chemistry of pallasites, described the structure of the metal, and summarized and reviewed the previous literature. Widmanstätten figures are developed in the larger patches of metal in many pallasites (Fig. 32). Mason (56) thoroughly reviewed the literature on pallasites and included excellent illustrations of the major textural types.

The nickel-iron in pallasites contains from about 8 to 15 percent Ni, so it is understandable that the larger patches of metal have Widmanstätten structure. Analyses of nickel-iron from a number of pallasites were reported by Lovering et al. (55, Table XIII). The metal generally seems to resemble that of the octahedrites.

The olivine from most of the more than 40 known pallasites has been analyzed recently by Buseck and Goldstein (57). They find that in the vast majority of pallasites the olivine crystals are compositionally homogeneous

Table XIII   Partial Analyses of Nickel-Iron from Eight Pallasites[a]

Meteorite	Ni, Weight Percent	Co, Weight Percent	Cu (ppm)
Admire	12.45	0.50	233
Albin	10.43	0.57	223
Bendock	9.20	0.58	143
Brenham	10.98	0.60	168
Glorieta Mountain	11.79	0.54	217
Imilac	11.32	0.47	190
Newport	10.83	0.58	240
Springwater	13.16	0.60	174

[a]After Lovering et al. (55).

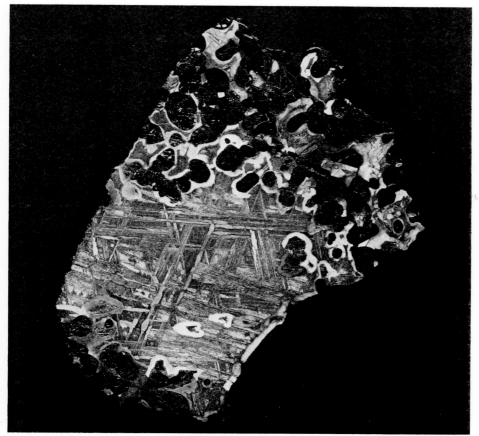

Figure 32   A polished and etched section of the Brenham pallasite showing Widmanstätten structure in a large patch of metal that is free of olivine crystals. Length of specimen is approximately 16 cm. (Courtesy of the Smithsonian Institution.)

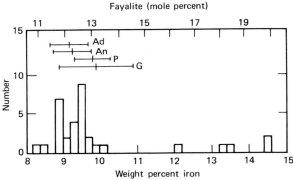

Figure 33   Distribution of olvine compositions in pallasites.
Observe that the total range of composition is small and
apparently bimodal, although the sample is small. Only four
pallasites contained olivine with measurable ranges of
composition (Ad = Admire, An = Anderson, G = Glorieta
Mountain, P = Pojoaque). (From Buseck and Goldstein 57;
Courtesy of The Geological Society of America.)

and appear to be bimodally distributed (Fig. 33) about Fa_{12} and $Fa_{18.5}$. Also,
the olivines from pallasites are depleted in Ni and Ca relative to terrestrial
olivines. The cooling rates of pallasites have been determined by using the
metallic phase of the pallasites in a manner similar to that used for
octahedrites (57). The results show that pallasites have cooled more slowly
than octahedrites. This has been interpreted to mean that pallasites have
come from deeper in the interiors of parent bodies, probably asteroids, than
have octahedrites.

MESOSIDERITES

Only 25 mesosiderites are known, and at least 7 of them are observed falls.
This is a much higher proportion of falls than has been observed for pallasites,
and the present absolute abundance of mesosiderites may be greater than that
of pallasites. The term "mesosiderites" was used by Brezina (7) to include
the stony-irons with both olivine and bronzite included in the metal
groundmass. A separate grouping was assigned to similar meteorites that also
contained plagioclase, but this separation was later abandoned by Prior (3).
Prior concluded that mesosiderites were mixtures of pallasitic and eucritic
materials; more specifically, he suggested that the mixture of pallasitic and
eucritic magmas led to the formation of the mesosiderites. The mesosiderites
were restudied and the previous literature was reviewed by Lovering (58), but
he concluded that Prior's suggestions were essentially correct. Oxygen
isotopes in eucrites, howardites, and the stony portions of mesosiderities

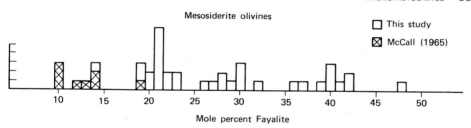

Figure 34   Composition of olivines in mesosiderites. Notice the much wider range of composition as compared with that of the pallasites (Fig. 33). (From Powell 60; reprinted by permission of Pergamon Press Ltd.)

were analyzed by Taylor et al. (59), who demonstrated that these meteorite classes form a chemical group that is distinct from the chondrites. It was also concluded that the eucrites, howardites, and mesosiderites are genetically related on the basis of chemical, mineralogical, and textural evidence by Duke and Silver (34). They further suggested that mesosiderites might be lunar fragments, but stipulated that they could not exclude an asteroidal origin.

The most comprehensive study of the mesosiderites is that of Powell (60), who concludes that although the mesosiderites, eucrites, and howardites appear to be genetically related, the model involving mixing of pallasitic and achondritic material is untenable. This conclusion is based mainly on the nickel content of mesosiderite metal, which is less than 8.9 percent as compared with Ni values that are mostly greater than 9 percent in the pallasites, and differences in olivine compositions between pallasites and mesosiderites (Fig. 34). Powell proposes a complex origin for the mesosiderites that requires magmatic differentiation, brecciation, metal-silicate mixing, burial and insulation, metamorphism, and removal from the parent body. The requirement for burial and insulation is based mostly on the very low cooling rates that have been calculated for the mesosiderites, 0.1°C per million years between 500°C and 350°C, which is the slowest cooling rate found for any group of meteorites. Many of the mesosiderites contain silicate portions that are brecciated (Fig. 35). The brecciation may have resulted from impact metamorphism, but this has not definitely been established.

MICROMETEORITES

This name has been applied to a number of different kinds of particles. The term micrometeorite or micrometeoroid has been applied to small particles ablated from the surface of larger meteoroids, interstellar and interplanetary cosmic dust, primary meteoritic material of small size (from various possible sources), and particles of other origins. Recondensed and

Figure 35 Polished and etched slab of the Mt. Padbury, Australia mesosiderite (USNM collection) showing the brecciated texture of the silicates. Also observe the octahedrite structure of the large metal grains. Width of field of view is approximately 15 cm.

melted material from the impacting body has been collected around several terrestrial impact craters (61), and this type of material has been called "micrometeorite" by some investigators.

Extraterrestrial components of dust recovered from deep sea sediment cores (62) and in polar cap ice (63) have been observed by the presence of cosmogenic isotope anomalies, but for the most part these particles probably are ablation droplets or dust from larger meteorites or both.

Attempts to collect primary micrometeorites directly by airborne collectors and suborbital rocket flights have been rather unsuccessful. Contamination by small particles from numerous sources has been a difficult problem to overcome. High altitude aircraft flights along the trajectory or downwind from the trajectory of observed large meteoroids have collected particles that have been ablated from the larger body (64). Several micrometeoroid impact craters on spacecraft windows and equipment have been observed, and residue deposited around the crater has been analyzed, but the data have been rather poor. The micrometeorite environment (mass-flux distribution) near the Earth has been the subject of many investigations and flight experiments. These data have been summarized by Zook et al. (65), and a composite curve for the mass-flux distribution is included (Fig. 36). Primary micrometeorites

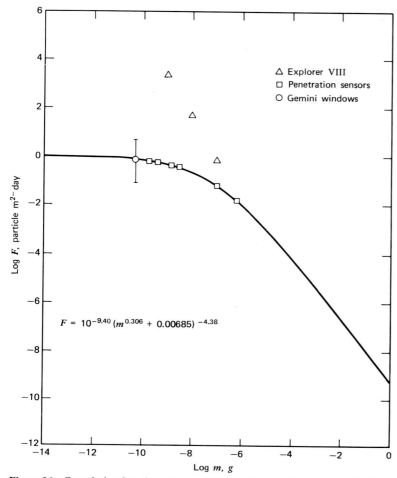

Figure 36   Cumulative flux (log F) versus meteoroid mass (log m), after Zook et al. (65). No acoustic detector data have been used in the construction of the curve. (Courtesy of H. Zook.)

may originate from a number of sources including the asteroids, cometary dust, and as secondary lunar dust ejecta. The collection and analysis of micrometeorites to resolve compositional questions may have to await the flight and recovery of large surface area collectors from long duration Earth-orbital space flights.

ORIGINS OF METEORITES

AGES

The ages of formation of all of the classes of meteorites apparently are very old, approximately 4.6×10^9 yr, the presently accepted age of the Solar System. In fact, our present estimate of the age of the Solar System is based on lead isotope measurements from troilite separated from iron meteorites (36).

Ages of formation may be interpreted from several different types of analyses. Most commonly rubidium-strontium, potassium-argon, argon 39-argon 40, and uranium-thorium-helium ages are used for this measurement. However, there are some inherent limitations in the use of these dating methods to determine the "age of formation." These limitations chiefly derive from the fact that the ages are affected to a greater or lesser degree by thermal metamorphic events. These metamorphic events mostly cause the age to appear younger than the actual age of formation. However, the possibility remains that the measured age may be the actual age of formation (cooling) of the meteorite, a younger age that records the time of an important thermal metamorphic event in the meteorite's history (such as the residual temperature from shock accompanying an impact), or some intermediate age between the two. Also, the meteorites that are clastic rocks, such as the chondrites and the brecciated achondrites, almost certainly contain individual clasts and particles of different ages that predate the formation of the meteorite as a solid body. Thus, there is the potential for additional confusion in the observed ages. However, in spite of all of these possible problems, the rubidium-strontium ages of chondrites and achondrites tend to cluster around 4.6×10^9 yr (66). The potassium-argon ages tend to be somewhat younger, but are consistent with small amounts of radiogenic argon loss (34). A number of workers have identified individual meteorites or subgroups of meteorites that have anomalously young ages or metamorphic events (67), but most of these can be correlated with petrographic evidence of shock in the meteorites or other anomalous features.

Some of the basaltic achondrites may have crystallized somewhat later, perhaps $200\text{-}400 \times 10^6$ yr (68). These observations in meteorites with igneous textures have led some researchers to conclude that magma generation was initiated very early in Solar System history. However, there always is the possibility that some or all of these achondrites with igneous textures may be crystallized impact melts and thus may not represent primary magmas.

The cosmic ray exposure ages of most stony meteorites are only a few tens of millions of years or less, ranging from unmeasurably short (or perhaps well shielded) to as much as 2.2×10^8 yr. Most stony meteorites have cosmic-ray exposure ages of less than 40 million years (69, Fig. 37), which is substantially less than the average cosmic-ray exposure age of iron meteorites (70). There tend to be distinct modes in the distribution of cosmic-ray exposure ages in

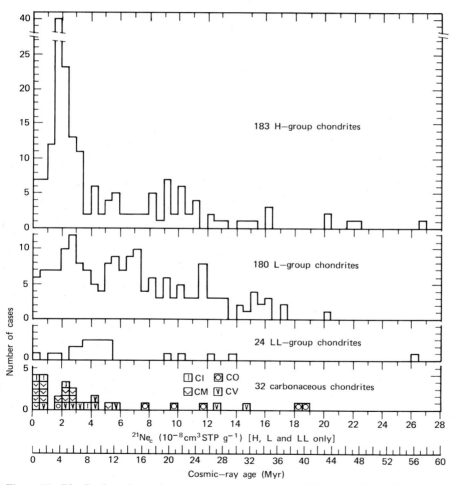

Figure 37   Distribution of cosmic-ray exposure ages in chrondritic meteorites. (After Mazor et al., 69; reprinted by permission of Pergamon Press Ltd.)

stony meteorites (Fig. 37) and also possibly in the iron meteorites. These modes may indicate collisions or shock events that produced Earth-crossing orbits for a large number of fragments, or the time of breakup of larger objects already in Earth-crossing orbits.

ORBITS

Two observations of stony meteorite falls were sufficiently well photographed to allow precise calculations of the pre-entry orbits. One meteorite, Pribram, fell in Czechoslovakia in 1959, and was photographed simultaneously from two different points (71). The Lost City, Oklahoma meteorite fell within the coverage of the Prairie Network (72) on January 3, 1970 (Fig. 1), and the photographic coverage was excellent. In addition, many meteors have been photographed, but no fragments have been recovered. These observations and calculations almost uniformly result in elliptical orbits for the observed bodies. These orbits have low inclinations to the plane of the ecliptic and many have their aphelia between Mars and Jupiter, in or near the "asteroid belt" (73). This is a strong argument for the origin of at least some meteorites as asteroid belt fragments that have been perturbed into Earth-crossing orbits by Jupiter or from ejecta derived from Apollo asteroids and short period comets. LaPaz (74) claimed that meteoroids traveled in hyperbolic orbits and, hence, are extra-Solar System material. However, later more accurate observations of meteors have shown that this is not true, and there are no modern observations of primary hyperbolic meteors.

Arnold (75) published a Monte Carlo simulation method that makes it possible to determine the parent body from which a meteorite comes, if the observations of the fall are sufficiently quantitative. This method has been used and refined by other workers (76), and from the data currently available, various families of asteroids seem to be the most likely possibilities.

Some meteorites should come to us as secondary ejecta from the Moon. The Moon has a low escape velocity (approximately 2.3 km/sec, avg), and primary meteorites that strike the lunar surface must accelerate some ejecta from the cratering event into the velocity spectrum such that material escapes the Moon but not the Earth-Moon System. The Earth will sweep up most of this material as meteorites. This mechanism, as well as chemical arguments, led Urey (77) to suggest that many of the stony meteorites might be this secondary lunar ejecta. O'Keefe (78) and others argued that tektites represented impact fused lunar surface material. We know now, from the direct investigations of lunar samples, that neither of these ideas is correct. However, the mechanism seems to be a perfectly valid one, and somewhere in the collections of meteorites there should be pieces of the Moon, provided that our collections are sufficiently complete. No lunar fragment, as yet, has been identified from presently known meteorites.

Thus we have evidence that some meteorites are very likely to be fragments of asteroids, and also we have plausible mechanisms that should produce meteorites, but no known examples. Another unknown is whether or not any of the meteorites have been derived from cometary sources. Fragments from one of the annual cometary meteor showers have never (knowingly) been recovered. These meteors mostly are small and possibly are fragile bodies, but it is intriguing to speculate that larger stony bodies may exist within comets. If this is true, perhaps there should be cometary fragments somewhere among the meteorite collections — they are not yet positively recognized.

Meteorites are inexpensive samples of portions of our Solar System, and many of them apparently come from beyond the present reach of manned or sophisticated unmanned spacecraft. The current decline of sample returns from space missions probably will result in an even more intensive study of meteorites in coming years than in the past.

REFERENCES AND NOTES

1. Ball, R. S. (1879) Speculations on the source of meteorites: Nature, vol. 19, p. 493-495.

2. Mason, B. (1972) The mineralogy of meteorites: Meteoritics, vol. 7, no. 3, p. 309-326.

3. Prior, G. T. (1920) The classification of meteorites: Mineralog. Mag., vol. 19, p. 51-63.

4. Mason, B. (1962) Meteorites, Wiley, New York and London, 274 p.

5. Rose, G. (1863) Beschreibung and eintheilung der meteoriten auf grund der sammlung im mineralogischen Museum zu Berlin: Physik. Abhandl. Akas. Wiss., Berlin, p. 23-161.

6. Tschermak, G. (1883) Beitrag zur classification der meteoriten: Sitzber. Akad. Wiss. Wein, Math.-naturw. Kl., Abt. I, vol. 88, p. 347-371, and other papers.

7. Brezina, A. (1904) The arrangement of collections of meteorites: Proc. Am. Philos. Soc., vol. 43, p. 211-247.

8. Urey, H. C., and H. Craig (1953) The composition of the stone meteorites and the origin of the meteorites: Geochim. et Cosmochim. Acta, vol. 4, p. 36-82.

9. Van Schmus, W. R., and J. A. Wood (1967) A chemical-petrologic classification for the chondritic meteorites: Geochim. et. Cosmochim. Acta., vol. 31, p. 747-765.

10. Wood, J. A. (1962) Metamorphism in chondrites: Geochim. et Cosmochim. Acta, vol. 26, p. 739-749.

11. King, E. A., Jr., E. Schonfeld, K. A. Richardson, and J. S. Eldridge (1969) Meteorite fall at Pueblito de Allende, Chihuahua, Mexico: Preliminary information: Science, vol. 163, p. 928-929, see also R. S. Clarke, Jr., E. Jarosewich, B. Mason, J. Nelen, M. Gomez, and J. R. Hyde (1970) The Allende, Mexico, meteorite shower: Smithson. Contr. Earth Sci., no. 5, Feb. 1971, 53 p.

12. Folinsbee, R. E., J. A. V. Douglas, and J. A. Maxwell (1967) Revelstoke, a new Type I carbonaceous chondrite: Geochim. et Cosmochim. Acta, vol. 31, p. 1625-1635.

13. Wiik, H. B. (1956) The chemical composition of some stony meteorites: Geochim. et Cosmochim. Acta, vol. 9, p. 279-289. See also Van Schmus, W. R., and J. M. Hayes (1974) Chemical and petrographic correlations among carbonaceous chondrites: Geochim. et Cosmochim. Acta, vol. 38, p. 47-64.

14. Sztrokay, K. L., V. Tolnay and M. Földvari-Vogl (1961) Mineralogical and chemical properties of the carbonaceous meteorite from Kaba, Hungary: Acta Geologica, vol. 7, p. 57-103.

15. Marvin, U. B., J. A. Wood, and J. S. Dickey, Jr. (1970) Ca-Al rich phases in the Allende Meteorite: Earth and Planetary Sci. Letters, vol. 7, p. 346-350.

16. Gray, C. M., D. A. Papanastassiou, and G. J. Wasserburg (1973) The identification of early condensates from the Solar Nebula: Icarus, vol. 20, no. 2, p. 213-239.

17. Oro, J. (1972) Extraterrestrial organic analyses: Space Life Sci., vol. 3, p. 507-550. See also Lawless, J. G. (1973) Amino acids in the Murchison meteorite: Geochim. et Cosmochim. Acta, vol. 37, p. 2207-2212.

18. Mason, B. (1963) Olivine composition in chondrites: Geochim. et Cosmochim. Acta, vol. 27, p. 1011-1023; also, Mason B. (1962) The classification of chondritic meteorites: Am. Mus. Nat. Hist., Novitates, no. 2085, 20 p. See also ref. 2, p. 81.

19. Borgström, L. H. (1903) Die Meteoriten von Hvittis und Marjalahti: Bull. Comm. Geol. Finlande, vol. 14, p. 1-80.

20. Ringwood, A. E. (1961) Silicon in the metal phase of enstatite chondrites and some geochemical implications: Geochim. et Cosmochim. Acta, vol. 25, p. 1-13; see also Ringwood, A. E. (1961) Chemical and genetic relationships among meteorites: Geochim. et Cosmochim. Acta, vol. 24, p. 159-197.

21. Mason, B. (1966) The enstatite chondrites: Geochim. et Cosmochim. Acta, vol. 30, p. 23-39.

22. Grant, R. W. (1967) The occurrence of silica minerals in meteorites: Ph.D. Thesis, Harvard Univ., 150 p.

23. Brezina, A. (1885) Die Meteoritensammlung des k.k. mineralogischen Hofkabinettes in Wien: Jahrb. k. k. Geol. Reichsanstalt, vol. 35, p. 151-276.

24. Roy, S. K. (1957) The problems of the origin and structure of chondrules in stony meteorites: Fieldiana, Geol., vol. 10, p. 383-396. This is an excellent review of the previous ideas on the origin of chondrules.

25. Ringwood, A. E. (1959) On the chemical evolution and densities of the planets: Geochim. et Cosmochim. Acta, vol. 15, p. 257-283. See also Ringwood (1961) Chemical and genetic relationships among meteorites: Geochim. et Cosmochim. Acta, vol. 24, p. 159-197.

26. Mason, B. (1960) The origin of meteorites: Jour. Geophys. Res., vol. 65, p. 2965-2970. See also ref. 18.

27. Whipple, F. L. (1966) Chondrules: Suggestion concerning the origin: Science, vol. 153, p. 54-56. Also, Cameron, A. G. W. (1966) The accumulation of chondritic material: Earth and Planet. Sci. Letters vol. 1, p. 93-96.

28. Nelson, L. S., M. Blander, S. R. Skaggs, and K. Keil (1972) Use of a CO_2 laser to prepare chondrule-like spherules from supercooled molten oxide and silicate droplets: Earth and Planet. Sci. Letters, vol. 41, p. 338-344. See also Englund, E. J. (1969) Experimental studies of the origin and thermal metamorphism of chondrules in chondritic meteorites: Master's Thesis, Univ. of Vermont, 33 p. plus appendices, Feb., 1969.

29. King, E. A., Jr., M. F. Carman, and J. C. Butler (1972) Chondrules in Apollo 14 samples: Implications for the origin of chondritic meteorites: Science, vol. 175, p. 55-56; also, King, E. A., J. C. Butler and M. F. Carman (1972) Chondrules in Apollo 14 samples and size analyses of Apollo 14 and 15 fines: Proc. Third Lunar Sci. Conf., Geochim. et Cosmochim. Acta, Supp. 3, vol. 1, E. A. King, Jr., ed., p. 673-686.

30. Wahl, W. (1907) Die enstatitaugit: Tschermak's mineral., petrogr. Mitt., vol. 26, p. 1-131.

31. Michel, H. (1912) Die Feldspate der Meteoriten: Tschermak's mineral. petrog. Mitt., vol. 31, p. 563-658.

32. Wahl, W. (1952) The brecciated stony meteorites and meteorites containing foreign fragments: Geochim. et Cosmochim. Acta, vol. 2, p. 91-117.

33. Moore, C. B. (1962) The petrochemistry of the achondrites: in Researches in Meteorites, C. B. Moore, ed., Wiley, New York, p. 164-178.

34. Duke, M. B., and L. T. Silver (1967) Petrology of eucrites, howardites and mesosiderites: Geochim. et Cosmochim. Acta, vol. 31, p. 1637-1665.

35. Lonsdale, J. T. (1947) The Peña Blanca Spring Meteorite, Brewster County, Texas: Am. Mineralogist, vol. 32, p. 354-364.

36. Patterson, C. (1956) Age of meteorites and the Earth: Geochim. et Cosmochim. Acta, vol. 10, p. 230-237.

37. Duke, M. B. (1966) The Shergotty Meteorite: Magmatic and shock metamorphism features: Am. Geophys. Union, Transactions, vol. 47, p. 481, *abstract*. See also Milton, D. J., and P. S. DeCarli (1963) Maskelynite: Formation by explosive shock: Science, vol. 140, p. 670-671.

38. Hintenberger, H., H. König, L. Schultz, and H. Wänke (1964) Radiogene, spallogene und primordiale Edelgase in Steinmeteoriten: Zeit. Naturforsch., vol. 19a, p. 327-341. Also, Megrue, G. H. (1966) Rare-gas chronology of calcium-rich achondrites: Jour. Geophys. Res., vol. 71, p. 4021-4027.

39. Gordon, S. G. (1931) The Grootfontein, Southwest Africa, meteoritic iron: Acad. Nat. Sci. Phila., Proc., vol. 83, 251-255.

40. For examples of analyses of irons see Moore, C. B., C. F. Lewis, and D. Nava (1969) Superior analyses of iron meteorites: Meteorite Res., Symp. 1968, Proc., p. 738-748. Also Lewis, C. F. and C. B. Moore (1971) Chemical analyses of thirty-eight iron meteorites: Meteoritics, vol. 6, no. 3, p. 195-205.

41. Wasserburg, G. J., H. G. Sanz, and A. E. Bence (1968) Potassium-feldspar phenocrysts in the surface of Colomera, an iron meteorite: Science, vol. 161, p. 684-687.

42. Olsen, E. and E. Jarosewich (1971) Chondrules: First occurrence in an iron

meteorite: Science, vol. 174, p. 583-585. See also Bunch, T. E., K. Keil and E. Olsen (1969) Mineralogy and petrology of silicate inclusions in iron meteorites: Contrib. Mineral. Petrol., vol. 25, p. 297-340.

43. Perry, S. H. (1944) The metallography of meteoritic iron: U. S. Natl. Mus., Bull., vol. 184, 206 p.

44. This fact has long been recognized, but for a thorough treatment of the subject see Buchwald, V. F. (1968) The austenite-ferrite transformation: Tables relating the Widmanstätten angles in iron meteorites to the plane section: Center for Meteorite Studies, Ariz. State Univ., Publ. No. 7, May, 1968, 13 p. plus tables.

45. Uhlig, H. H. (1954) Contribution of metallurgy to the study of meteorites. Part I — Structure of metallic meteorites, their composition and the effect of pressure: Geochim. et Cosmochim. Acta, vol. 6, p. 282-301. See also Goldstein, J. I. and R. E. Ogilvie (1965) Fe-Ni Phase Diagram: NASA Pub. X-640-65-117, Goddard Space Flight Center, Greenbelt, Md., March, 1965, 10 p.

46. Goldberg, E., A. Uchiyama and H. Brown (1951) The distribution of nickel, cobalt, gallium, palladium and gold in iron meteorites: Geochim. et Cosmochim. Acta, vol. 2, p. 1-25; Wasson, J. T. (1972) Parent body models for the formation of iron meteorites: Proc. 24th, Int. Geol. Cong., Montreal, sec. 15, p. 161-168; Wasson, J. T. (1970) The chemical classification of iron meteorites — IV. Irons with Ge concentrations greater than 190 ppm and other meteorites associated with group 1: Icarus, vol. 12, p. 407-423. See also Wasson, J. T. (1967) The chemical classification of iron meteorites: 1. A study of iron meteorites with low concentrations of gallium and germanium: Geochim. et Cosmochim. Acta, vol. 31, p. 161-180, also several other papers in the intervening period, mostly in Geochim. et Cosmochim. Acta and Jour. Geophys. Res.

47. Goldstein, J. I., and J. M. Short (1967) The iron meteorites, their thermal history and parent bodies: Geochim. et Cosmochim. Acta, vol. 31, p. 1733-1770. See also Wood, J. A. (1964) The cooling rates and parent planets of several iron meteorites: Icarus, vol. 3, p. 429-459. Also Goldstein, J., and R. E. Ogilvie (1965) The growth of the Widmanstätten pattern in metallic meteorites: Geochim. et Cosmochim. Acta, vol. 29, p. 893-920.

48. Uhlig, H. H. (1955) Contribution of metallurgy to the origin of meteorites, Part II – The significance of Neumann bands in meteorites: Geochim. et Cosmochim. Acta, vol. 7, p. 34-42.

49. Heymann, D., M. E. Lipschutz, B. Nielson, and E. Anders (1966) Canyon Diablo meteorite: Metallographic and mass spectrometric study of 56 fragments: Jour. Geophys. Res., vol. 71, p. 619-641. Lipschutz, M. E., and R. R. Jaeger (1966) X-ray diffraction study of minerals from shocked iron meteorites: Science, vol. 152, p. 1055-1057. Jaeger, R. R., and M. E. Lipschutz (1968) X-ray diffraction study of kamacite from iron meteorites: Geochim. et Cosmochim. Acta, vol. 32, p. 773-779.

50. Carter, N. L., and G. C. Kennedy (1964) Origin of diamonds in the Canyon Diablo and Novo Urei meteorites. Jour. Geophys. Res., vol. 69, p. 2403-2421. Also Carter, N. L., and G. C. Kennedy (1966) Origin of diamonds in the Canyon Diablo and Novo Urei Meteorites — a reply: Jour. Geophys. Res., vol. 71, p. 663-672.

51. Anders, E., and M. E. Lipschutz (1966) Critique of paper by N. L. Carter and G. C. Kennedy, "Origin of diamonds in the Canyon Diablo and Novo Urei meteorites": Jour. Geophys. Res., vol. 71, p. 643-661. Also Anders E., and M. E. Lipschutz (1966) Reply: Jour. Geophys. Res., vol. 71, p. 673-674.

52. Comerford, M. F. (1969) Meteorites: An X-ray analysis of deformed kamacite: Jour. Geophys. Res., vol. 74, p. 6675-6678.

53. Chladni, E. F. F. (1794) Uber den Ursprung der von Pallas gefundenen und anderer Eisenmassen, and Uber einiger damit in Vergindung stehende Naturerscheiningen: Riga, J. F. Hardknoch, 63 p.

54. Prior, G. T. (1916) On the genetic relationship and classification of meteorites: Mineralog. Mag., vol. 238, p. 56-60.

55. Lovering, J. F., W. Nichiporuk, A. Chodos, and H. Brown (1957) The distribution of gallium, germanium, cobalt, chromium, and copper in iron and stony-iron meteorites in relation to nickel content and structure: Geochim. et Cosmochim. Acta, vol. 11, p. 263-278.

56. Mason, B. (1963) The pallasites: Am. Mus. Nat. Hist., Novitates, no. 2163, 19 p.

57. Buseck, P. R., and J. I. Goldstein (1969) Olivine compositions and cooling rates of pallasitic meteorites: Geol. Soc. Amer., Bull., vol. 80, p. 2141-2158. See also Buseck, P. R., and J. I. Goldstein (1968) Pallasitic meteorites: Implications regarding the deep structure of asteroids: Science, vol. 159, p. 300-302.

58. Lovering, J. F. (1962) The evolution of the meteorites — evidence for the co-existence of chondritic, achondritic, and iron meteorites in a typical parent meteorite body: in Researches on Meteorites, C. B. Moore, ed., Wiley, New York, p. 179-198.

59. Taylor, H. P., M. B. Duke, L. T. Silver, and S. Epstein (1965) Oxygen isotope studies of stone meteorites: Geochim. et Cosmochim. Acta, vol. 29, p. 498-512.

60. Powell, B. N. (1969) Petrology and chemistry of mesosiderites I. Textures and composition of nickel-iron: Geochim. et Cosmochim. Acta, vol. 33, p. 789-810. Also, Powell, B. N. (1971) Petrology and chemistry of mesosiderites II. Silicate textures and compositions and metal-silicate relationships: Geochim. et Cosmochim. Acta, vol. 35, p. 5-34.

61. Hodge, P. W., and F. W. Wright (1970) Meteoritic spherules in the soil surrounding terrestrial impact craters: Nature, vol. 225, p. 717-718.

62. Merrihue, C. (1964) Rare gas evidence for cosmic dust in modern Pacific red clay: New York Acad. Sci., Annals, vol. 119, p. 351-367. See also Tilles, D. (1965) Anomalous argon isotope ratios in particles from Greenland ice and Pacific Ocean sediments: Am. Geophys. Union, Trans., vol. 46, p. 117, abstract.

63. Fireman, E. L. (1967) Evidence for extraterrestrial particles in polar ice: Meteor Orbits and Dust, Proc. Symp. 1965, G. S. Hawkins, ed., NASA SP-135, SCA vol. II., p. 373-379, also vol. 11, Smithson. Contr. Astrophys.

64. Carr, M. H. (1970) Atmospheric Collection of debris from the Revelstoke and Allende fireballs: Geochim. et Cosmochim. Acta, vol. 34, p. 689-700.

65. Zook, H. A., R. E. Flaherty and D. J. Kessler (1970) Meteoroid impacts on the Gemini windows: Planet. Space Sci., vol. 18, p. 953-964. See also Kerridge, J. F.

(1970) Micrometeorite environment at the Earth's orbit: Nature, vol. 228, p. 616-619.

66. Golopan, K., and G. W. Wetherill (1969) Rubidium-strontium age of amphoterite (LL) chondrites: Jour. Geophys. Res., vol. 74, p. 4349-4358; (1970) Rubidium-strontium studies on enstatite chondrites: whole meteorite and mineral isochrons: Jour. Geophys. Res., vol. 75, p. 3457-3467; (1971) Rubidium-strontium studies on black hypersthene chondrites: Effects of shock and reheating: Jour. Geophys. Res., vol. 76, 8484-8492; also, Kaushal, S. K., and G. W. Wetherill (1969) Rb^{87}-Sr^{87} age of bronzite (H group) chondrites; Jour. Geophys. Res., vol. 74, p. 2717-2726; (1970) Rubidium-87 — Strontium-87 age of carbonaceous chondrites: Jour. Geophys. Res., vol. 75, p. 463-468.

67. For example see Heymann, D. (1967) The origin of hypersthene chondrites: ages and shock effects of black chondrites: Icarus, vol. 6, p. 189-221.

68. Hohenberg, C. M., M. N. Munk, and J. H. Reynolds (1967) Spallation and Fissiogenic Xenon and Krypton from stepwise heating of the Pasamonte Achondrite; the case for extinct plutonium 244 in meteorites; relative ages of chondrites and achondrites: Jour. Geophys. Res., vol. 72, p. 3139-3177.

69. Mazor, E., D. Heymann, and E. Anders (1970) Noble gases in carbonaceous chondrites: Geochim. et Cosmochim. Acta, vol. 34, p. 781-824.

70. Voshage, H. (1967) Bestrahlungsalter und Herkunft der Eisenmeteorite: Zeit, Naturforsch, vol. 22a, p. 477-506.

71. Ceplecha, Z. (1961) Multiple fall of Pribram meteorites photographed: Astron. Inst. Czech., Bull., vol. 12, p. 21-47.

72. McCrosky, R. E. and H. Boeschenstein (1965) The Prairie Meteorite Network: Smith. Astrophys. Obs., Spec. Rept., no. 173, 23 p.

73. McCrosky, R. E., A. Posen, G. Schwartz, and C. Y. Shao (1971) The Lost City Meteorite: Its recovery and a comparison with other fireballs: Smith. Astrophys. Obs., Spec. Rept., no. 336, 41 p.

74. LaPaz, L. (1958) The effects of meteorites upon the Earth (including its inhabitants, atmosphere, and satellites): Adv. Geophys., Vol. 4, p. 217-350.

75. Arnold, J. R. (1965) The origin of meteorites as small bodies. II. The model: Astrophys. Jour., vol. 141, p. 1536-1547.

76. Wetherill, G. W. (1968) Dynamical studies of asteroidal and cometary orbits and their relation to the origin of meteorites: in Origin and Distribution of the Elements, L. H. Ahrens, ed., Pergamon Press, Oxford, p. 423-443. Also, Wetherill, G. W. (1969) Relationships between orbits and sources of chondritic meteorites: in Meteorite Research, P. M. Millman, ed., D. Reidel, Dordrecht, Chap. 48, p. 573-589.

77. Urey, H. C. (1965) Meteorites and the Moon: Science, vol. 147, p. 1262-1265. Also, Urey, H. C. (1959) Primary and secondary objects: Jour. Geophys. Res., vol. 64, p. 1721-1737.

78. O'Keefe, J. A. (1960) The origin of tektites: Tech. Note D-490, NASA, 26 p. and later papers.

SUGGESTED READING AND GENERAL REFERENCES

Mason, B. (1962) Meteorites: Wiley, New York and London, 274 p. A comprehensive review of meteorites.

Wood, J. (1968) Meteorites and the Origin of Planets: McGraw-Hill, Earth and Planetary Sci. Ser., 117 p.

Wasson, J. T. (1974) Meteorites; Classification and Properties: vol. 10, Rocks and Minerals, Springer-Verlag, New York, Heidelberg, Berlin, 316 p. A modern, rather chemically oriented review of meteorites that includes excellent lists of classified meteorites.

Hey, M. H. (1966) Catalogue of Meteorites, third revised and enlarged edition, British Museum, Pub. No. 464, 637 p.

Tschermak, G. (1885) The Microscopic Properties of Meteorites: *translation* and reproduction of original German, John A. Wood, and E. Mathilde Wood, Smithsonian Contr. to Astrophys., Smithsonian Astrophys. Obs., vol. 4, no. 6, 239 p. Excellent photomicrographs and descriptions of stony meteorites.

To anyone who has worked with them, tektites are probably the most frustrating stones ever found on earth.   Henry Faul, 1966

2. Tektites

INTRODUCTION

The origin of tektites has been one of the most vigorously debated scientific subjects of the past 30 or more years. F. E. Suess first brought tektites to international scientific attention about 1900, and first named them tektites after the Greek "τηχτος," meaning melted or molten (1). Suess concluded that tektites were a glassy variety of meteorite, after studies of the Czechoslovakian tektites (Fig. 1) and a thorough review of the previous literature. In 1933, L. J. Spencer concluded that tektites were impact glass formed by the fusion of rocks at the Earth's surface from the heat generated by impacting meteorites (2). This conclusion was based on observed similarities of tektites with impact-produced glass at the Henbury, Australia and Wabar, Saudi Arabia meteorite craters. Barnes (3) brought tektites to attention in the recent Western Hemisphere literature with his report of the first North American tektites (Fig. 2) from Grimes County, Texas, and his comprehensive review of the previous literature. In his 1939 article, Barnes concluded that tektites are a variety of fulgurite (lightning fusions), but stated that Spencer's previous hypothesis could not be easily dismissed (4). Barnes later became a strong advocate of the terrestrial impact origin of tektites.

OCCURRENCES, CHEMISTRY, AND PHYSICAL PROPERTIES

Tektites currently are known from four major occurrences. These can be grouped as shown in Table I. Tektites are high silica glass (Table II) with rare

69

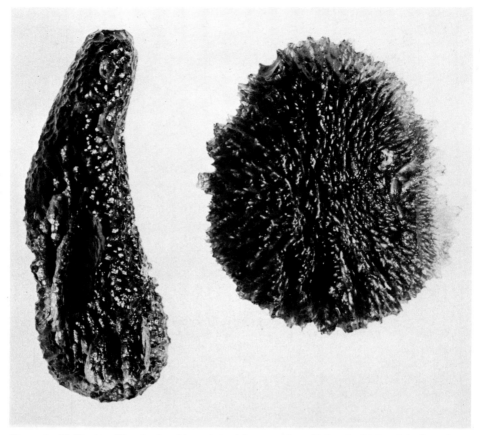

Figure 1 Moldavites (Czechoslovakian tektites) from southern Bohemia. Right: disk-shaped, transparent green specimen; left: dark green, transparent, bent drop-shaped tektite. Fine surface sculpture is caused by etching in soil acids and groundwater. The drop-shaped specimen is approximately 6 cm long.

mineral inclusions. Their shapes are commonly fluidal (Fig. 3) or are those of abraded pebbles that have been stream worn together with other detritus. Tektites range in size from less than 0.01 mm in diameter (microtektites) to large masses weighing more than 10 kg. Some of the Australian specimens, called "buttons," have extremely well preserved aerodynamically sculptured shapes (Fig. 4) that are interpreted to be the result of two periods of melting: (a) the initial fusion or melting that formed the glass from preexisting rock, and (b) a remelted layer with flow structure and morphological detail, such as

Figure 2   Large bediastie (Texas tektite) from near Bedias, Texas. Approximate diameter of the specimen is 5.5 cm. The glass is dark brown, nearly opaque in thick pieces.

ring waves on the leading surface, resulting from ablation during hypervelocity passage through the Earth's atmosphere. These interpretations and supporting experimental work that reproduces closely the forms of the "buttons" have been developed extensively by Chapman and his co-workers (5).

ORIGIN

Chapman concluded, on the basis of aerodynamic arguments, that tektites could originate *only* from the Moon. He further refined the argument based on the apparent distribution of tektite subvarieties in Australia to conclude that some Australian tektites could have originated only as ejecta from the lunar

Table I   The Four Major Tektite Groups

Name[a]	Localities	K-Ar Age	Approximate Number of Specimens[b]
Australasian tektites	Australia, Philippines, Billiton, Indochina, Thailand, Sumatra, and other southeast Asian sites	0.7 ± 0.1 my.	$n \times 10^5$
Ivory Coast tektites	Ivory Coast	1.3 ± 0.2 my.	$n \times 10^2$
Moldavites	Czechoslovakia (Bohemia, Moravia)	15 ± 0.5 my.	$n \times 10^4$
North American tektites	United States (Texas, Georgia, and Martha's Vineyard)	34 ± 1 my.	$n \times 10^4$

[a]In addition to the major group names, many groups of tektites from more local areas have been named, for example, bediasites (Texas), indochinites, billitonites, philippinites, etc.
[b]Number of specimens preserved in collections only, the actual number of tektites that occur in the different groups is undoubtedly much larger.

Table II   Examples of the Major Element Compositions of Tektites in Weight Percent

Oxide	Australites[a]	Bediasites[b]	Moldavites[c]	Ivory Coast[d]
SiO_2	73.45	76.37	80.07	71.05
TiO_2	0.70	0.76	0.80	0.70
Al_2O_3	11.53	13.78	10.56	14.60
MgO	2.05	0.63	1.46	3.29
CaO	3.50	0.65	1.87	1.67
Na_2O	1.28	1.54	0.51	1.71
K_2O	2.28	2.08	2.95	1.53
FeO	4.54	3.81	2.29	5.51
P_2O_5	—	0.04	—	—
Fe_2O_3	—	0.19	0.15	0.18
MnO	—	0.04	0.11	0.08

[a]Average of 17 Australite analyses from Taylor, S. R. (1962) The Chemical Composition of Australites: Geochim. et Cosmochim. Acta, vol. 26, p. 685-722.
[b]Average of 21 Bediasites analyzed by F. Cuttitta and M. K. Carron from Chao, E. C. T. (1963) The petrographic and chemical characteristics of tektites: in Tektites, J. O'Keefe, ed., Univ. of Chicago Press, Chap. 3, p. 51-94.
[c]Average of 8 analyses from Chao, E.C.T. (1963) Ibid.
[d]Average of 3 analyses from Barnes, V. E. (3).

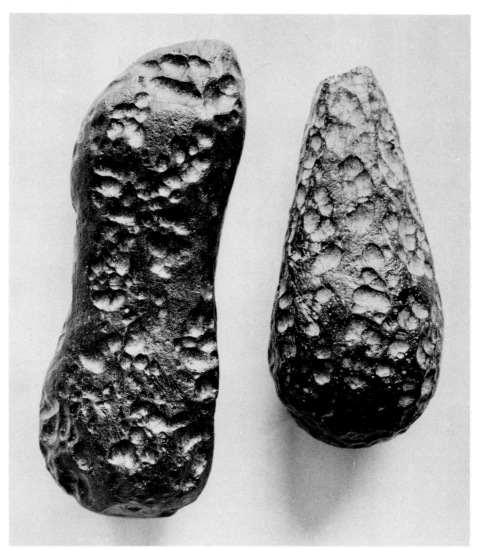

Figure 3 Indochinites showing the dumbbell and tear-drop fluidal shapes. These tektites are very dark brown glass, virtually opaque. The dumbbell is approximately 10 cm long.

(a)

Figure 4 Australites showing the "button" shape that results from aerodynamic sculpturing and ablation during hypervelocity passage through the atmosphere. *(a)* anterior surface, *(b)* posterior surface, *(c)* side view of 25 mm diameter specimen from approximately 13 km north of Princetown, Victoria.

(b)

(c)

74

(d)

Figure 4 (Cont'd)   *(d)* anterior
surface, *(e)* posterior surface, *(f)*
side view of a 26.1 mm diameter
tektite from 10 km east of Port
Campbell, Victoria. (Courtesy of
George Baker.) Color is medium
brown in strong transmitted light,
black in reflected light.

(e)

(f)

75

crater Tycho. O'Keefe vigorously supported the hypothesis of the lunar origin of tektites on the basis of theoretical and chemical arguments (6).

Several observations have lead to the generally accepted conclusion that tektites are a variety of impact glass. Walter identified coesite in southeast Asian tektites (7). Coesite is produced at the Earth's surface only at impact crater sites, and this observation links tektites to high pressure, probably impact events. Baddeleyite, a high temperature phase of ZrO_2, was identified in North American tektites (8). Baddeleyite is known to be one of the impact metamorphism products of zircon ($ZrSiO_4$), and this occurrence together with the previously identified lechatelierite (3, Fig. 5) is evidence of a very high temperature history for tektites. Chao's discovery of iron-nickel spherules in Philippine tektites (9) as well as Spencer's previous report of metal particles in tektites, linked tektites to meteoritic material. All of these observations taken together with the general petrography of tektites are compelling evidence for the origin of tektites as impact glass (10).

Chemically, tektites are known to be very similar to crustal terrestrial rocks such as granite and terriginous sediments except for minor differences. The iron in tektites is almost entirely reduced to ferrous iron, and tektites are extremely dry, with water contents approximately 1×10^3 less than analyzed terrestrial rocks and glass with generally similar compositions. Numerous similarities in major and trace element composition as well as similarities in isotopic chemistry between tektites and terrestrial rocks have been well documented (11). However, prior to the summer of 1969, virtually nothing was known of the chemistry of lunar rocks, and no lunar samples were available for detailed chemical analysis and comparison with tektites.

During the height of the debate concerning the origin of tektites, two large impact craters were recognized that are close to two of the major tektite occurrences. The Ries Crater, a 24 km diameter feature in southern Germany, was identified as an impact crater (page 112). The Ries Crater is only about 300 km from the nearest moldavites in southern Bohemia. Also, the Bosumtwi Crater (sometimes called Ashanti), an 11 km diameter, circular, lake-filled feature in Ghana at a similar distance from the Ivory Coast tektites, was recognized as a large impact site. K-Ar dating of impact glass from the two craters (12) indicates that the craters have the same age of formation as the nearby tektites. This relation is suggestive that the tektites formed as impact glass at the two craters, but still leaves room for other remote possibilities. The Sr/Rb and strontium isotope composition of the crater glass at Bosuntwi and the Ivory Coast tektites was investigated by Schnetzler et al. (13, Fig. 6), who concluded that "The evidence available at present suggests that the Ivory Coast tektites are most probably the fusion products of meteoritic impact at the Bosumtwi crater site." This article left little doubt in the minds of most members of the scientific community that tektites are, in fact, *terrestrial* impact glass (10). The subsequent direct analyses of the lunar surface by the Surveyor alpha backscattering experiment and the later return

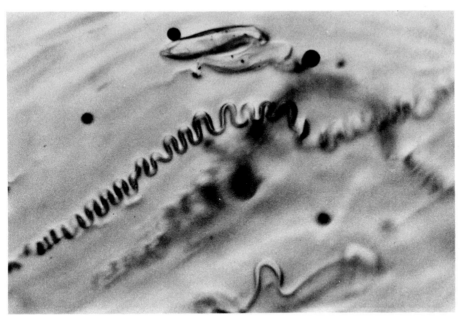

Figure 5 Lechatelierite (silica glass) inclusions in a tektite from Dodge County, Georgia. Top: irregular lechatelierite grain with numerous bubbles at the margin of the grain; bottom: sinuous lechatelierite grain elongated parallel to flow structure. Lechatelierite inclusions are numerous in all tektites. Lengths of fields of view are approximately 2 mm, plane polarized light.

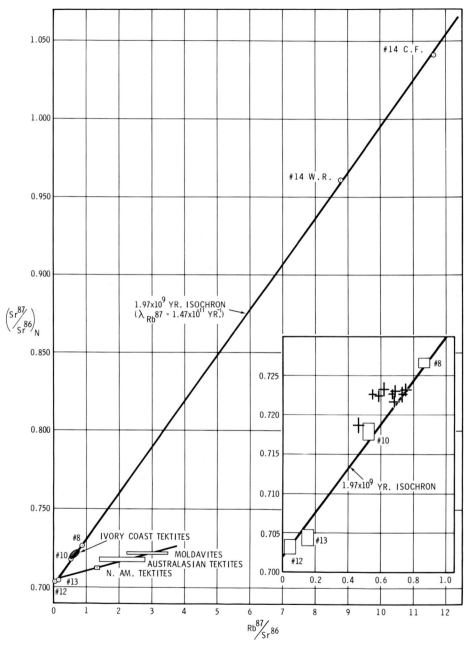

Figure 6   Sr-Rb isochron for the Ivory Coast tektites and Bosumtwi Crater rocks (sample numbers) showing that both materials lie on the same isochron if slight volatilization of Rb relative to Sr occurs during fusion. (Courtesy of C. C. Schnetzler, 13; copyright 1966 by the American Association for the Advancement of Science.)

and analysis of lunar samples from the Apollo and Luna missions demonstrated that lunar rocks are not suitable parent materials for tektites.

Although the place of origin now seems certain, a number of questions are left unanswered or are only vaguely answered concerning the genesis of tektites. Foremost is the mechanism by which tektites can escape from the atmosphere during their initial fusion and ejection from a terrestrial crater so that, at least, some specimens can obtain the "button" or other aerodynamically sculptured shapes. Also, the nature of the impacting bodies that have produced the tektites is not known. The impacting bodies may be cometary (14), asteroidal, or both, and it is particularly frustrating that no large impact crater has yet been recognized that is associated with the most recent and largest tektite strewnfield — the Australasian tektites.

REFERENCES AND NOTES

1. Suess, F. E. (1898) Uber die Herkunft der Moldavite aus dem Weltraume: Akad. Wiss. Wien Sitzungsber. Anzeiger., Nr 24, p. 2; (1900) Die Herkunft der Moldavite und verwandter Gläser: K. geol. Reichsanstalt, Wien Jahrb., Band 50, Heft 2, p. 193; and later papers.

2. Spencer, L. J. (1933) Origin of tektites: Nature, vol. 131, p. 117-118, 876.

3. Barnes, V. E. (1939) North American tektites: Univ. of Texas Publ. 3945, part 2, p. 477-583. The tektites were first identified by H. B. Stenzel, a stratigrapher and paleontologist with the Univ. of Texas, Bureau of Economic Geology.

4. Some years later, Barnes concluded that tektities were the products of large terrestrial impacts, in general agreement with Spencer's hypothesis. See Barnes, V. E. (1961) Tektites: Sci. Am., vol. 205, no. 5, p. 58.

5. Chapman, D. R., and H. K. Larson (1963) The lunar origin of tektites: Tech. Note D-1556, NASA, Feb., 1963, 66 p., and other papers.

6. O'Keefe, J. A. (1960) The origin of tektites: Tech. Note D-490, NASA, 26 p., and later papers.

7. Walter, L. W. (1965) Coesite discovered in tektites: Science, vol. 147, p. 1029-1032.

8. Clarke, R. S., Jr., and J. F. Wosinski (1963) The ZrO_2 inclusion in the Martha's Vineyard tektite: Prog. with Abstrs., Sec. Int'l. Tektite Symp., Pittsburgh, Pa., p. 27, title; and King, E. A., Jr. (1966) Baddeleyite inclusion in a Georgia tektite: Am. Geophys. Union, Trans., vol. 47, no. 1, p. 145, abstract.

9. Chao, E. C. T., I. Adler, E. J. Dwornik, and Janet Littler (1962) Metallic spherules in tektites from Isabela, Philippine Islands: Science, vol. 135, p. 97. It is interesting to note that in spite of close similarities of these spherules with those that they examined in terrestrial impact glass Chao et al. state ". . . we favor the lunar or extraterrestrial origin advocated by previous workers. . ."

10. King, E. A., Jr. (1968) Recent information on the origin of tektites: Shock Metamorphism of Natural Materials, B. M. French and N. M. Short, eds., Mono Book Corp., Baltimore, Md., 1968, p. 626, abstract. See also Faul, Henry (1966) Tektites are terrestrial: Science, vol. 152, p. 1341-1345.

11. Taylor, S. R. (1962) The chemical composition of australites: Geochim. et Cosmochim. Acta, vol. 26, p. 685; King, E. A., Jr. (1962) Possible relation of tuff in the Jackson Group (Eocene) to bediasites: Nature, vol. 196, no. 4854, p. 569; Tilles, David (1964) Stable silicon isotope ratios in tektites: Geochim. et Cosmochim. Acta, vol. 28, p. 1015, and many others.

12. Gentner, W., H. J. Lippolt, and O. Müller (1963) Kalium-Argon-Alter des Bosumtwi-Kraters in Ghana und die chemische Beschaffenheit seiner Gläser: Max-Planck-Institut für Kernphysik, Heidelberg, 1963/V/6, 7 p., also Zeitschr. Naturforschung, vol. 19a, No. 1, p. 150-153, and other papers by the Heidelberg group.

13. Schnetzler, C. C., W. H. Pinson, and P. M. Hurley (1966) Rubidium-Strontium age of the Bosumtwi Crater Area, Ghana, compared with the age of the Ivory Coast tektites: Science, vol. 151, p. 817.

14. Urey, H. C. (1963) Cometary collisions and tektites: Nature, vol. 197, no. 4864, p. 228.

SUGGESTED READING AND GENERAL REFERENCES

Tektites, V. E. Barnes, and M. A. Barnes, eds., Benchmark Papers in Geology; Dowden, Hutchinson and Ross, Stroudsburg, Pa., 1973, 445 p.

Faul, Henry (1966) Tektites are terrestrial: Science, vol. 152, p. 1341.

Barnes, V. E. (1961) Tektites: Sci. Am., vol. 205, no. 5, p. 58.

Tektites, John O'Keefe, Ed., Univ. of Chicago Press, Chicago, 1963, 228 p.

Craters are the most basic topographic form of planetary surfaces in the inner portion of the Solar System.   A. Woronow, 1972

3. Craters

INTRODUCTION

Craters are one of the most fundamental and widely distributed topographic forms on the surfaces of planets and moons. They probably also are abundant on the surfaces of many asteroids and meteoroids, but this observation has not yet been made, except for the moons of Mars, which may be captured asteroids.

On large planetary bodies, approximately the size of the Moon and larger, craters can result from many different processes, including volcanism, impact, subsidence, secondary impacts, and collapse. The two most important of these processes are impact and volcanism. The formation of craters associated with volcanic eruptions, such as calderas, maars, and the summit craters of cinder cones, is well documented in the classic geologic literature (1, 2, 3) and will not be considered here. Volcanic craters have been identified on the Moon and Mars, and illustrations and discussion of them are found in Chapters 6 and 7.

On smaller planetary bodies, asteroids and moons, which are not large enough to have (or to have had) hot, active interiors in which exothermic radioactivity could provide the heating necessary to produce magma, volcanic craters should not be present. However, *all* of the bodies in the Solar System must collide with meteoroids, asteroids and comets, so that we may anticipate that impact craters will be ubiquitous features of stony and metallic surfaces everywhere in the Solar System.

IMPACT CRATERS

CRATERING MECHANICS

Most of the phenomena that accompany the impact of a meteoroid or other hypervelocity projectile are of such short duration that they are difficult to study. Also, the energy of the interaction of the projectile and the target is quite great. Many meteoroids, asteroids, and comets move through the Solar System with high velocities. For example, large meteoroids might encounter the Earth with velocities that range from approximately 11 km/sec to as much as 72 km/sec. From the simple kinetic energy equation, $E = 1/2\,mv^2$, where E = kinetic energy, m = mass, and v = velocity, it is apparent that the impact of even a relatively small hypervelocity mass will produce a very high energy event. For example, the formation of Meteor Crater, Arizona required an estimated 6×10^{22} ergs (4, see also footnote 1 below), and the cratering event that formed the Imbrium Basin on the Moon may have required a total energy in excess of 1×10^{32} ergs. We can make useful analogies of atomic bomb craters at the Atomic Energy Commission Nevada Test Site (Fig. 1) in the estimation of the energy expended to form some impact craters, particularly because some of the Nevada Test Site craters are approximately the same size as some natural impact craters. The nuclear blasts were highly instrumented and documented and the total energy yields of the devices used are well known.

High velocity missile test impacts have been used to study impact craters of the order of feet to tens of feet in diameter (5). Numerous laboratory methods have been used to study the impacts of very small projectiles (6), including two-stage light gas guns, Van de Graaff accelerators, and high velocity rifles. The best of these techniques can achieve projectile velocities of approximately 30 km/sec; however, most methods can obtain only 10 to 15 km/sec with reliability. Even so, these methods have been most useful in understanding the basic phenomena associated with hypervelocity impacts. In natural impacts only the crater usually is observed, whose volume and dimensions can lead to an estimate of the total energy release, but both the mass and velocity of the impacted meteoroid are unknown.

Sequence of Events. The development of many experimental hypervelocity craters has been observed by ultra-high speed photography and instrumental techniques. Gault and co-workers (7) have recognized three stages of natural impact crater formation based on observations of natural craters and of experimental hypervelocity impacts, mostly from the light gas gun accelerated projectiles (Table I).

The *Compression Stage* is initiated when the projectile contacts the surface of the target. Shock waves are generated immediately in both the target and

---

[1]Various estimates range from about 3.3×10^{21} to 2.7×10^{24} ergs; for further discussion see R. B. Baldwin, (1963) The measure of the moon: Univ. of Chicago Press, Chicago, p. 178.

Figure 1 The Sedan Crater in an oblique aerial view. The crater is 390 m in diameter and 98 m deep. It was formed in alluvium at the Atomic Energy Commission Nevada test site on July 6, 1962, by the explosion of a 100-kiloton thermonuclear device buried 194 m below the surface. Notice the hummocky ejecta immediately surrounding the crater as well as the crude rays and groups of secondary craters. (Courtesy of the U.S. Atomic Energy Commission, Nevada Test Site.)

Table I   Sequence of Events in the Formation of an Impact Crater[a]

1. Compression Stage a. Initial contact b. Jetting c. Terminal engulfment	Transfer of projectile kinetic energy to target and establishment of intense shock waves
2. Excavation Stage a. Radial expansion b. Lateral flow c. Ejection	Cratering process as a manifestation of shock waves and attendant rarefaction waves
3. Modification a. Slumping b. Isostatic adjustment c. Erosion and infill	Potential post-cratering alterations *not* attributable to shock waves

[a]After Gault et al. (7).

the projectile. The shock waves in the target are propagated radially away from the point of contact along a nearly hemispherical front, and pressures in the compressed target material commonly are as much as several megabars for a small fraction of a second. Even though the highest pressures obtain only for a very short time, this environment is sufficient to cause drastic changes of state in the target materials. Similarly, a shock wave is generated in the impacting projectile, which continues along its original velocity vector; however, the projectile deforms, begins to break up and, in most hypervelocity impacts, vaporizes at a rapid rate. During the impact of the projectile, the complex interaction of shock waves with the free surface of the projectile generates rarefaction waves behind the shock waves. The first occurrence of these rarefaction waves is marked by ''jetting'' (8), when high velocity particles from the projectile-target interface are hydrodynamically ejected from the impact site. The jetted material includes incandescent and molten fragments of target and projectile, as well as ionized gas. Ejection velocities of material in jets is as much as several times the initial velocity of the impacting projectile (7). The ''impact flash'' described by other workers (9) is associated with the onset of jetting. The Compression Stage is terminated by the reflection of the shock wave from the back side of the projectile. By this time the projectile is highly deformed in parallel with the developing crater in the target (Fig. 2).

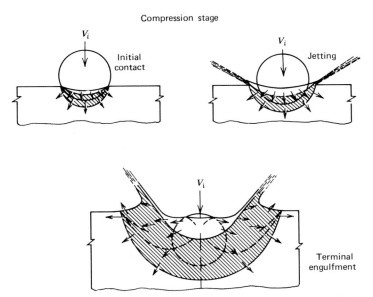

Figure 2   Schematic representations of the impact phenomena associated with the compression stage of crater development, from Gault et al. (7, Courtesy of Mono Book Corp., Baltimore, Md.)

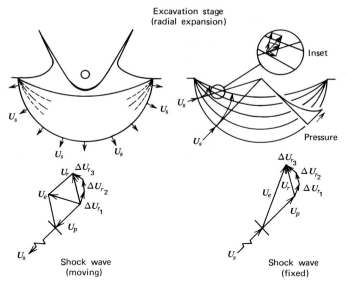

Figure 3 Schematic representation of the radial expansion of
material behind a shock wave produced by a hypervelocity impact,
from Gault et al. (7, Courtesy of Mono Book Corp., Baltimore, Md.)

During the *Excavation Stage,* the greatest mass of material is ejected from
the embryonic crater at relatively low stresses and intermediate to low
velocities. The shock front and compression of material expands radially
away from the impact point, thereby distributing the kinetic energy from the
projectile over a rapidly increasing volume and mass of target. Thus the peak
pressure in the target material decreases very rapidly away from the impact
point (Fig. 3). The shock wave imparts motion to the target material that
initially is parallel to the motion of the wave front, but which quickly becomes
parallel to the developing crater wall. This results in lateral flow of ejecta,
which escapes the crater at or near the rim with little, if any, material
escaping from the central portion of the crater (Figs. 4 and 5). There is a
marked tendency for the early ejecta to be thrown out at very shallow angles,
and for the later ejecta to come out at higher angles, reflecting the
development of crater depth and the steepness of the walls during the later
stages of crater development. The crater assumes its final shape while much
ejecta is still in ballistic trajectories. Some of this ejecta will be in very high
angle trajectories over the crater and will fall back to form a loose debris layer
blanketing the crater. Some of this material may even cascade back down the
crater walls.

The *Modification Stage* includes post-cratering modifications of crater
morphology. Erosion and infilling are the dominant processes of

Excavation stage
Lateral flow and ejection

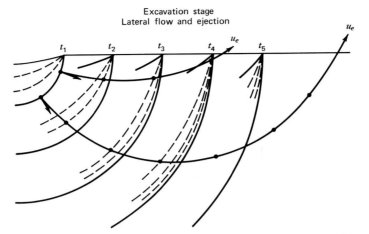

Figure 4   Schematic diagram of the lateral flow and ejection of material from an impact crater, from Gault et al. (7, Courtesy of Mono Book Corp., Baltimore, Md.)

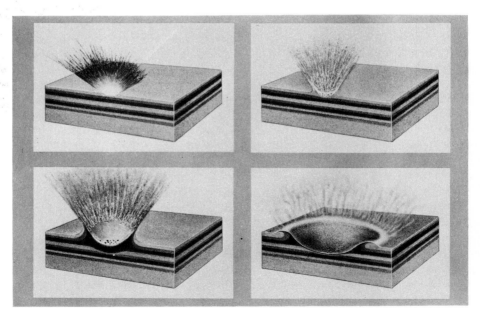

Figure 5   Representation of hypervelocity impact crater development at four different points during the Compression and Evacuation Stages, from Gault et al. (7, Courtesy of Mono Book Corp., Baltimore, Md.)

post-cratering modification on the Earth, Moon, Mars, and probably on most other planetary surfaces. Only very recent craters, such as Meteor Crater, Arizona, show little effect of these processes. Very old craters may have their morphologies drastically affected, for example, the impact craters on the Canadian Shield.

Slumping of the rim and walls commonly is observed on large craters (10), especially those of more than 10 km diameter. Although a fresh impact crater may have a depth to diameter ratio of 1:5 or 1:6, slumping soon after the event may change this ratio considerably, that is, 1:8 or 1:7. The terraced walls so commonly observed on large lunar craters almost certainly are caused by slumping.

Central peaks are tilted and jumbled blocks that apparently have been torn from the crater floor during the terminal stages of crater formation or just after the crater has formed. The formation of central peaks may result from inelastic rebound of the rocks that are subjacent to the crater floor. This should occur immediately after the shock front passes through and highly compresses the rocks beneath the crater. Central peaks rarely are observed in small craters, but seem to be present in virtually all impact craters larger than approximately 10 km diameter. Reflected shock waves from deep layers, and velocity interfaces also have been cited as a possible cause of central peak formation.

Isostatic adjustments to the morphology of large craters is certainly an important process of crater modification. The central portion and deepest part of the crater gradually is uplifted and may cause additional structural complications in the crater. Apparent depth to diameter ratios of 1:20 or even less may result.

Any normal geologic process such as faulting, volcanism, glaciation, and the like can modify an impact crater substantially but, of course, these are not related specifically to the presence of the crater. A *possible* exception is volcanism. Tuffaceous ejecta deposits occur around many terrestrial impact craters and, in fact, have been described as volcanic rock units prior to the recognition of shock effects. Several terrestrial impact craters, for example Manicouagan, are floored with substantial amounts of impact melt, apparently derived from near surface rocks. However, there has been some speculation that large impacts may produce melting of already warm rocks in the interior of a planet. Certainly, large impacts could produce fractures in the surface or crust that could lower the pressure of hot subsurface rocks and act as channelways for the ascent of magma, but whether or not a large impact can raise the temperature of a portion of a planetary crust sufficiently to achieve a significant amount of partial melting is not demonstrated at the present time. This possibility should be carefully investigated. It has been especially fashionable to speculate about the origins of the lava flows that filled the mare basins on the Moon, but it now appears that the time interval between the giant impacts that formed the mare basins and the eruption of the uppermost lavas that filled them is too great for the magmas to be a direct result of the impact itself. However, the ages of the first lava flows to erupt into the newly formed mare basins are not known.

*Rankine-Hugoniot Equations.*Theoretical work with shock waves dates from the middle of the last century with the papers of Stokes (11) and Rankine (12), who established the basic concepts and mathematical models still in use today. Experimental work with shock effects based on the Rankine-Hugoniot equations began much later with the advent of important industrial and military applications (13). An excellent summary of the development of shock wave theory, experimentation, and shock effects on rocks and minerals has been published by Stöffler (14).

If the free surface of a solid body is rapidly accelerated by the impact of a projectile or by the compressional waves accompanying an explosion, a stress wave propagates through the body of the solid with supersonic velocity. The compressibility of solids decreases with increasing pressures, and the stress wave steepens to become a shock wave (or "shock front", 14). The shock wave is immediately followed by a rarefaction wave which travels at a velocity faster than the shock front. This situation derives from the fact that behind the shock front there is a positive particle velocity that increases the sound velocity in the compressed solid. Therefore, the rarefaction wave gradually overtakes the shock wave and the peak pressure is decreased. Thus the shock wave rapidly broadens away from the site of wave generation.

The following derivation of the Rankine-Hugoniot relation is adapted from Stöffler (14), and assumes that the solid material transmitting the shock wave behaves hydrodynamically under compression. It also is assumed that conservation of momentum, mass, and energy can be applied across the shock front.

If we consider a plane shock wave, traveling with velocity U, into a solid material initially at rest and then accelerated to a particle velocity u, the density ρ_0, the pressure P_0, and the internal energy E_0 rise to ρ_1, P_1, and E_1. From the conservation of mass, we obtain

$$\rho_0 U = \rho_1(U - u) \tag{a}$$

From the conservation of momentum we obtain

$$P_1 - P_0 = \rho_0 U u \tag{b}$$

The conservation of energy can be expressed by

$$P_1 u = \tfrac{1}{2} \rho_0 U u^2 + \rho_0 U (E_1 - E_0) \tag{c}$$

Eliminating U and u by the combination of equations (a) and (c):

$$E_1 - E_0 = \tfrac{1}{2}(P_1 + P_0)(V_0 - V_1) \tag{d}$$

where V is the specific volume $1/\rho$. This equation is termed the

Rankine-Hugoniot relation and describes the locus of all shock states (P_1, V_1, E_1) that may obtain from the passage of shock waves in a solid of the initial state P_0, V_0, E_0. For a more detailed and rigorous derivation of the above equations see Duvall and Fowles (15) or McQueen et al. (16).

The compression phase of a shock event is a nonisentropic process, whereas the adiabatic expansion of the target by the rarefaction wave is isentropic. Therefore, the passage of a shock wave through a solid material results in an increase in entropy. If the Rankine-Hugoniot relation is represented in the P-V plane, the resultant line is termed the "Hugoniot curve" (Fig. 6). In shock events, it is apparent that a certain amount of irreversible work is done, and that most of this work is represented by post-shock heat. Thus the melting and vaporization apparent in shocked target materials is easily accounted for from inspection of the Hugoniot curve. These changes in target materials are discussed in the chapter on impact metamorphism.

A much simplified relation has been used (16) to approximate the Hugoniot curves of many materials. This relation involves only one pair of variables; the shock wave velocity (Us) and the particle velocity (Up) and are approximately linear if combined with the weak-wave speed (C_0 = bulk dilational wave speed), and entropy (S) in the following manner:

$$Us = C_0 + SUp \qquad \text{(e)}$$

Materials that depart from linearity in this relation generally have large void spaces or porosity, large elastic waves, or phase transitions.

Much of the greatly heated materials in the target will be ejected from the crater and will cool rapidly. A relation to describe the time required for cooling of filimentous ejecta from an impact event has been developed by Cavarretta et al. (17) as follows:

$$t = \frac{mc_v\alpha}{S\sigma} \left(\frac{1}{T_2^3} - \frac{1}{T_1^3} \right) \qquad \text{(f)}$$

where m is the mass and S is the surface area of the filament, c_v is the heat capacity for the ejected material, α is the opacity, σ is the Stephan constant, and T_1 and T_2 are the initial and final temperatures, respectively.

Energy Partitioning.   Hypervelocity impacts are, for all practical purposes, explosive events. Most of the energy conversion in the impact is expended in the acceleration and excavation of particles from the volume that eventually becomes the crater. However, several other phenomena that require significant fractions of the total energy of the event are observed.

One of the most impressive phenomena accompanying a hypervelocity impact is the destruction of the projectile. In most cases, only small fragments comprising an exceedingly small fraction of the total projectile mass survive as relatively unchanged material. For example, at Meteor Crater, Arizona

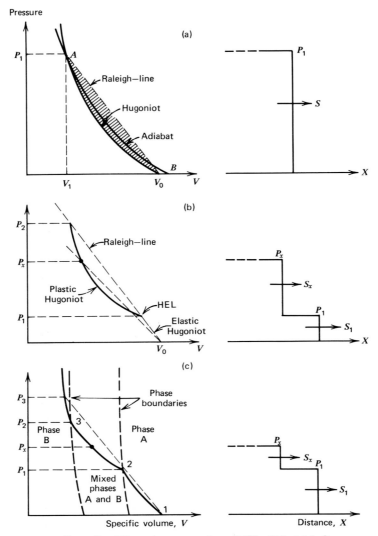

Figure 6   Generalized Hugoniot curves, from Stöffler (14). *(a)* Left:
Hugoniot curve and expansion adiabat for shock state $A(P_1, V_1)$ for
hydrodynamic conditions. Right: shock profile for a single shock with an
amplitude P_1. *(b)* Left: Hugoniot curve characteristic of solids with a
dynamic elastic limit. Right: shock profile for a two wave shock
transition with an elastic precursor of the amplitude P_1 and a following
plastic wave of the amplitude P_1 and a following plastic wave of
amplitude P_x; S_1 = elastic shock, S_x = plastic shock. *(c)* Left: Hugoniot
curve characteristics of solids with an exothermic phase transition.
Right: shock profile for a two wave shock transition with a first shock
wave of amplitude P_1 and a second transformational shock wave of
amplitude P_x; S_1 = plastic shock, S_x = transformational shock.

approximately 100,000 to 300,000 tons[2] of nickel-iron impacted the surface of the Earth, but only a few tens of tons of large iron fragments are known to have been recovered from the crater and nearby plains (18). It is remarkable that *any* large fragments survive. These pieces probably were spalled from the rear surface of the meteoroid by the shock wave at impact, or were stripped from the surface of the meteoroid by atmospheric pressure immediately before the impact. However, many thousands of tons of the impacting body can now be found as nickel-iron and iron oxide spherules, mostly 0.1 to 0.3 mm diameter, scattered about the plains surrounding the crater (19). These tiny spherules probably represent melted and recondensed material from the vaporized body of the projectile. The minimum energy required to vaporize the impacting meteoroid can be calculated precisely, except that the mass is not known. The energy involved in vaporizing the projectile is only a small fraction of the total energy of the event, even if rather generous estimates of the size of the body are used.

There may be a bright flash of light accompanying a hypervelocity impact. This phenomenon has been observed in a number of experimental impacts, and probably results from the ionization of gases at the instant of impact, as well as the ejection of incandescent molten material from the interface between the target and the projectile. The "flash" is of exceedingly short duration, as the name implies.

The amount of heat generated in an impact is fairly large, but arises from several sources. Rapid compression raises the temperature of the target and the projectile, but only a small fraction of the ejecta from an impact crater actually is melted. Most of the ejecta from large terrestrial impacts does not appear to have been heated to any significant degree.

The energy budgets of impacts are not well known and deserve a great deal more experimental work (20). The actual partitioning of energy probably is a function of velocity and mass of the impacting object, composition and physical state of the target and projectile, relative densities and viscosities of the target and projectile, and a number of other parameters.

REFERENCES AND NOTES

1. MacDonald, G. A. (1972) Volcanoes: Prentice-Hall, Englewood Cliffs, N. J., 510 p.
2. Bullard, F. M. (1962) Volcanoes: in history, in theory, in eruption: Univ. of Texas Press, Austin, Tex., 441 p.
3. Jaggar, T. A. (1947) Origin and development of craters: Geol. Soc. Am. Mem. 21, 508 p.
4. Shoemaker, E. M. (1963) Impact mechanics at Meteor Crater, Arizona: *in* The

[2]There is a large range in the estimates of the mass of the impacting body at Meteor Crater because the velocity of the body at the time of impact also is unknown.

Solar System, vol. 4, The moon, meteorites and comets, B. M. Middlehurst and G. P. Kuiper, eds., Univ. Chicago Press, Chicago, p. 301-336.

5. Moore, H. J. (1971) Craters produced by missile impacts: Jour. Geophys. Res., vol. 76, p. 5750-5755.

6. Cable, A. J. (1970) Hypervelocity accelerators: *in* High-Velocity Impact Phenomena, R. Kinslow, ed., Academic Press, New York, p. 1-21.

7. Gault, D. E., W. L. Quaide, and V. R. Oberbeck (1968) Impact cratering mechanics and structures: *in* Shock Metamorphism of Natural Materials, B. M. French and N. M. Short, eds., Mono Book Corp., Baltimore, Md., p. 87-99.

8. Gault, D. E., E. M. Shoemaker, and H. J. Moore (1963) Spray ejected from the lunar surface by meteoroid impact: NASA Pub., TN D-1767, 39 p.

9. Gehring, J. W., A. C. Charters, and R. L. Warnica (1964) Meteoroid impact on the lunar surface: *in* The Lunar Surface Layer, J. W. Salisbury and P. E. Glaser, eds., Academic Press, New York, p. 215-264.

10. Quaide, W. L., D. E. Gault, and R. A. Schmidt (1965) Gravitative effects on lunar impact structures: Annals N. Y. Acad. Sci., vol. 123, p. 563-572.

11. Stokes, G. G. (1848) On a difficulty in the theory of sound: Phil. Mag., Ser. 3, vol. 33, p. 349-356.

12. Rankine, W. J. M. (1870) On the theory of waves of finite disturbance: Trans. Roy. Soc., London, vol. 160, p. 277-288.

13. Schall, R. (1944) cited by W. Döring and H. Schardin (1948) Hydro- and Aerodynamics: Fiat review of German Science, 1939-1946.

14. Stöffler, D. (1972) Deformation and transformation of rock-forming minerals by natural and experimental shock processes: I. Behavior of minerals under shock compression: Fortschr. Mineral., vol. 49, p. 50-113.

15. Duvall, G. E. and C. R. Fowles (1963) Shock waves: *in* High pressure physics and chemistry, R. S. Bradley, ed., Academic Press, New York, vol. 2, p. 209-291.

16. McQueen, R. G., S. P. Marsh, J. W. Taylor, J. N. Fritz, and W. J. Carter (1970) The equation of state of solids from shock wave studies: *in* High Velocity Impact Phenomena, R. Kinslow, ed., Academic Press, New York, p. 293-417.

17. Cavarretta, G., A. Coradini, R. Funiciello, M. Fulchignoni, A. Taddeucci, and R. Trigila (1972) Glassy particles in Apollo 14 soil 14163,88: Peculiarities and genetic considerations: *in* Proc. Third Lunar Sci. Conf., vol. 1, E. A. King, Jr., ed., Geochim. et Cosmochim. Acta, Suppl. 3, MIT Press, Cambridge, Mass., p. 1085-1094.

18. Mason, B. (1962) Meteorites: Wiley, New York, 228 p.

19. Nininger, H. H. (1951) Condensation globules at Meteor Crater: Science, vol. 113, p. 755-756. Also, Baldwin, R. B. (1963) The Measure of the Moon: Univ. of Chicago Press, Chicago, p. 14-16, and Rinehart, J. S. (1957) A soil survey around the Barringer Crater: Sky and Telescope, vol. 16, p. 8. Also, Blau, P. J., H. J. Axon, and J. I. Goldstein (1973) Investigation of the Canyon Diablo metallic spheroids and their relationship to the breakup of the Canyon Diablo Meteorite: Jour. Geophys. Res., vol. 78, p. 363-374.

20. Pond, R. B., and C. M. Glass (1970) Metallurgical observations and energy partitioning: *in* High-Velocity Impact Phenomena, R. Kinslow, ed., Academic Press, New York, p. 419-461.

SUGGESTED READING AND GENERAL REFERENCES

Kinslow, Ray, ed. (1970) High-Velocity Impact Phenomena: Academic Press, New York, 579 p.

Gault, D. E., W. L. Quaide, and V. R. Oberbeck (1968) Impact cratering mechanics and structures: *in* Shock Metamorphism of Natural Materials, B. M. French and N. M. Short, eds., Mono Book Corp., Baltimore, Md., p. 87-99.

Shoemaker, E. M. (1963) Impact mechanics at Meteor Crater, Arizona: *in* The Solar System, vol. 4, The moon, meteorites and comets, B. M. Middlehurst and G. P. Kuiper, eds., Univ. Chicago Press, Chicago, p. 301-336.

The kinetic energy of a large mass of iron travelling at a high velocity, was suddenly transformed into heat, vaporizing a large part of the meteorite and some of the earth's crust, so producing a violent gaseous explosion, which formed the crater. . . .   L. J. Spencer, 1937

4. Terrestrial Impact Craters

INTRODUCTION

The scientific community has been very slow to accept impact as an important and common geological process on the surface of the Earth. As recently as 1953 (1), some scientists have argued against an impact origin for Meteor Crater, Arizona. Bucher (2), Snyder and Gerdemann (3), Currie (4), and others have made eloquent pleas for the origins of specific structures by other processes such as "cryptoexplosions"; however, nothing is known of this kind of process if, in fact, it exists. The opposing arguments, summarizing the faults with the "cryptovolcanic" or "cryptoexplosion" arguments and supporting the impact hypothesis, have been well made by Dietz (5) and French (6). Well before this time, several authors such as L. J. Spencer and R. A. Daly had attributed both small craters and large structures to meteoroid or asteroid impact (e.g., 7, 8, 9). Two investigators, as early as 1938 (8), even had attempted theoretically to deduce the structures in layered rocks that would result from impact of a large meteorite (Fig. 1).

The number of well documented impact craters on the surface of the Earth has risen astonishingly in the past 20 years, and new craters are reported annually. Systematic searches of old shield areas and desert regions by careful inspection of aerial photographs have located dozens of new impact

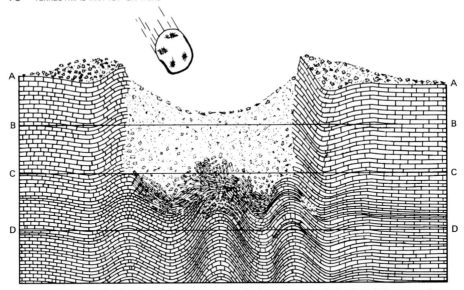

Figure 1   Hypothesized section through a meteorite impact crater in horizontally layered sedimentary rocks. (After Boon and Albritton, 1937; reprinted by permission of Southern Methodist University Press.)

craters that have been confirmed by later surface and sample investigations. Petrographic investigations and reinvestigations of a number of supposedly volcanic explosion features have provided evidence of shock metamorphism and structures unique to impact craters. The number of impact craters currently recognized is variously estimated at between 50 and 120 (e.g., 6, 10, 11). Table I lists most of the currently accepted terrestrial impact craters or impact sites, although definitive criteria of shock metamorphism may not have

Table I   Known and Probable Terrestrial Impact Craters and Structures[a]

Name	*Latitude*	*Longitude*	*Diameter*[b] *(km)*
Aouelloul, Mauritania	20°15'N	012°41'W	250 m
Bosumtwi, Ghana	06°32'N	001°23'W	10.5
Boxhole, N. T., Australia	22°37'S	135°12'E	175 m
Brent, Ontario	46°05'N	078°29'W	4
Campo del Cielo, Argentina	27°28'S	061°30'W	70 m
Carswell, Saskatchewan	58°27'N	109°30'W	30
Charlevoix, Quebec	47°32'N	070°18'W	35
Clearwater Lake East, Quebec	56°05'N	074°07'W	15
Clearwater Lake West, Quebec	56°13'N	074°30'W	30
Crooked Creek, Missouri	37°50'N	091°23'W	5
Dalgaranga, Western Australia	27°45'S	117°05'E	21 m

Table I (Cont'd)   Known and Probable Terrestrial Impact Craters and Structures[a]

Name	Latitude	Longitude	Diameter[b] (km)
Decaturville, Missouri	37°54'N	092°43'W	6
Deep Bay, Saskatchewan	56°24'N	102°59'W	9
Dellen, Sweden	61°50'N	016°45'E	12
Flynn Creek, Tennessee	36°16'N	085°37'W	3.6
Gosses Bluff, Northern Territory, Australia	23°48'S	132°18'E	22
Haviland, Kansas	37°37'N	099°05'W	11 m
Henbury, Northern Territory, Australia	24°34'S	133°10'E	150 m
Holleford, Ontario	44°28'N	076°38'W	2
Kaalijarvi, Estonian SSR	58°24'N	022°40'E	110 m
Kentland, Indiana	40°45'N	087°24'W	6
Köfels, Austria	47°13'N	010°58'E	5
Lac Couture, Quebec	60°08'N	075°18'W	10
Lappajarvi, Finland	63°10'N	023°40'E	10
Liverpool, Northern Territory, Australia	12°24'S	134°03'E	1.6
Manicouagan, Quebec	51°23'N	068°42'W	65
Manson, Iowa	42°35'N	094°31'W	30
Meteor Crater, Arizona	35°02'N	111°01'W	1220 m
Mien Lake, Sweden	56°25'N	014°55'E	5
Middlesboro, Kentucky	36°37'N	083°44'W	7
Mistastin, Labrador	55°53'N	063°18'W	20
Monturaqui, Chile	23°56'S	068°17'W	0.48
New Quebec, Quebec	61°17'N	073°40'W	3.2
Nicholson, Northwest Territories	62°40'N	102°41'W	12.5
Odessa, Texas	31°48'N	102°30'W	168 m
Pilot Lake, Northwest Territories	60°17'N	111°01'W	5
Ries, Germany	48°53'N	010°37'E	24
Rochechouart, France	45°50'N	000°56'E	15
St. Martin, Manitoba	51°47'N	098°33'E	24
Serpent Mound, Ohio	39°02'N	083°25'W	6.4
Sierra Madera, Texas	30°36'N	102°55'W	13
Sikhote Alin, Primorye, Terr. Siberia, USSR.	46°07'N	134°40'W	26.5 m
Siljan, Sweden	61°05'N	015°00'E	45
Steen River, Alberta	59°31'N	117°38'W	25
Steinheim, Germany	48°02'N	010°04'E	3
Strangways, N.T., Australia	15°12'S	133°35'E	16
Sudbury, Ontario	46°36'N	081°11'W	100
Tenoumer, Mauritania	22°55'N	010°24'W	1.8
Vredefort, South Africa	27°00'S	027°30'E	100
Wabar, Saudi Arabia	21°30'N	050°28'E	90 m
Wanapitei, Ontario	46°44'N	080°44'W	8.5
Wells Creek, Tennessee	36°23'N	087°40'W	14
West Hawk Lake, Manitoba	49°46'N	095°11'W	2.7
Wolf Creek, Western Australia	19°18'S	127°47'E	850 m

[a]From Dence (76).   [b]Diameter of largest crater if more than one.

been found at all of the locations included. Only a *few* examples of impact craters, illustrating the typical forms, structures, and deformation, are discussed in the pages that follow.

METEOR CRATER, ARIZONA

This feature is one of the most recent natural hypervelocity impact craters known on the surface of the Earth (Fig. 2), and it is among the most thoroughly studied. The crater is located in northern Arizona, south of Interstate 40 and Route 66, about 22 miles west of Winslow, Arizona. In the older literature, the names Canyon Diablo Crater, Coon Butte Crater, Crater Mound, Coon Mountain Crater, and Barringer Crater also are used. Meteor Crater is approximately 1220 m in diameter and 180 m deep, and it is formed in the flat lying Permian and Triassic rocks of the southern Colorado Plateau. the deformation caused by the cratering event is apparent and easily mappable (13). However, even for this crater, there has been a great deal of discussion concerning its origin.

G. K. Gilbert and Marcus Baker (14) first drew attention to the crater in 1891 with the publication of their preliminary observations of the crater and the associated meteoritic iron. They concluded that a large meteorite had caused the crater. Further attention was directed toward Meteor Crater in 1891 and 1892 by A. E. Foote, the well known Philadelphia mineral dealer and natural historian, who visited the site to view the large concentration of iron meteorites (15). He noted that many of the masses of iron were found surrounding an elevation with a large cavity in the center. Foote searched for volcanic materials but found none and concluded that he could not explain the cause of "this remarkable geological phenomenon." A few years later, Gilbert *changed* his opinion and supported the view that the crater had been caused by a steam explosion of volcanic origin (16). Few new data were added to the discussion for a number of years, until the efforts of D. M. Barringer and his associates (17), during attempts to develop the crater as a commercial enterprise, made a strong case for the origin of the crater by the impact and explosion of a large iron meteorite. Some workers have disputed the impact origin of the crater recently (1), but no doubt remains.

Unusual fracturing and metamorphism in samples of Coconino Sandstone from Meteor Crater were described by Merrill (18), an observation that has been duplicated at many other impact sites (see Impact Metamorphism). Chao and his co-workers first described natural occurrences of coesite and stishovite from Meteor Crater (19). Thus Meteor Crater is not only a fine example of an impact crater, but this crater has played a central role in the development of recent ideas concerning the mechanics of impact and the structures that impact features should display (13).

Meteor Crater is not perfectly circular; in fact, it appears to be rather square in plan view. Shoemaker attributes this outline to two prominent sets

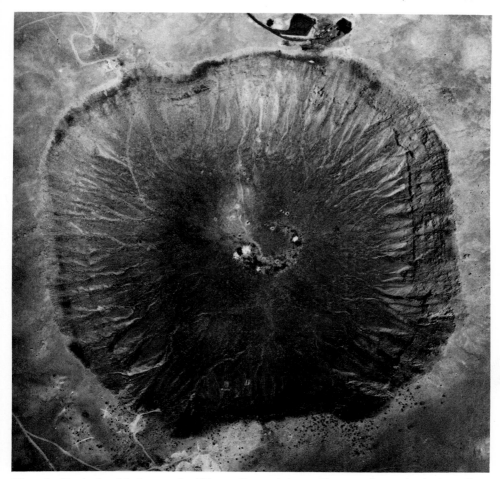

Figure 2   Vertical aerial photograph of Meteor Crater, Arizona. The crater is not circular in outline, but is square with rounded corners because of a preexisting joint system in the sedimentary country rocks. The light spots in the bottom of the crater are dumps from shafts and holes that were dug in search for a buried mass of nickel-iron. The crater is approximately 1220 m in diameter and almost 180 m deep. (Courtesy of Arizona State University.)

of faults and joints in the underlying rocks (13). The crater has a prominent raised rim, 30 to 60 m above the surrounding plain (Fig. 3), and the remnants of an extensive ejecta blanket are well preserved. Some of the ejecta still can be mapped at a distance of one crater diameter from the rim (Fig. 2). One of the critical relations that is observed in the rim is the overturning of beds at some places. Thus small sections of the rim stratigraphy are inverted (Fig. 4) in an "overturned flap." A number of drill holes and a few mine

Figure 3 View of the raised north rim of Meteor Crater, Arizona, taken about 2 km away. The low building on the northeast rim is the tourist center and museum.

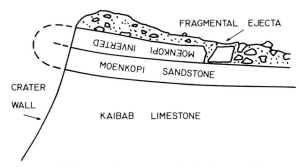

Figure 4 Schematic diagram showing the inverted rim stratigraphy or "overturned flap" as it occurs at Meteor Crater, Arizona. Such structures, as well as low angle thrust faults, rim anticlines, vertical beds, and slightly overturned beds, are common in the rims of well preserved impact craters.

shafts have been driven into the floor and rim of the crater, in a vain search for a large nickel-iron body, such that the basic profile of the crater is fairly well established. However, rather poor records were kept for most of the excavations and drill holes, and much important information cannot now be recovered. A small quarry operation in the ejecta on the south and southwest sides of the crater has exposed a good cross section through the ejecta near the rim.

The ejecta appears to be chaotic at first glance, but it is surprisingly well ordered. Base surge, ballistic ejecta, and fallback deposits can be identified. The ejecta is very poorly sorted, containing much very fine material, and also blocks as much as 30 m in diameter. The inverted stratigraphy of the rocks in which the crater is formed is clearly visible in the distribution of the ejecta. The dark red Moenkopi Formation, which was the uppermost stratigraphic unit at the site at the time of the impact, is the farthest ejecta from the crater rim. Moenkopi debris is overlain nearer the crater by Kaibab Limestone ejecta, which is overlain by Coconino Sandstone ejecta even closer to the crater rim. The ejecta contains many small fragments of the meteorite, but most of the iron is completely oxidized. Although most of the ejecta does not appear shocked or metamorphosed in any way, there are abundant fragments of fused and partially fused rock, sheared rocks, and rock fragments coated with glass.

The age of Meteor Crater is not precisely known, but it obviously is rather young. Quaternary and Recent lake and playa beds partially have filled the floor of the crater, and Pleistocene talus mantles the lowest parts of the crater walls (13). Although some stories about the formation of the crater are told by the local Indians, these tales are thought to be of recent origin. No organic debris has been recovered from beneath the ejecta, which would be excellent material to date by the C^{14} method. Estimates of the age range from about 20,000 to 40,000 years.

It is interesting to note that shatter cones are very rare at Meteor Crater, although the lithologies of some of the underlying rocks apparently are well suited for their formation.

There has been much discussion about the direction from which the meteorite struck the surface (13, 20, 21). Most of this discussion has been based on the symmetries of the crater, ejecta distribution, and distribution of iron fragments. However, the answer is not clear from the available evidence. Estimates of the mass of the impacting nickel-iron body range from about 30,000 to more than 200,000 tons.

Access to Meteor Crater is good and can be obtained for the payment of a small fee. A private museum and concession is situated on the north rim, and trails may be followed to descend onto the crater floor and around the rim. It is to the credit of the Barringer Crater Company that the best preserved impact crater now known is easily available to scientists and students as well as to the general public.

ODESSA CRATERS, TEXAS

In 1921, S. R. McKinney picked up a 2.7 kg piece of iron on the rim of a large shallow depression near Odessa, Texas. The specimen subsequently was identified as an iron meteorite, and the depression as the largest of a group of at least five meteorite craters. The largest crater (Fig. 5) is 170 m in diameter and now has a maximum depth of approximately 5 m (23). The next largest crater is 21 m in diameter, and the remaining craters are considerably smaller.

E. H. Sellards, a geologist with the University of Texas, began a field reconnaissance in the area of the craters in the mid-1920s. However, Sellards was slow to be convinced of the meteoritic origin of the craters, but eventually published material on the craters and the meteorites, including a structural cross section of the main crater (24, 25). The object of much of the work at the craters was simply the recovery of fragments of the meteorite and the building of a display at the Texas Memorial Museum in Austin.

D. M. Barringer, Jr., visited the Odessa Craters and compared the largest one to Meteor Crater, Arizona (26). H. H. Nininger first visited the Odessa Craters in 1932, and in 1933 he searched the area of the craters with a magnetic balance and a magnetic plow, recovering more than 1000 small meteorite fragments.

The meteorite was described by Merrill in 1922, and it was found to be a coarse octahedrite, *very* similar to the specimens from Meteor Crater, Arizona. It was later determined that many chemical and petrographic similarities existed between specimens from the two localities, so much so that Nininger suggested that the two cratering events might be related[1].

[1]Nininger later discarded this idea. Detailed chemical analyses also have documented differences in the two irons.

Figure 5   Aerial photograph of the Odessa Crater taken in 1934 by Otto Roach. The rim to rim diameter of the crater is approximately 168 m. (Courtesy of H. H. Nininger.)

The main crater at Odessa has a rim that ranges from an average of approximately 2 m to more than 3.6 m above the adjacent plain (23). Sellards and Evans (25) report that the original depth of the crater was more than 25 m below the level of the plain. The local rocks are flat-lying Cretaceous limestones and sandstones. However, around the rim of the crater the rocks are tilted up, dipping away from the crater as much as 60 deg. The rocks have been pushed away and lifted up from the central part of the crater forming a ring anticline. In a section shown by Sellards (24), the rim anticline is slightly overturned and thrust faulted at one locality within the present crater rim.

The mass of the impacting iron at the site of the main crater has been estimated to be 315 tons (21). However, the recovered mass from all of the Odessa Craters is much less, perhaps only one ton.

Evans continued to work on the craters intermittently until 1961 (27) and published new maps and surveys; however, very little has been done with the petrography of shocked rocks or impactite. The main Odessa Crater is fairly well exposed and, although Sellards and Evans dug several trenches and made a number of other excavations, the Odessa Craters clearly deserve restudy in light of recent information on shock metamorphism and crater mechanics.

HAVILAND CRATER, KANSAS

A large number of pallasites had been gathered on a farm near Haviland, Kansas between 1885 and 1900. Nininger visited the farm in 1925 with the hope of finding an overlooked specimen of the well known Brenham meteorite. A small "buffalo wallow" attracted Nininger's attention during his visit. The depression had a small raised rim that had been almost obliterated by wheat farming. After several years of negotiating with the owners of the property, Nininger was allowed to excavate the "buffalo wallow" in 1933, in collaboration with the Colorado Museum of Natural History. He recovered a number of large meteorite fragments from the little crater, some weighing as much as 40 kg, together with a large number of oxidized and weathered small pieces (28). The largest specimens were found at a depth of 3.3 m.

This structure is not an explosion crater, as are Meteor Crater and the large Odessa Crater, but resulted from a much lower velocity impact. The mass of meteorite was sufficiently small that it was slowed down by atmospheric friction prior to impact. Such low velocity craters commonly are termed "penetration funnels" or "penetration pits." This particular example was 17 m long and 11 m wide.

A number of craters of this type occur at Campo del Cielo, Argentina, associated with numerous and large masses of nickel-iron (29), although a few of the larger craters may be explosion craters or intermediate types of craters.

The fall of the Sikhote-Alin meteorites on the western side of the Sikhote mountain range in Russia was witnessed by hundreds of persons on February

12, 1947 (30). A large nickel-iron meteorite broke up in the atmosphere, and the fragments produced many small penetration pits and low velocity impact craters ranging up to 26.5 m in diameter.

Numerous other examples of these small meteorite impact craters and penetration pits are cited in the literature (e.g., 31, 21).

THE CANADIAN CRATERS

Many large impact structures and eroded craters have been discovered or recognized in Canada during the past 20 years (Figs. 6, 7), primarily through the search program initiated by the Dominion Observatory, and the work of

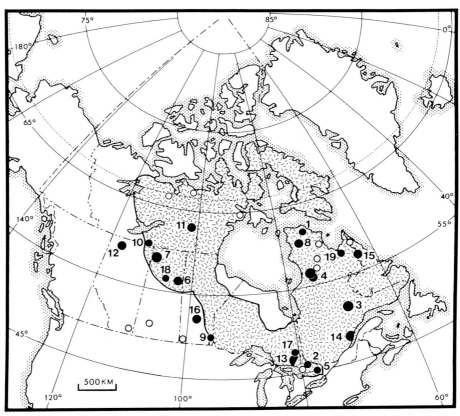

Figure 6   Map of Canada showing the locations of known impact craters (closed circles) and a number of possible impact structures (open circles). See Table II for identifications and approximate ages. Figure 7 shows the relative sizes and approximate map outlines. Notice the high density of craters on the old Canadian Shield (stippled area). (Courtesy of the Earth Physics Branch, Department of Energy, Mines and Resources, Ottawa, Canada.)

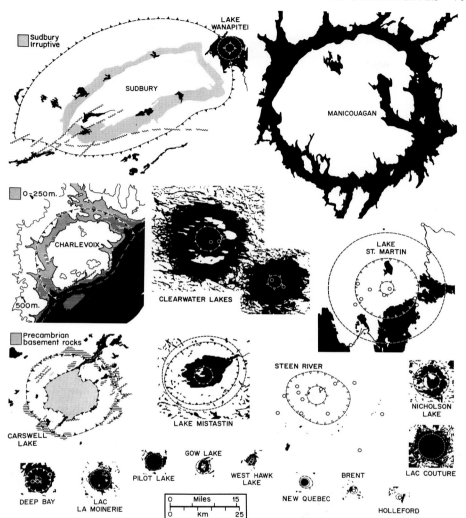

Figure 7   Map views of the 20 known Canadian impact structures, illustrating the relative sizes and outlines. For locations see Fig. 6, additional information in Table II. (Courtesy of the Earth Physics Branch, Department of Energy, Mines and Resources, Ottawa, Canada.)

the Department of Energy, Mines and Resources, Ottawa. The discovery of the impact origin of the New Quebec Crater (also called the Chubb Crater or Ungava Crater) by Meen (32) in 1950 and of the Brent Crater by Millman et al. in 1951 (33) sparked a systematic search by C. S. Beals (34) and his associates of circular features with raised rims shown on aerial photographs. They concentrated especially on the old rocks of the Canadian Shield. Many of the structures that were discovered are relatively inaccessible, but a determined program of geophysical surveys, mapping, drilling, and sample examinations have produced dramatic results. M. R. Dence and his co-workers have been particularly persistent in examining numerous structures in the field and in working with samples from the suspected craters to establish their modes of origin (35).

Among the Canadian craters and structures now known or strongly suspected to be of impact origin are the following: New Quebec Crater, Brent Crater (Fig. 8), Holleford Crater (Fig. 9), Deep Bay (Figs. 10 and 11), Carswell, West Clearwater Lake (Fig. 12), East Clearwater Lake (Fig. 12), West Hawk Lake (Fig. 13), Manicouagan (Fig. 7), the Sudbury Structure (Fig. 7), Lac Couture (Fig. 14), Pilot Lake and Nicholson Lake (Fig. 7, Table II). These features range from approximately 2 to more than 65 km in diameter. The evidence for the meteoritic origins of these features is mainly petrographic, based on shock metamorphism visible in thin sections from surface outcrops or samples taken from core drill holes. Supporting evidence comes chiefly from field observations and geophysical surveys. A number of

Figure 8   The Brent impact structure, Canada. Recognized in 1951. The approximate diameter of the structure is 4 km. (Courtesy of the Earth Physics Branch, Department of Energy, Mines and Resources, Ottawa, Canada.)

Figure 9 The Holleford impact structure, Canada, which has been filled in with many later sedimentary rocks. (Courtesy of the Earth Physics Branch, Department of Energy, Mines and Resources, Ottawa, Canada.)

Figure 10 Oblique aerial photograph of Deep Bay Crater looking toward the south. (Courtesy of the Earth Physics Branch, Department of Energy, Mines and Resources, Ottawa, Canada.)

Figure 11   Vertical aerial photomosaic of Deep Bay Crater, Canada. Diameter of the structure is approximately 9 km. Sharp breaks in tone and lighting are mosaic edges. (Courtesy of the Earth Physics Branch, Department of Energy, Mines and Resources, Ottawa, Canada.)

other large circular features in Canada *may* be the remnants of impact craters, but definitive evidence is lacking.

The recognition of the Canadian craters has been hampered by their old geologic ages. As a result, some of them have been filled in by sediments, some have been involved in regional metamorphism, and most have experienced great changes in their morphologies from the action of continental glaciers. The ages of most of the craters can be restricted to Paleozoic or Precambrian. These craters are much too old to expect to find weathered meteoritic iron or other meteoritic debris associated with them.

One Canadian impact site that deserves special attention is the Sudbury Structure, Ontario, not only because of its large size (originally, a *crater* 50 km in diameter) but also because of the large amount of metallic ores associated with it. The mines in this area have produced several billion dollars

Figure 12 Vertical aerial photomosaic of Clearwater Lakes, Canada. The ages of these structures are identical within experimental error, and they apparently represent a doublet cratering event, possibly from the breakup of a large incoming body within the Roche limit of the Earth. The larger structure is approximately 25 km in diameter. (Courtesy of the Earth Physics Branch, Department of Energy, Mines and Resources, Ottawa, Canada.)

worth of nickel, iron, copper, and other heavy metals since the discovery of ore in 1883.

The meteoritic or asteroidal impact origin for the Sudbury Structure was first proposed by R. S. Dietz (36). Support for this view has come from Bray (37), who determined that the distribution and orientation of shatter cones (page 146) in the Sudbury Structure were compatible with an impact origin.[2] Shatter cones are particularly large and well formed in the Mississagi Quartzite south and southwest of the Sudbury Structure (See Impact Metamorphism, Fig. 11; 38).

A number of petrographic features indicative of shock and impact metamorphism have been described by French (39) from thin sections of the Onaping Formation (also called Onaping Tuff or Tuff-Breccia). These observations include multiple sets of planar features (page 139) and partially annealed planar features in grains of both quartz and feldspar included in lithic fragments, unusually intense deformation of granite inclusions, glass-rimmed crystalline rock fragments, and impactite glass. It should be emphasized that the planar features have a strong preference for the

[2]Bray was not entirely convinced of the impact origin of the Sudbury Structure, however, and suggested that final judgment be withheld until additional evidence was available.

Figure 13 Vertical aerial photograph of the West Hawk Lake impact structure, Canada. This structure is only approximately 3 km in diameter. (Courtesy of the Earth Physics Branch, Department of Energy, Mines and Resources, Ottawa, Canada.)

crystallographic orientations noted for planar features from other impact sites. French concludes that the Onaping Formation is the welded fallback unit from the impact that took place 1.7 billion years ago.

Dietz suggested that the impacting metal body might be responsible for the ore deposits, especially because they are rich in nickel as are common types of iron meteorites. However, this idea neither is supported by field observations nor by isotopic comparisons of the ore bodies with meteoritic

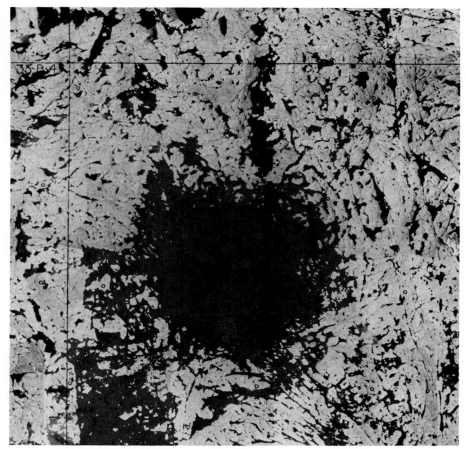

Figure 14   The Lac Couture impact structure, Canada, as shown in a vertical aerial photograph mosaic. The approximate diameter of the structure is 10 km. (Courtesy of the Earth Physics Branch, Department of Energy, Mines and Resources, Ottawa, Canada.)

materials. Also, it is very difficult to explain the high proportions of copper and sulfur, which are not found in these proportions in known varieties of meteorites.

The present form of the Sudbury Structure is caused by deformation and metamorphism subsequent to the impact. If these are accounted for, then the original shape of the structure is much more circular. Dietz estimates that the original crater was roughly circular, 50 km in diameter, 3 km deep, and

Table II   Probable Meteorite Craters and Impact Structures in Canada, 1972[a]

Name	Figure 6 Map Reference	Diameter, (km)	Age ($\times 10^6$ Years)	Year Confirmed
New Quebec Crater	1	3	less than 1	1950
Brent	2	4	450±40	1951
Manicouagan	3	65	210±4	1954
West Clearwater Lake	4	25	285±30	1955
East Clearwater Lake	4	14.5	285±30	1955
Holleford	5	2	550±50	1955
Deep Bay	6	9	100±50	1956
Carswell	7	30	485±50	1960
Lac Couture	8	10	300±150	1959
West Hawk Lake	9	3	150±50	1962
Pilot Lake	10	5	300±150	1965
Nicholson Lake	11	12.5	300±150	1965
Steen River	12	13.5	95±7	1965
Sudbury Structure	13	100	1700±200	1966
Charlevoix	14	35	350±25	1966
Lake Mistastin	15	20	202±25	1966
Lake St. Martin	16	24	225±25	1969
Lake Wanapitei	17	8.5	300±150	1969
Gow Lake	18	5	more than 150	1972[b]
Lac La Moinerie	19	8	more than 150	1972[b]

[a]Compiled by the Earth Physics Branch, Department of Energy, Mines and Resources, Ottawa, Canada.
[b]Confirmed after preparation of Table I by Dence (76).

formed by an explosive event estimated at 3×10^{29} ergs. This size of crater and magnitude of event could be produced by the impact of a small asteroid or comet.

RIES KESSEL, STEINHEIM BASIN AND STOPFENHEIM KUPPEL, GERMANY

The Ries Kessel (Ries Crater) is an impact site 24 km in diameter in southern Germany (Fig. 15). The old walled town of Nördlingen is situated within the structure. The argument concerning the origin of this structure, which traditionally has been described as a volcanic or "cryptoexplosion" structure (40), has been a European parallel to the debate concerning the origin of Meteor Crater in this country.

Figure 15 Oblique aerial photograph of the Ries (Nördlinger-Ries) Crater, Germany. The largest community visible is the old walled city of Nördlingen. The rim of the structure is marked by forests on the higher elevations. (By permission of Luftbild Albrecht Brugger, Stuttgart, Germany.)

The first suggestion that the Ries structure might be an impact site was made by E. Werner (41) in 1904. This paper set off a long exchange of views about the origin of the structure (e.g., 42, 43, 44), with approximately 100 papers published concerning the structure, its origin, the associated "tuffs," etc. between 1900 and 1970. Detailed mineralogical and petrographic works describing high pressure minerals (coesite and stishovite) and shock metamorphism in rocks from the Ries (45) together with geophysical studies (46), and reexaminations of the field relationships have left no doubt that it is a large impact crater. Virtually all of the features that are attributable to shock metamorphism have been found in Ries rocks. The structure is dated at approximately 15 my, apparently identical in age to the Czechoslovakian tektites (moldavites), which has prompted Gentner et al. (47) to suggest the Ries Crater might be the source of the moldavites.

The fallback unit from the crater is well exposed in a quarry at Otting and at many other localities inside and outside the crater. It can be divided easily into recrystallized and unrecrystallized zones. This breccia is termed "suevite." The suevite contains both shock metamorphosed rock fragments (see Impact Metamorphism, Fig. 10) and glassy inclusions. Many of the glassy inclusions are aerodynamically sculptured bombs, termed "Fladen" (after the German for pancake). Suevite has gained general use as a petrologic term. The cathedral at Nördlingen is constructed of suevite, thereby beautifully displaying the texture of the breccia. This material could be mistaken for volcanic tuff by inexperienced observers; however, in detail the textures are not at all similar to ash flow tuffs.

The Steinheim Basin is located approximately 30 km southwest of the Ries Crater, but it is a much smaller structure only a few kilometers in diameter near the village of Heidenheim. Numerous shatter cones and large breccias have been found there in the Jurassic limestone, and the age of the structure apparently is identical to the Ries. Evidence for the impact origin of the Steinheim Basin has been summarized by von Engelhardt et al. (48).

The Stopfenheim Kuppel is a nearly circular structure located approximately 20 km northeast of the Ries Crater. It is a radially faulted, uplifted area of Mesozoic sedimentary rocks approximately 9 km in diameter. This structure has been interpreted as a "cryptovolcanic" feature related to the "volcanic" Ries structure by Picard (49). However, the structure was brought to the attention of the Heidelberg research group by M. R. Bloch. An investigation of the structure produced evidence of shock metamorphism in quartz grains derived from the structure, so that it seems that this feature also is an impact site (50).

All three of these structures lie on a straight line, and they appear to be of the same age; thus, it is probable that they are genetically related (Storzer et al., 50). The model proposed is that the three structures represent a triplet cratering event by means of a single body breaking up under tidal forces as it neared the Earth. A quantitative treatment of this mechanism has been put forth by Sekiguchi, and by Aggarwal and Oberbeck (51).

THE AUSTRALIAN CRATERS

WOLF CREEK CRATER

The Wolf Creek Crater (Fig. 16) is located in Western Australia in a remote region. It is approximately 117 km south of Hall's Creek, and is not easily accessible. The crater was discovered in 1937 by observation from an airplane, but it was not visited on the ground until 1948 when an Australian expedition reached the locality and collected a number of oxidized meteoritic fragments (52). These fragments were especially abundant on the southwest rim of the crater. Cassidy visited the site with some difficulty in 1953, but was not successful in finding unoxidized meteoritic iron, although he did recover more than 630 kg of oxidized meteoritic material (53).

Wolf Creek Crater is slightly elongated with a north-south rim to rim dimension of 945 m and an east-west dimension of 838 m. The crater has a

Figure 16 Wolf Creek Crater, Western Australia; photograph taken from a low altitude aircraft. Observe the apparent asymmetry of the ejecta blanket. Rim to rim diameter is approximately 884 m. (Courtesy of S. R. Taylor.)

prominent raised rim that rises approximately 25 to 33 m above the surrounding plain and as much as 55 m above the crater floor (54). The age of the crater is not precisely known, but it is estimated to be Pleistocene to Recent (55). The country rock is mostly quartzite of probable Precambrian age. There are many blocks and large rocks on and outside the rim. It clearly is a meteorite explosion crater. However, as recently as 1965, McCall (56) has concluded that the available evidence favors a ''cryptovolcanic'' origin for the crater. This crater is in need of detailed mapping and petrographic study.

BOXHOLE CRATER

The Boxhole Crater is located near Boxhole Station, Central Australia, and was discovered in 1937 (57). Numerous iron meteorites, including one weighing 82 kg have been recovered in and around the crater together with weathered and oxidized fragments. The crater is perfectly circular, 166 m in diameter rim to rim, with a raised rim that is 14 m above the crater floor on the north and 6 m above the floor on the south (54). The meteorites are similar to the Henbury iron, medium octahedrite. The age of the crater is not known, and no shock metamorphosed rocks have been recognized at this site.

DALGARANGA CRATER

On Dalgaranga Station, Western Australia, G. E. Willard observed a small crater in 1923, approximately 25 km northeast of Dalgaranga Mountain. This crater together with the associated meteorites was reported from Willard's description by Simpson, in 1938 (58) in a survey article on Western Australia meteorites. Nininger and Huss visited the site in 1959 and collected a number of small meteorite fragments (the largest weighing 57 gm) and also oxidized meteorite material. The crater is only 21 m in diameter and is approximately 3.2 m deep (59), similar to some of the smaller craters at Odessa. Many of the small meteorite specimens show evidence of shock metamorphism. The impacting body was either a mesosiderite or an iron with numerous silicate and sulfide inclusions. The age of the crater has not been determined, but Nininger and Huss estimate that it may be 25,000 years old.

GOSSES BLUFF

Gosses Bluff is a roughly circular raised structure located on Missionary Plain between the MacDonnell and James ranges approximately 160 km west of Alice Springs (Fig. 17 and 18). Although the structure resembles a slightly deformed crater, this is fortuitous because of the lithologies of the rocks and the erosion that has occurred in relatively recent times. The origin of the Gosses Bluff structure has been attributed variously to ''cryptovolcanism,''

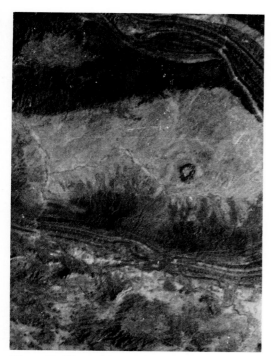

Figure 17   View of the Gosses Bluff impact structure taken from Earth orbit by the crew of Gemini V. A portion of NASA photograph S-65-45568. (Courtesy of NASA, Johnson Space Center.)

the action of a "mud volcano," igneous intrusion, and salt diapirism (e.g., 60, 61). However, other investigators recently have concluded that the structure is the result of a large impact.

The available data on the origin of the Gosses Bluff structure and a report of new drilling and mapping of the area were presented by P. J. Cook (62). He concluded that there was more evidence at that time to support an impact origin than any internal origin. He cited the abundant and well developed shatter cones, abundant glassy material and breccia, and the strain and fracturing found in quartz grains from rocks in the structure.

Extensive mapping and field work at the site, including seismic and gravity surveys, as well as petrographic work on collected specimens were reported by Milton et al. (63). They measured the orientations of numerous shatter cones and reconstructed them into their initial positions. They concluded that the focus of the cone apices and axes was approximately 600 m or less below

Figure 18   Aerial photograph of the ring-shaped central uplift zone of the Gosses Bluff impact structure. The actual size of the original crater was much larger, possibly as much as 20 km. Photograph by C. Zawartko. (Courtesy of P. R. Brett.)

the pre-event ground surface, although there were some variations in the elevation as determined from different stratigraphic units. However, this could be due to the somewhat nonhemispherical advance of the shock front. Virtually all of their determinations of shatter cone orientations indicated a focus near the center of the structure.

Many samples of melt breccia and rocks with obvious evidence of shock metamorphism were investigated by Milton et al. Abundant quartz-bearing samples with planar features in individual grains oriented parallel to $\{10\bar{1}1\}$, $\{0001\}$, and $\{10\bar{1}3\}$ were found. Breccia consisting of sintered shock-melted fragments very similar to the well known suevite from the Ries Crater also was found. This material is particularly well exposed, together with a section of rocks that exhibit a gradation in shock metamorphism, at Mt. Pyroclast, a 40 m high hill located 5 km south of the center of the structure. This suevitelike breccia may be a portion of the fallback material from the cratering event.

It is apparent that the raised portion of the structure is only the central uplift area of a much larger crater that has been deeply eroded. Milton et al. estimate the diameter of the original crater was 20 km. The age of the event has been measured as earliest Cretaceous (approximately 130 my) by both potassium-argon and fission track dating (63).

HENBURY CRATERS

A cluster of 13 or more craters in Australia's Northern Territory, near Henbury, was discovered in 1931 and has received attention in the scientific literature since 1932, when both the craters and the iron meteorites associated with them were first described (64, 65, 66). Alderman (67) reported the results of the first scientific expedition to the Henbury Craters and described both the craters and the meteorites. Furthermore, he compared the craters to Meteor Crater, Arizona.

A magnetometer survey of 12 craters in the Henbury group was made by Rayner (68), who did not find any large magnetic anomalies associated with any of the craters. Rayner concluded that the craters must have been formed by explosive impact because of the lack of any indication that there were large buried masses of nickel-iron. P. W. Hodge reviewed the work on the Henbury Craters until 1965 and published a well illustrated description of the craters, including a number of low altitude aerial photographs (Fig. 19) and a report of a previously undiscovered crater associated with the group (69).

Most of the recent work at this site has been done by D. J. Milton (70), who began field work there in 1963. Milton mapped a number of the craters and

Figure 19   Aerial photograph of the largest Henbury Craters, Australia. The craters are numbered in accordance with the identification by Alderman in 1932. The largest crater in the group has a rim to rim diameter of approximately 145 m. (Courtesy of P. W. Hodge.)

identified a crudely looped ray of ejecta[3] associated with one of the smaller craters. One particularly interesting observation was the intact, narrow wall between two closely adjacent craters. Such observations on the lunar surface had led to the interpretation that one or both of the lunar craters must be volcanic (collapse) in origin or the narrow septum would not be preserved.

The largest crater in this group has an average diameter of approximately 145 m, and together with the next three largest craters (ranging down to 70 m diameter) have been studied extensively by Milton. One of the chief goals of this work was to determine what structures might occur in the crater walls and rims that might be unique to impact craters. Tangential folds (overturned away from the crater), shear folds (formed by displacements parallel to the crater walls), overturned flaps, thrust slices, and fault bounded folds of several types are among the structural features described after careful mapping of the crater walls and rims. Many of these features are known from other impact craters, but never had so many different types of structures been observed in the same site.

The abundant glassy impactite associated with the larger craters has been analyzed extensively by Taylor (71), who reports that the composition of the impactite shows few differences in chemical composition from the unaltered crater parent rocks.

ARABIAN CRATERS

The Wabar Craters, Saudi Arabia, were discovered in 1932 by Philby (72) who was in search of the lost city of Wabar.[4] Philby detoured to this locality to see a piece of iron "as large as a camel." However, he found only smaller pieces which he returned to England for study. There was never any doubt concerning the origin of the craters. Spencer (7) quickly confirmed the impact origin of the Wabar Craters and studied the abundant glassy impactite from the site. Spencer concluded that the large Wabar Crater (there are certainly two craters and possibly three), which is approximately 100 m in diameter, was an explosion crater and was not produced by simple percussion. Spencer (73) further concluded as follows:

> The kinetic energy of a large mass of iron (nickel-iron) travelling at a high velocity, was suddenly transformed into heat, vaporizing a large part of the meteorite and some of the earth's crust, so producing a violent gaseous explosion, which formed the crater and back-fired the remnants of the meteorite. Such an explosion crater will be circular in outline whatever be the angle of approach of the projectile. The materials

[3]Looped rays have long been observed on the Moon associated with the bright, relatively recent rayed craters such as Copernicus. However, until this observation, none were known to be associated with terrestrial impact craters.

[4]These craters are called Al-Hadidah Craters by the Arabs and also in some of the literature.

collected at the Wabar craters afford the clearest evidence that very high temperatures prevailed. The desert sand was not only melted, yielding a silica-glass, but this boiled (b.p. about 3,500° + C) and was vaporized. The meteoritic iron was also in large part vaporized, afterwards condensing as a fine drizzle. Minute spheres of nickel-iron, of the same composition as the meteorite, are preserved in the bubbly silica glass. In some portions these are present to the extent of about two million per cubic centimetre of the glass. The fact that these minute spheres show a bright and highly polished surface suggests that the earth's atmosphere was blown aside in the fiery blast.

This is a classic, early, qualitative statement of the major events associated with the terrestrial impact of a large meteorite.

The craters are partially covered with drifting sand (Fig. 20) and are not well exposed. In fact, the target material impacted by the meteorite may have been largely unconsolidated sand. However, the largest crater does have a blocky rim (Fig. 21), indicating that coherent material from a few meters or

Figure 20   Aerial photograph of the largest Wabar Crater, Saudi Arabia, showing the blocky rim and partial sand cover. Diameter of the crater is approximately 100 m. (Courtesy of Arabian American Oil Company.)

Figure 21   View across a portion of the largest Wabar Crater, Saudi Arabia, showing the blocky rim of reflective, shock welded sandsone and excavated blocks. (Courtesy of V. E. Barnes.)

tens of meters below the surface has been broken up and ejected from the crater or that shock lithification of loose sand has occurred (74).

There are abundant small fresh and weathered nickel-iron fragments at the impact site both within and outside of the crater. However, one fresh metal fragment weighing more than 2180 kg has been recovered recently (75). This may well be the Al Hadidah or large iron about which Philby had been told, but which had been covered by shifting sands at the time of his visit. The meteorite is a medium octahedrite and, as is common at Meteor Crater and other iron meteorite impact sites, many pieces show petrographic evidence of shock damage.

Although the craters are located in a remote portion of the Rub-al-Khali (empty quarter), they have been visited frequently by geophysical survey crews in the course of oil exploration in recent years. The Wabar Craters are one of the few landmarks in the area.

CRITERIA FOR RECOGNITION OF IMPACT STRUCTURES

The recognition of very recent impact structures and craters is not difficult, because of the associated meteorite fragments, obvious shock metamorphism, and explosion features. However, the recognition of ancient and badly deformed craters may be much more difficult. This has been a particular problem with many of the old craters on the Canadian Shield. Dence (76) recently has summarized the most usable criteria (Table III).

It seems obvious that many terrestrial impact craters have escaped identification by geologists. Numerous new examples certainly will be recognized during the coming decades, especially in areas where very old rocks are exposed.

Table III   Criteria for the Identification of Terrestrial Impact Craters[a]

Criterion	Nature and Status	Examples
1. Presence of meteorites	Rare except in ejecta of young craters	Barringer, Henbury
2. Circular plan	Distinctly circular near center. Modified at margins by: (a) Preexisting structures (b) Erosion, slight to moderate	Brent Barringer, Manicouagan New Quebec, Deep Bay, West Hawk Lake
	Obscured by: (a) Deep erosion (b) Later cover (c) Later tectonic events	Nicholson Lake, Dellen Lake Holleford, Lake St. Martin Charlevoix, Sudbury
3. Rim structure	Raised, overturned rim only apparent in young simple craters. In complex craters rim has dropped to form: (i) Subdued uplift, or (ii) Disturbed zone, or (iii) Peripheral trough	Deep Bay Ries Manicouagan
4. Central structure	Bowl of breccia in simple craters. Central uplift in complex craters either: (i) Single peak, or (ii) Ring Structure	Barringer, Brent Steinheim Gosses Bluff, Clearwater Lake West
5. Gravity anomaly	Generally negative. May be enhanced by sedimentary fill. Most clearly developed in craters of moderate size. In complex large craters may be obscured by: (a) Central uplift of heavy rocks (b) Erosion (c) Regional gravity variations	Deep Bay, Lake Wanapitei Clearwater Lake West Nicholson Lake Carswell, Manicouagan

Table III (Cont'd) Criteria for the Identification of Terrestrial Impact Craters[a]

Criterion	Nature and Status	Examples
6. Magnetic field	Variable, commonly subdued, merge with regional field. Distinct anomalies may be present over suevite and melt rock concentrations	Clearwater Lake craters, Deep Bay, Brent Ries
7. Seismic velocities	Crater rocks show lower seismic velocities than country rocks. Craters in stratified rocks have central region of chaotic structure	Deep Bay, Brent Gosses Bluff, Sierra Madera
8. Brecciation	Observed in surface samples and drill core. Rim rocks show mainly monomict breccias overlain by mixed ejecta, if preserved. Mixed breccias within crater interlayered with melt rock concentrations	Brent, Ries Clearwater Lake West Brent, West Hawk Lake
	Country rocks in central uplift cut by pseudotachylites and by veins of mixed breccia and melt rocks	Vredefort, Manicouagan
9. Shock metamorphism	Main criterion for hypervelocity impact. Includes shatter coning, planar elements in minerals, glassy solid states, high-pressure phases, complete melting to form mixed breccias, glasses and pools or sheets of melt rocks. Present in ejecta breccias or in mixed breccias within crater, also in country rocks underlying central region of crater. Not in rim rocks. May be obscured by annealing, hydrothermal alteration (zeolites, etc.), later regional metamorphism	Barringer, Ries Brent, Clearwater Lakes Charlevoix Manicouagan, Sudbury

[a]From Dence (76).

REFERENCES AND NOTES

1. Hager, D. (1953) Crater Mound (Meteor Crater), Arizona, a Geologic Feature: Bull., Am. Assoc. Petrol. Geol., vol. 37, p. 821-857.

2. Bucher, W. H. (1963) Cryptoexplosion structures caused from without or from

within the earth? ("astroblemes" or "geoblemes"?): Am. Jour. Sci., vol. 261, p. 597-649.

3. Snyder, F. G., and P. E. Gerdemann (1965) Explosive igneous activity along an Illinois-Missouri-Kansas axis: Am. Jour. Sci., vol. 263, p. 465-493.

4. Currie, K. L. (1965) Analogues of lunar craters on the Canadian shield: Annals, N.Y. Acad. Sci., vol. 123, p. 915-940. See also Currie, K. L. (1968) A note on shock metamorphism in the Carswell circular structure, Saskatchewan, Canada: *in* Shock Metamorphism of Natural Materials, B. M. French and N. M. Short, eds., Mono Book Corp., Baltimore, Md., p. 379-381.

5. Dietz, R. S. (1963) Cryptoexplosion structures: a discussion: Am. Jour. Sci., vol. 261, p. 650-664.

6. French, B. M. (1968) Shock metamorphism as a geological process: *in* Shock Metamorphism of Natural Materials, B. M. French and N. M. Short, eds., Mono Book Corp., Baltimore, Md., p. 1-17.

7. Spencer, L. J. (1932) Meteorite craters: Nature, vol. 129, p. 781-784.

8. Boon, J. D., and C. C. Albritton, Jr. (1938) Established and supposed examples of meteoritic craters and structures: Field and Laboratory, vol. 6, no. 2, p. 44-56. See also earlier papers in Field and Laboratory.

9. Daly, R. A. (1947) The Vredefort ring-structure of South Africa: Jour. Geol., vol. 55, p. 125-145.

10. Short, N. M. (1968) A worldwide inventory of features characteristic of rocks associated with presumed meteorite impact structures: *in* Shock Metamorphism of Natural Materials, B. M. French and N. M. Short, eds., Mono Book Corp., Baltimore, Md., p. 255-266.

11. Freeberg, J. H. (1966) Terrestrial impact structures — a bibliography: U. S. Geol. Survey, Bull. 1220, 91 p.; also (1969) Terrestrial impact structures — a bibliography 1965-1968, U. S. Geol. Survey, Bull. 1320, 39 p.

12. McKee, E. D. (1954) Stratigraphy and history of the Moenkopi Formation of Triassic Age: Geol. Soc. Am., Memoir 61, 133 p. See also earlier papers by the same author.

13. Shoemaker, E. M. (1963) Impact mechanics at Meteor Crater, Arizona: *in* The Solar System, Vol. IV, The Moon, Meteorites and Comets, B. M. Middlehurst and G. P. Kuiper, eds., The Univ. of Chicago Press, Chicago, p. 301-336.

14. Gilbert, G. K., and M. Baker (1891) A meteoric crater: Astron. Soc. Pacific Pub., vol. 4, no. 21, p. 37.

15. Foote, A. E. (1891) Geologic Features of the Meteoritic locality in Arizona: Acad. Nat. Sci. Phila., Proc., vol. 43, p. 407; also Foote, A. E. (1892). A new locality for meteoritic iron with a preliminary notice of the discovery of diamonds in the iron: Proc. Am. Assoc. Adv. Sci., vol. 40, p. 279-283.

16. Gilbert, G. K. (1896) Presidential Address, Geol. Soc. Wash., March, 1896, see also (1896) Science, N. S., vol. 3, p. 1-13.

17. Barringer, D. M., (1905) Coon Mountain and its Crater: Acad. Nat. Sci. Phila., Proc. vol. 57, p. 861-886; see also later papers, e.g., Barringer, D. M. (1926) Exploration at Meteor Crater: Eng. Mining Jour. Press, vol. 121, No. 2, p. 59, No. 11, p. 450-451.

18. Merrill, G. P., (1907) On a peculiar form of metamorphism in siliceous sandstone: Proc. U. S. Natl. Mus., vol. 32, p. 547-551.

19. Chao, E. C. T., E. M. Shoemaker, and B. M. Madsen (1960) First natural occurrence of coesite: Science, vol. 132, p. 220. Also, Chao, E. C. T., J. J. Fahey, J. Littler, and D. J. Milton (1962) Stishovite, SiO_2, a very high pressure mineral from Meteor Crater, Arizona: Jour. Geophys. Res., vol. 67, p. 419-421.

20. Rinehart, J. S. (1957) A soil survey around the Barringer Crater: Sky and Teles., vol. 16, no. 8, p. 366-369.

21. Baldwin, R. B. (1963) The measure of the Moon: Univ. of Chicago Press, Chicago, 488 p.

22. Bjork, R. L. (1961) Analysis of the formation of Meteor Crater, Arizona, a preliminary report: Jour. Geophys. Res., vol. 66, p. 3379-3387.

23. Monnig, O. E., and R. Brown (1935) The Odessa, Texas Meteorite Crater: Pop. Astron., vol. 43, p. 34-37.

24. Sellards, E. H., and G. Evans (1941) Statement of progress of investigation at Odessa Meteor Craters: Univ. of Texas, Bur. Eco. Geol., Sept. 1, 1941, 12 p.

25. Sellards, E. H., and G. Evans (1944) Odessa Meteor Craters, views in Texas Memorial Museum: Mus. Notes, vol. 6, p. 13, July 1944.

26. Barringer, D. M., Jr. (1928) A new meteor crater: Acad. Nat. Sci. Phila., Proc., vol. 80, p. 307-311.

27. Evans, G. L. (1961) Investigations at the Odessa meteor craters: in Proc. Geophys. Lab./Lawrence Rad. Lab. Cratering Symp., Wash., D.C., March, 1961; Univ. Cal., Livermore, Rad. Lab. Rept. UCRL-6438, pt. 1, paper D, 11 p. (Rept. for AEC).

28. Nininger, H. H., and J. D., Figgins (1934) The excavation of a meteorite crater near Haviland, Kansas: Am. Jour. Sci., vol. 28, p. 312-313, abstract. See also the Proceedings of the Colorado Museum of Natural History, vol. 7, no. 3; vol. 12, p. 13-14.

29. Spencer, L. J. (1933) Meteorite craters as topographical features of the Earth's surface: Geog. Jour., vol. 81, p. 227-243. For a detailed investigation of these craters see Renard, M. L., and W. A. Cassidy (1971) Entry trajectory and orbital calculations for the Crater 9 meteorite, Campo del Cielo, Argentina: Jour. Geophys. Lab./Lawrence Rad. Lab. Cratering Symp., Wash., D. C., March, 1961;

30. Krinov, E. L. (1963) The Tunguska and Sikvote-Alin meteorites: in The Solar System IV, The Moon, Meteorites and Comets, B. M. Middlehurst and G. P. Kuiper, eds., Univ. of Chicago Press, Chicago, p. 208-234.

31. Krinov, E. L. (1963) Meteorite craters on the Earth's surface: in The Solar System IV, The Moon, Meteorites and Comets, B. M. Middlehurst and G. P. Kuiper, eds., Univ. of Chicago Press, Chicago, p. 183-207.

32. Meen, V. B. (1950) Chubb Crater, Ungava, Quebec: Jour. Roy. Astron. Soc. Canada, vol. 44, p. 169-180.

33. Millman, P. M., B. A. Liberty, J. F. Clark, P. L. Willmore, and M. J. S. Innes (1960) The Brent Crater: Pub. Dominion Obs., vol. 24, no. 1, p. 1-43.

34. Beals, C. S., M. J. S. Innes, and J. A. Rottenberg (1963) Fossil meteorite crater

in The Solar System IV, The Moon, Meteorites and Comets, B. M. Middlehurst and G. P. Kuiper, eds., Univ. of Chicago Press, Chicago, p. 235-284.

35. Dence, M. R. (1965) The extraterrestrial origin of Canadian craters: Ann. New York Acad. Sci., vol. 123, p. 941-969 (also reprinted as Dominion Obs., Contributions, vol. 6, no 11). Also, Dence, M. R., M. J. S. Innes, and P. B. Robertson (1968) Recent geological and geophysical studies of Canadian craters: *in* Shock Metamorphism of Natural Materials, B. M. French and N. M. Short, eds., Mono Book Corp., Baltimore, Md., p. 339-362, and numerous other papers.

36. Dietz, R. S. (1962) Sudbury structure as an astrobleme: Am. Geophys. Union, Trans., vol. 43, no. 4, p. 445-446, *abstract:* also Dietz, R. S. (1964) Sudbury structure as an astrobleme: Jour. Geol., vol. 72, no. 4, p. 412-434.

37. Bray, J. G. and geological staff (1966) Shatter cones at Sudbury: Jour. Geol., vol. 74, no. 2, p. 243-245.

38. Dietz, R. S., and L. W. Butler (1964) Shatter-cone orientation at Sudbury, Canada: Nature, vol. 204, no. 4955, p. 280-281. Also, Dietz, R. S. (1968) Shatter cones in cryptoexplosion structures: *in* Shock Metamorphism of Natural Materials, B. M. French and N. M. Short, eds., Mono Book Corp., Baltimore, Md., p. 267-285.

39. French, B. M. (1967) Sudbury structure, Ontario — some petrographic evidence for origin by meteorite impact: Science, vol. 156, no. 3778, p. 1094-1098; a somewhat more extensive version of the same paper appears as NASA, Goddard Space Flight Center, Pub. X-641-67-67, February 1967, 56 p.

40. Gümbel, C. W. (1870) Uber den Riesvulkan und uber vulkanisch Erhscheinungen im Rieskessel: Akad. Wiss. München Sitzungsber., Abt. 1, p. 153-200.

41. Werner, E. (1904) Das Ries in der schwäbisch-fränkischen Alb: Blätter der Schwab., Albvereins, vol. 16, p. 153-167.

42. Kranz, W. (1928) Vulkanexplosionen, Sprengtechnik, praktische Geologie und Ballistik: Deutsche Geol. Gesell. Zeitschr., vol. 80, p. 257-307. See also other papers by this author from 1911 to 1952.

43. Ahrens, W. (1929) Die Tuffe des Nördlinger Rieses und ihre Bedeutung für das Gesamtproblem: Deutsche Geol. Gesell. Zeitschr., vol. 81, p. 94-99.

44. Seeman, R. (1939) Versuch einer vorwiegend tektonischem Erklarung des Nördlinger Rieses: Neues Jahrb. Mineral. Geol. and Paläo., Beilage- Bank 81, Abt. B, no. 1, p. 70-166; no. 2, p. 169-214.

45. Shoemaker, E. M., and E. C. T. Chao (1961) New evidence for the impact origin of the Ries basin, Bavaria, Germany: Jour. Geophys. Res., vol. 66, p. 3371-3378; also, Engelhardt, W. v., and D. Stöffler (1968) Stages of shock metamorphism in

46. Kahle, H.-G. (1969) Abschätzung der Störungsmasse in Nördlinger Ries: Zeitschr. für Geophys., Band 35, p. 317-345, also (1970) Deutung der Schwereanomalien in Nördlinger Ries: Zeitschr. für Geophys., Band, 36, p. 601-606, and earlier papers by K. Jung and others.

47. Gentner, W., H. J. Lippolt, and O. A. Schaeffer (1963) Argonbestimmung am Kaliummineralien, XI — Die Kalium-Argon-Alter des Gläser der Nördlinger Rieses und der böhmischmahrischen Tektite: Geochim. et Cosmoschim. Acta, vol. 27, no. 2, p. 191-200.

crystalline rocks of the Ries Basin, Germany: *in* Shock Metamorphism of Natural Materials, B. M. French and N. M. Short, eds., Mono Book Corp., Baltimore, Md., p. 159-168, and numerous other papers by von Engelhardt, his students, and colleagues at the Univ. of Tübingen.

48. Engelhardt, W. v., W. Bertsch, D. Stöffler, P. Groschopf, and W. Reiff (1967) Anzeichen für den meteoritischen Ursprung des Beckens von Steinheim: Naturwissenschaften, vol. 54, no. 8, p. 198-199. See also, Groschopf, P., and W. Reiff (1966) Ergebnisse neuerer Untersuchungen in Steinheimer Becken: Ver. Vaterländ. Naturkunde Wurttemberg Jahreshefte, vol. 121, p. 155-168.

49. Picard, L. (1923) Die Frankische Alb von Weissenburg i.B. und Umgebung: Dissertation, Univ. of Freiburg, Germany.

50. Storzer, D., W. Gentner, and F. Steinbrunn (1971) Stopfenheim Kuppel, Ries Kessel and Steinheim Basin: A triplet cratering event: Earth Planet. Sci. Letters, vol. 13, no. 1, p. 76-78.

51. Sekiguchi, N. (1970) On the fission of a solid body under the influence of tidal forces; with application to the problem of twin craters on the moon: The Moon, vol. 1, p. 429-439; Aggarwal, H. A. and V. R. Oberbeck (1974) Roche limit of a solid body: Astrophys. Jour., vol. 191, p. 577-588.

52. Reeves, F., and R. O. Chalmers (1949) Wolf Creek crater: Austral. Jour. Sci., vol. 11, p. 154-156.

53. Cassidy, W. A. (1954) The Wolf Creek, Western Australia, meteorite crater (CN=-1278,192): Meteoritics, vol. 1, no. 2, p. 197-199.

54. Cassidy, W. A. (1968) Descriptions and topographic maps of the Wolf Creek and Boxhole Craters, Australia: *in* Shock Metamorphism of Natural Materials, B. M. French and N. M. Short, eds., Mono Book Corp., Baltimore, Md., p. 623, *abstract*.

55. Guppy, D. J., and R. S. Matheson (1950) Wolf Creek meteorite crater, Western Australia: Jour. Geol., vol. 58, p. 30-36.

56. McCall, G. J. H. (1965) Possible meteorite craters — Wolf Creek, Australia: Annals, New York Acad. Sci., vol. 123, art. 2, p. 970-998.

57. Madigan, C. T. (1937) The Boxhole crater and the Huckitta meteorite (central Australia): Roy. Soc. South Austral., Trans. and Proc., vol. 61, p. 187-190. See also Madigan, C. T. (1940) The Boxhole meteoritic iron, central Australia: Min. Mag., vol. 25, no. 168, p. 481-486.

58. Simpson, E. S. (1938) Some new and little-known meteorites found in Western Australia: Min. Mag., vol. 25, no. 163, p. 157-171.

59. Nininger, H. H., and G. I. Huss (1960) The unique meteorite crater at Dalgaranga, Western Australia: Min. Mag., vol. 32, no. 251, p. 619-639.

60. McNaughton, D. A., T. Quinlan, R. M. Hopkins, and A. T. Wells (1968) Evolution of salt anticlines and salt in the Amadeus Basin, Central Australia: Geol. Soc. Am. Spec. Paper 88, p. 229-247.

61. Ranneft, T. S. M. (1970) Gosses Bluff, Central Australia, as fossil mud volcano: Am. Assoc. Petrol. Geol., Bull. vol. 54, p. 417-427.

62. Cook, P. J. (1968) The Gosses Bluff cryptoexplosion structure: Jour. Geol., vol. 76, no. 2, p. 123-139.

63. Milton, D. J., B. C. Barlow, Robin Brett, A. R. Brown, A. Y. Glikson, E. A. Manwaring, F. J. Moss, E. C. E. Sedmik, J. Van Son, and G. A. Young (1972) Gosses Bluff impact structure, Australia: Science, vol. 175, p. 1199-1207.

64. Alderman, A. R. (1932) The Henbury (central Australia) meteoritic iron: South Austral. Mus. Rec., vol. 4, no. 4, p. 555-563.

65. Spencer, L. J. (1932) Meteoric iron and silica-glass from the meteorite craters of Henbury (Central Australia) and Wabar (Arabia), with chemical analysis by M. H. Hey: Mineralog., Mag., vol. 23, no. 142, p. 387-404.

66. Bartrum, C. O. (1932) The meteorite craters at Henbury, central Australia: Brit. Astron. Assoc. Jour., vol. 41, no. 4, p. 263-264.

67. Alderman, A. R. (1932) The meteorite craters at Henbury, central Australia, with addendum by L. J. Spencer: Mineralog. Mag., vol. 23, no. 136, p. 19-32.

68. Rayner, J. M. (1939) Examination of the Henbury meteorite craters by the methods of applied geophysics: Austral. and New Zeal. Assoc. Adv. Sci., Rept., vol. 24, p. 72-78.

69. Hodge, P. W. (1965) The Henbury Meteorite Craters: Smithsonian Contr. to Astrophysics, vol. 8, no. 8, p. 199-203. See also Hodge, P. W., and F. W. Wright (1971) Meteoritic particles in the soil surrounding the Henbury Meteorite Craters: Jour. Geophys. Res., vol. 76, no. 17, p. 3880-3895.

70. Milton, D. J. (1968) Structural geology of the Henbury meteorite craters, Northern Territory, Australia: U. S. Geol. Survey, Prof. Paper 599-C, p. C1-C17. Also see Milton, D. J. and F. C. Michel (1965) Structure of a ray crater at Henbury, Northern Territory, Australia: U. S. Geol. Survey, Prof. Paper 525-C, p. C5-C11.

71. Taylor, S. R. (1966) Australites, Henbury impact glass and subgreywacke — a comparison of the abundances of 51 elements: Geochim. et Cosmochim. Acta, vol. 30, no. 11, p. 1121-1136; see also Taylor, S. R. (1967) Composition of meteorite impact glass across the Henbury strewnfield: Geochim. et Cosmochim. Acta, vol. 31, no. 6, p. 961-968.

72. Philby, H. St. J. (1932) Wabar craters: Nature, vol. 129, p. 932.

73. Spencer, L. J. (1937) Meteorites and the craters on the Moon: Nature, vol. 139, p. 655-657.

74. Short, N. M. (1966) Shock-lithification of unconsolidated rock materials: Science, vol. 154, no. 3748, p. 382-384. See also the cover on this journal issue.

75. Abercrombie, T. J. (1966) Saudi Arabia, beyond the sands of Mecca: Natl. Geog., vol. 129, no. 1, p. 1-53; see photograph on page 35.

76. Dence, M. R. (1972) The nature and significance of terrestrial impact structures: Proc. 24th Internat. Geol. Cong., Montreal, Sec. 15, p. 77-89.

SUGGESTED READING AND GENERAL REFERENCES

Krinov. E. L. (1963) Meteorite craters on the Earth's surface: in The Solar System. B. M. Middlehurst and G. P. Kuiper, eds., Vol. IV. The Moon, Meteorites and Comets, Univ. of Chicago Press. Chicago, p. 183-207.

Baldwin, R. B. (1963) Modern terrestrial meteoritic craters; Probable and possible terrestrial meteoritic craters; Ancient meteoritic craters and cryptovolcanic structures: *in* The Measure of the Moon, Univ. of Chicago Press, Chicago, Chaps. 2, 3 and 4, p. 6-105.

Dence, M. R. (1972) The nature and significance of terrestrial impact structures: Proc. 24th Internat. Geol. Cong., Montreal, Sec. 15, p. 77-89.

Natural shock metamorphism is produced by the nearly-instantaneous transfer of the kinetic energy of an impacting meteorite into the surrounding rock by means of intense shock waves. Such a process is quite distinct from normal geological metamorphism and it becomes necessary to think in terms of microseconds instead of megayears.   Bevan French, 1966

5. Impact Metamorphism

INTRODUCTION

The hypervelocity impacts of large meteorites, asteroids, and comets generally cause considerable metamorphism in the rocks that formerly occupied the volume of the resultant crater, and in the rocks immediately adjacent to the crater. This was recognized in 1907 by Merrill (1) who carefully described unusual metamorphism in rocks at Meteor Crater, Arizona. The research work of many geoscientists during the past 20 years has been devoted to documenting the effects of impact-produced shock waves in rocks and minerals. The chief objective of this research has been to establish mineralogic, petrologic, and field criteria for the recognition of impact craters, shocked rocks, and geologically old impact sites on the Earth.

A forcing function in this loosely coordinated work was the Apollo Program. It was widely recognized that some portion of the lunar samples would be shock metamorphosed, perhaps a very large fraction. This metamorphism might increase the difficulty involved in making the usual sorts of petrographic, mineralogical, and geochemical observations and interpretations, thus obscuring lunar history and processes. With the presidential announcement of the United States' goal of returning men (and

131

samples) from the Moon by 1970, this research took on a special urgency, and a large volume of data and results were generated about a virtually new topic in a short period of time (2). A number of more or less diagnostic criteria for the recognition of shock metamorphism gradually were established.

MINERALOGIC AND PETROLOGIC CRITERIA

The most useful mineralogic and petrologic criteria for shock metamorphism are those that can be applied to the common rock types and to common rock-forming minerals. The following petrologic features indicative of shock are those that are most easily recognized in most shock metamorphosed rocks.

COESITE

Coesite is a dense, high pressure polymorph of silica that was first synthesized by Coes in 1953 (3). This phase had never been recognized in nature. Nininger first suggested that coesite, if it occurred naturally anywhere on the surface of the Earth, might be found at Meteor Crater, Arizona (4), because of the high pressures generated in the local silica-rich rocks by the meteorite impact. In 1960, Chao and co-workers examined pieces of highly shocked Coconino Sandstone from Meteor Crater and discovered the first natural occurrence of coesite (5). Subsequently, coesite has been identified at a number of other impact sites such as the Ries and Wabar Craters. Impact sites remain the only common natural occurrences of coesite that have been reported.

Coesite generally occurs as small grains in other silica phases in shocked rocks (Fig. 1). It is easily converted to tridymite or cristobalite by heating for a short time, and some highly shocked rocks that may have contained coesite initially now only have high temperature phases of silica preserved because the residual temperatures of the rocks were more than several hundred degrees. Coesite may be very difficult to find petrographically, even in rocks that are known to contain substantial amounts, because of the very small grain size. However, the X-ray diffraction pattern of coesite has two strong peaks that make it easy to identify, even in samples in which coesite is present in small amounts (Table I). Some samples of the shocked Coconino Sandstone from Meteor Crater contain sufficient coesite that it can be detected in diffractograms of bulk samples of the sandstone. More commonly, the coesite must be concentrated or separated from the other mineral phases in the shocked rock before its presence can be established. Two methods have been used successfully: (a) HF treatment of pulverized and sized samples (6), and (b) centrifugation of pulverized and sized samples in liquids of appropriate density.

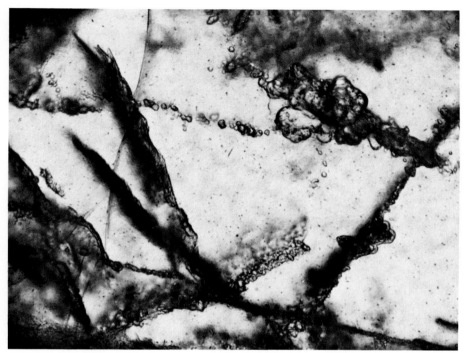

Figure 1 Coesite in trails and clusters along fractures in quartz in a shocked rock from the Ries Crater, Germany; plane polarized light, length of field of view is approximately 500 μm. (Courtesy of Dieter Stöffler.)

The HF technique is easy and is especially useful for quartz sandstone or other quartz-rich rocks. This technique depends on the lesser solubility of the coesite to concentrate it in the acid treated residue. However, in aluminous rocks, HF dissolution may lead to the formation of aluminum fluoride trihydrate and other aluminum-fluorine compounds that have some X-ray peaks in their diffractograms which could be (and have been) mistaken for

Table I Comparison of Physical and Optical Properties of Coesite and Stishovite[a]

Mineral	Specific Gravity	Optical Character	Refractive Indices Range	Strong X-ray Diffraction Peaks (Å)
Coesite	2.93	Biaxial (+)	1.59–1.61	$d = 3.09, 3.44$
Stishovite	4.35	Uniaxial (+)	1.79–1.83	$d = 2.96, 1.53$

[a]After Frondel (1962) Dana's The System of Mineralogy, 7th ed., vol. III, Silica Minerals, Wiley, New York, 334 p.

coesite by unwary investigators. Most of these compounds are water soluble, and thorough washing of the residue before X-ray analysis will obviate the problem.

The centrifugation technique involves placing the finely pulverized and sized rock powder into a centrifuge tube with an acetone-bromoform solution that is adjusted for a density slightly less than that of coesite. After a few minutes in the centrifuge, the coesite (if present) and probably a few other heavy minerals will be in the bottom of the tube. The quartz, feldspar, glass, and other less dense phases will be at the top of the tube.

Coesite can be identified in residues from either of these methods by standard X-ray and/or optical techniques (Table I). The presence of coesite is considered a reliable criterion for impact metamorphism because coesite is known naturally almost exclusively from impact metamorphosed rocks.[1]

STISHOVITE

Stishovite is an even more dense polymorph of SiO_2 that is known to occur naturally *only* in impact metamorphosed rocks. In 1952, J. B. Thompson predicted the existence of a silica polymorph in which silica would be in octahedral coordination with oxygen, and he further suggested that the phase would have rutile structure (7). This phase was later synthesized in crystals as much as 0.5 mm long by Stishov and Popova (8) at pressures of 160,000 to 180,000 kg/cm² and 1200°C to 1400°C. Stishovite is the stable phase of silica at pressures above 100 kbar (Fig. 2). In impact metamorphosed rocks, stishovite occurs mostly in minute grains associated with coesite and other silica phases. The first natural occurrence was recognized in 1962 by Chao and co-workers in shocked Coconino Sandstone at Meteor Crater, Arizona. Stishovite can be concentrated from shocked rocks by the same methods as coesite, and it is easily identified from its X-ray diffraction pattern and other properties (Table I), even if it is intimately mixed with coesite. Stishovite converts to cristobalite when it is heated to only a few hundred degrees; thus, it is probably no longer present in some shocked rocks that originally contained stishovite because of the thermal effects associated with strong shock.

The presence of stishovite is a reliable criterion for impact metamorphism.

BADDELEYITE

Baddeleyite is monoclinic ZrO_2 that commonly is formed in impact glass by the thermal decomposition of zircon as follows:

$$ZrSiO_4 \rightarrow ZrO_2 + SiO_2$$

[1]The one exception presently known is a coesite inclusion in a diamond.

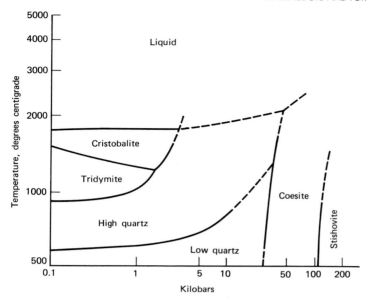

Figure 2   Silica phase diagram showing a portion of the stability fields of coesite and stishovite. (From *The Interpretation of Geological Phase Diagrams* by Ernest G. Ehlers. W. H. Freeman and Company, Copyright ©1972.)

El Goresy has described baddeleyite from a number of impact localities after discovering the occurrence of this mineral in samples from the Ries and Aouelloul Craters (9). The occurrence of baddeleyite in tektite glass also is well established (page 76). Baddeleyite can occur as irregular swarms of highly reflective rounded grains associated with amorphous silica, but it commonly occurs with silica and retains the euhedral crystal morphology of zircon (Fig. 3). Such inclusions in impact glass mostly are very small, and it is common to search a number of polished sections of shock-fused zircon-bearing rocks before finding an example. Baddeleyite resulting from the breakdown of zircon has not been reported in terrestrial volcanic or other igneous rocks. However, the mineral does occur rarely in vugs in terrestrial volcanic rocks and in some other geologic settings. Baddeleyite has been identified in a number of lunar samples in which it commonly occurs associated with ilmenite and ulvöspinel.

Experimental work indicates that the breakdown of zircon to baddeleyite and silica occurs at very high temperatures at one atmosphere pressure. Estimates of the temperature required for the breakdown of zircon (10) range from 1720°C to 1900°C in the simple system ZrO_2-SiO_2. No experimental work is available as yet that gives an indication of the required temperature at higher pressures in more complex multicomponent systems.

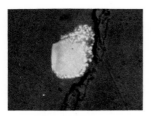

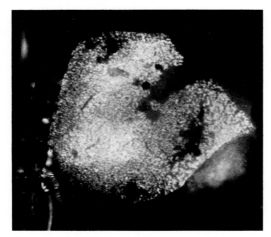

Figure 3 Baddeleyite resulting from the
breakdown of zircon in two impactite glasses. Top:
uncompleted breakdown of zircon to baddeleyite
plus silica from the Ries Crater, Germany, 600×.
Bottom: the outline of a former zircon grain now
composed of clusters of baddeleyite globules and
silica from the Aouelloul Crater, 600×. (Courtesy of
A. El Goresy.)

The occurrence of baddeleyite as a decomposition product of zircon in
glassy rocks is considered to be a reliable indicator of a very high temperature
history and a good criterion for shock metamorphism.

Baddeleyite as a breakdown product of zircon might also occur in
fulgurites, but it has not yet been reported. However, this would cause little
confusion because the geometry and occurrence of fulgurites as vitreous
crusts and hollow tubes could not easily be mistaken for fused or glassy rock
resulting from an impact.

LECHATELIERITE

Lechatelierite is silica glass. It is known in natural occurrences from a number
of impactites (11) and is common in tektites (page 76) and fulgurites.

Lechatelierite can be recognized easily by its very low refractive index, approximately 1.46, and chemical composition.

At surface temperature and pressure, lechatelierite exists metastably (Fig. 2) and most commonly is formed by the rapid melting and cooling of a preexisting silica phase. In the one-component system SiO_2, the temperature must exceed 1710°C. However, in multicomponent systems, such as common quartz-bearing rocks, the temperature required to melt silica may be somewhat lower. Nonetheless, the temperature required to melt quartz will exceed those temperatures normally achieved in near surface magmatic and volcanic processes.

One of the best-known occurrences of large pieces of lechatelierite is in the Libyan Desert and parts of the desert in western Egypt, where pieces (Fig. 4) of as much as a kilogram in weight are found scattered about the surface in a few areas. This "Libyan Desert Glass" probably is impactite, but conclusive proof is lacking at the present time.

The presence of lechatelierite in glassy, fused rocks is accepted as a good criterion for impact metamorphism, except for fulgurites. As indicated

Figure 4   Libyan Desert Glass (lechatelierite) in the Libyan Desert, Africa. The cobble has been faceted and polished by the windblown sand, and is translucent greenish yellow. The coin is approximately 2 cm in diameter. (Courtesy of V. E. Barnes.)

previously, fulgurites generally are easy to recognize because of their geometry and occurrence.

OTHER MINERALS

Other minerals can be used as possible indicators of shock history, but mostly because of their unusual textures in the surrounding glass or breakdown of euhedral crystals of one mineral phase to two or more others. Their presence alone is not a criterion for impact metamorphism. These modes of occurrence mostly indicate high temperatures and rapid cooling histories. Some particular textures may indicate strong subsolidus reduction. Minerals in this category include ilmenite, rutile, hematite, magnetite, chalcopyrite, spinel, and pseudobrookite (12).

CRYSTAL STRUCTURE DAMAGE

This kind of damage at a submicroscopic scale is characteristic of many minerals that have been shocked at high pressures. This phenomenon is easily seen in unrotated *single crystal* X-ray diffraction films of some minerals from highly shocked rocks (Fig. 5). Line broadening and "asterism" of X-ray

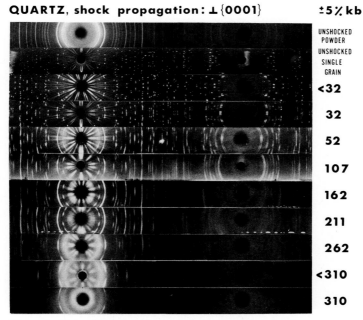

QUARTZ, shock propagation: ⊥ {0001}   **±5 ⁒ kb**

UNSHOCKED POWDER
UNSHOCKED SINGLE GRAIN
<32
32
52
107
162
211
262
<310
310

Figure 5 X-ray diffraction films showing the asterism and line broadening of quartz due to various intensities of shock (k bars, ±5% at right). The direction of shock propagation was perpendicular to {0001}. (Courtesy of Fred Hörz.)

diffraction spots indicates that there is a slight mosaicism of the crystal structure (13), or a breaking up of the structure into slightly nonparallel domains within what was formerly a single crystal with good long-range order. This effect can be observed in all of the major rock-forming minerals. Minerals that show this effect also commonly show highly undulatory extinction in crossed nicols. However, minerals with undulatory extinction induced by normal tectonic stresses do not show this effect.

DIAPLECTIC GLASS

Quartz and feldspar commonly are observed to have been converted to amorphous phases by the shock waves that accompany meteoritic impact. This is a solid state transition at relatively low temperatures. This phenomenon has been experimentally reproduced in nuclear explosions and in the laboratory (14). The term "diaplectic" is derived from the Greek word *diaplesso,* meaning to destroy by striking or beating, and was first proposed by von Engelhardt and Stöffler (15), who described this type of glass from the shock metamorphosed rocks of the Ries Crater. Diaplectic glass generally can be distinguished from ordinary glass by the absence of flow structure and vesicles, higher refractive index and density than ordinary glass (16) of the same composition, inclusions of coesite and/or stishovite in silica glass, and the euhedral shape of the previous crystalline phase that commonly is retained (Fig. 6). Diaplectic glass that retains the form of the previous crystalline phase is termed "thetomorphic" (17). Diaplectic glass is isotropic in crossed nicols and does not yield sharp X-ray diffraction lines, but only the single broad peak that is typical of most glasses.

The presence of diaplectic glass or of thetomorphs of silica or feldspar glass (thetomorphic or diaplectic plagioclase glass is *maskelynite*) is a reliable criterion of shock history. Such glasses are known to occur only in impact metamorphosed rocks.

PLANAR FEATURES

These features, also variously termed "planar elements," "shock lamellae," and the like, are known in quartz, feldspar, and other rock-forming minerals from a large number of impact structures and craters. They are easily visible in the optical microscope at intermediate magnification. Planar features were first brought to scientific attention in descriptions of rocks from Clearwater Lakes, Quebec, by McIntyre (18).

More than one set or orientation of planar features can occur in a single crystal grain, and as many as eight sets have been reported. Multiple sets of planar features have been produced in quartz-bearing rocks by experimental shock loading and by nuclear explosions. Short (19) first suggested that multiple sets of planar features might be a criterion for shock history. The orientations of planar features mostly are along rational crystallographic

Figure 6   Thetomorphic maskelynite (plagioclase glass) in the Shergotty eucrite. The light transparent plagioclase, *(a)* transmitted plane polarized light, has been transformed into glass that is isotropic in crossed polarizers *(b)* by shock. Width of fields of view is approximately 1.5 mm. (USNM 321.)

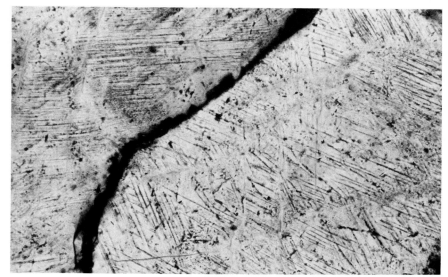

Figure 7   Multiple sets of planar features in quartz grains from shocked rocks at the Lake Mistastin impact structure. Such sets of planar features are highly suggestive of shock metamorphism. Plane polarized light, petrographic thin section, length of maximum dimension is approximately 0.5 mm. (Courtesy of Mike Dence.)

directions. For example, $\{10\bar{1}3\}$, $\{10\bar{1}1\}$, $\{11\bar{2}2\}$, $\{0001\}$ and $\{10\bar{1}0\}$ are some of the orientations of planar features that are observed commonly in shocked quartz. In shocked plagioclase, $\{001\}$, $\{010\}$, $\{100\}$, $\{1\bar{2}0\}$, $\{012\}$, and $\{130\}$ are among the orientations that are reported most frequently.

Planar features have been identified as microfractures, healed microfractures, slip planes, twin lamallae, and cleavage planes; however, most investigators now agree that the majority of planar features are glass that is less dense than the host phase.

Of course, not all of the planar features found in terrestrial quartz, feldspar, and other rock-forming minerals are the result of shock metamorphism. *Caution* must be used in applying this criterion. However, if multiple sets of planar features are observed that have the same general appearance as those shown here (Fig. 7) or in Short (20), and if the planar features occur in rocks that have any of the other mineralogic or petrologic criteria for the recognition of shock metamorphism, then shock metamorphism should be considered as a strong possibility.

KINK BANDS

Kink bands in biotite have been observed in rocks from a number of impact sites (21; Fig. 8), but this type of deformation can occur also as a result of

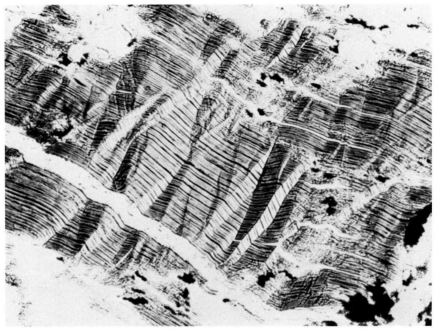

Figure 8   Kink bands in biotite from shocked granite gneiss, Ries Crater, Germany. Plane polarized light, length of field of view is approximately 0.5 mm.

tectonic forces. Some studies indicate that it may be possible to identify impact-induced kink banding by the crystallographic orientations and angles between adjacent segments (22). However, kink banding by itself generally is not accepted as a reliable criterion of shock metamorphism. Kink bands in biotite have been produced by nuclear explosions at the Atomic Energy Commission Nevada Test Site. Kink banding commonly occurs in quartzose and feldspar-rich biotite-bearing rocks in which the quartz and feldspar contain planar features.

UNDULATORY EXTINCTION

This may be extremely pronounced in many impact metamorphosed rocks and mineral grains. However, this is no criterion of shock metamorphism by itself. Strong undulatory extinction commonly results from tectonic forces in terrestrial rocks. Undulatory extinction must be considered together with other textural and mineralogic criteria. If the undulatory extinction is especially strong, or if there is other information that indicates that the rocks might be shocked, single grains should be x-rayed to search for line broadening or asterism (page 138).

FUSED ROCK GLASS

Fused rock glass or impactite (11) can be recognized on the basis of its mineral inclusions, heterogeneity, and general petrography. The glass may contain coesite, baddeleyite, lechatelierite, or other diagnostic inclusions, generally has strongly contorted schlieren or flow structure, and commonly contains mostly spherical or only slightly elongate vesicles (Fig. 9). Some of the mineral inclusions may show planar features, abundant microfractures, and strong undulatory extinction. Impactite rarely contains crystallites or microlites, which are so abundant in most volcanic glasses, unless the glass has been annealed in the lower portion of a hot fallback or base surge deposit. The most diagnostic criterion for the recognition of impact-melted glass is the presence of abundant, tiny nickel-iron spherules that are recondensed portions of the impacting body. However, cometary impact glass or the impact glass from stony meteorite craters might not be recognized by this criterion. The field occurrence is invaluable in the recognition of impactite, just as it is for many other geologic phenomena. A clear association of glass with a crater that contains abundant meteorite fragments, such as Meteor Crater, Arizona or the Wabar Craters of Saudi Arabia, presents a simplistic problem. However, the identification of impactite glass in a poorly exposed vein near a suspected impact structure or from an isolated surface fragment can be difficult. Rare igneous glasses, fulgurites, and synthetic glasses may show some of the same petrographic features. Tektites are a special variety of fused rock glass (see Chapter 2).

BRECCIA

Large volumes of breccia are associated with well preserved impact craters as interior fill or fallback deposits, as well as base surge and ejecta. The breccia tends to be chaotic with clasts of different lithologies mixed with little apparent order, depending on the lithologic variability of the target area.

The base surge and ejecta forms broad blanket deposits, which may tend to be glassy and more or less welded, that are superficially similar to some ash-flow tuffs. However, in detail the texture of the breccia is quite distinct from volcanic tuffs (Fig. 10). These breccias tend to contain abundant glassy impactite that has been fused or partially fused. Also, many of the individual rock fragments and mineral grains will show petrographic and mineralogic evidence of extreme shock.

If the crater that is the source of the breccias is large, individual fragments within the breccia may be quite large. For example, at Meteor Crater, Arizona, which is a relatively small impact crater, there are breccia blocks with their major dimensions as much as 30 m exposed on the crater rim. Breccia blocks of large size typically are found within craters along crater walls or immediately subjacent to the crater. However, in these occurrences

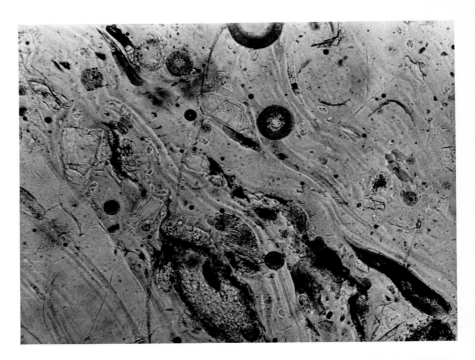

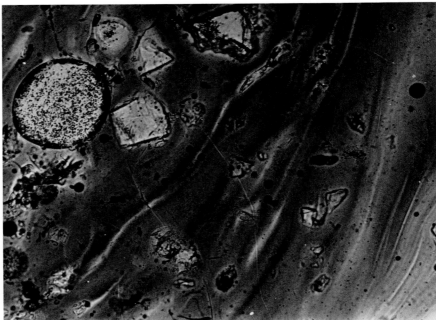

Figure 9 Two fields of view of impactite glass from the Ries Crater, Germany. Notice the many mineral inclusions, and strongly developed schlieren and spherical vesicles. Lengths of fields of view are approximately 750 μm. (Courtesy of D. Stöffler.)

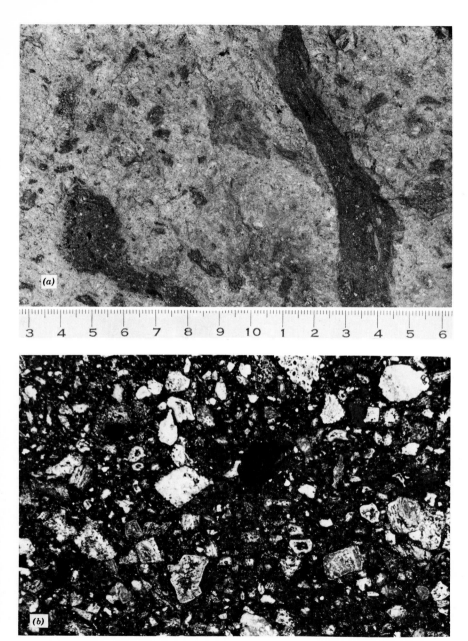

Figure 10 *(a)* Impact breccia (suevite) from the Ries Crater, Germany, collected at the quarry in Otting. Dark fragments are impactite glass (Fladen in part), matrix is fine grained rock, mineral, and glass fragments. The cathedral in Nördlingen is constructed from this rock, which is the fallback unit inside the Ries Crater. Scale is in centimeters, reflected light on fracture surface. *(b)* Polished slab of the Onaping Tuff from the Sudbury Structure, Ontario, Canada. Observe the diverse rock and mineral clasts and the densely welded glassy groundmass. Length of field of view is approximately 8 cm, reflected light.

the impact origin of the structure and the breccia usually is already known.

Some volcanic "vent agglomerates" may be texturally similar to some impact breccias. They do not, of course, contain clasts or mineral fragments that show shock damage.

If the crater is deeply eroded or totally eroded, the only breccia that remains may be pseudotachylite. This is a dark glassy rock with numerous clasts that commonly occurs in veins and dikes crosscutting the local country rocks adjacent to many impact sites (e.g., the Vredefort Structure, Africa). The included clasts and mineral grains may show evidence of shock damage, but thermal annealing caused by the enclosing molten rock that cools to form the matrix of the pseudotachylite may remove or reduce the evidence of shock. Much of the pseudotachylite that is abundant in many Precambrian areas of the Earth may have been impact produced, but definitive evidence is lacking.

STRUCTURAL CRITERIA

SHATTER CONES

The passage of shock waves in some rock types generates shatter cones, an easily recognizable structure that apparently is diagnostic of shock. These shatter cones range in size from more than 15 m long to less than 1 cm. Shatter cones have been recognized at many terrestrial impact sites including the Sudbury Structure (Fig. 11), Ries and Steinheim Basins, Sierra Madera, Flynn Creek, Gosses Bluff, and many others. R. S. Dietz has been the foremost advocate of the occurrence and orientations of shatter cones as indicators of shock history (23). Experimental work with shock waves has produced shatter cones in both natural and artificial materials, and crude cones may be formed at the bases of shot holes in quarries, mines, and road cuts.

True shatter cones are easy to recognize by their overall conical shapes and the braided striations that lie on the surface of the cone, which tend to have a "horsetail-like" appearance from the apex of the cone down the flanks. Shatter cones have been observed in a wide variety of rock types including limestone, dolomite, quartzite, gneiss, and shale.

Very large numbers of shatter cones occur at many impact sites, and if their orientations are measured and compensations are made for post-cone movements of the blocks in which they occur, the apices of the cones tend strongly to point toward the center of the structure. Presumably the apices point toward the advancing shock front and thus to the exact site of the impact. For deeply eroded structures, shatter cones may be found to point toward the center of the structure and also upward (e.g., Gosses Bluff). In these structures, the shatter cone orientations can be used to estimate the depth of erosion since the cratering event.

Figure 11 Large shatter cones in the Mississagi Quartzite on the south shore of Kelly Lake southwest of the Sudbury Structure. The cone axes are nearly parallel to the bedding of the steeply dipping rocks. When the beds are rotated back into their originally horizontal position, the apices and cone axes point to the approximate center of the Sudbury Structure. The hammer in the foreground is 41 cm long. (Courtesy of B. M. French.)

Other structural criteria for the recognition of impact sites are related to major crater structure and not to the features that might be found on a single hand specimen or a single rock outcrop (see Terrestrial Impact Craters).

INTERPRETATION OF SHOCK HISTORY

Pressure and temperature histories of individual samples are the objectives of most attempts to interpret shock history, other than simply to recognize that the sample has been shocked. Petrologic and mineralogic criteria for various pressures and temperatures have been recognized by comparison with experimental shock studies and nuclear explosion effects. The pressure-temperature history of an individual naturally shocked sample can only rarely be known within fine limits, but some limits *can* be determined. These interpretations are subject to the prodigious petrologic and mineralogic complexities of rocks and the seemingly infinite number of partially isolated microenvironments that can be found within them. It may be necessary to have the geometric relation of the sample in question to the cratering event and to the postcratering deposits, that is, good field information.

However, if a portion of the pressure-temperature history can be established, then it can be used to interpret the mineralogy, petrology, and even the chemistry of samples prior to the impact event. For example, if the maximum temperature experienced by a piece of impactite glass can be established, then proper consideration of alkali volatilization can be made in trying to reconstruct the chemistry of the parent rock. This kind of problem has proved to be important in the examination and interpretation of lunar samples. By the study of a number of samples from the same ejecta unit, or other stratigraphic unit in or around an impact crater or site, it may be possible to arrive at definite conclusions concerning the temperature of the unit at emplacement or the mode of emplacement of the unit itself.

Two noteworthy attempts to organize and quantify a portion of this information have been made (15, 24). These systems of classification (Tables II and III) are, at best, semiquantitative, but in most instances they can be used as a guide to the relative amount of shock metamorphism that an individual sample has experienced.

Table II   Classification of Stages of Shock Metamorphism[a]

Pressure (kbar)	Stage of Shock Metamorphism	Characteristic Deformations and Phase Transitions	Residual Temperature (°C)
ca. 100			ca. 100
	Stage I	Fracturing Plastic deformation (diaplectic quartz and feldspar)	
250–300			200–300
	Stage II	Phase transitions (diaplectic glasses of quartz and feldspar, high-pressure phases of SiO_2)	
500–550			1200–1500
	Stage III	Selective melting (normal glasses of quartz and feldspar, high-pressure phases of SiO_2)	
600–650			2000–3000
	Stage IV	Melting of all main rock-forming minerals (inhomogeneous rock melts, Fladen)	
ca. 1000			ca. 5000
		Volatization	

[a]From (15), Engelhardt, W. v., and D. Stöffler, Stages of Shock Metamorphism in the Crystalline Rocks of the Ries Basin, Germany, in *Shock Metamorphism of Natural Materials,* B. M. French and N. M. Short, Mono Book Corp., Baltimore, Md. (1968).

Table III Classification of Shock Metamorphosed Rocks and Shock Damage.[a]

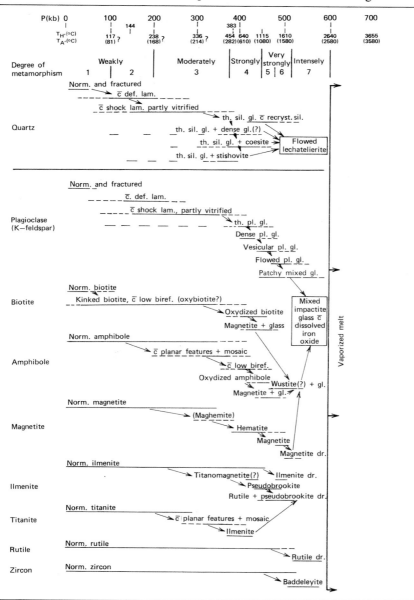

[a] T_H' is the peak temperature behind a shock wave associated with a given compression, T_A' is the residual temperature after the shock is relieved from a given Hugoniot temperature. Abbreviations: biref.–birefringence, c–with, def. lam.–deformation lamellae, dr.–droplets, gl.–glass, norm.–normal, pl.–plagioclase, qtz.–quartz, sil.–silica, th.–thetomorphic, recryst.–recrystallized. From (24), Chao, E. C. T., Pressure and Temperature Histories of Impact Metamorphosed Rocks — Based on Petrographic Observations, in *Shock Metamorphism of Natural Materials*, B. M. French and N. M. Short, Mono Book Corp., Baltimore, Md. (1968).

149

REFERENCES AND NOTES

1. Merrill, G. P. (1907) On a peculiar form of metamorphism in siliceous sandstone: Proc. U. S. Natl. Mus., vol. 32, p. 547-551.

2. For an example see Shock Metamorphism of Natural Materials, B. M. French and N. M. Short, eds., Mono Book Corp., Baltimore, Md., 1968, 644 p. This volume is the proceedings of a conference held at the NASA, Goddard Space Flight Center, April, 1966.

3. Coes, L., Jr. (1953) A new dense silica: Science, vol. 118, p. 131-132.

4. Nininger, H. H. (1956) Arizona's Meteorite Crater: Am. Meteorite Mus., Sedona, Ariz. 232 p. See footnote p. 50; also p. 154.

5. Chao, E. C. T., E. M. Shoemaker, and B. M. Madsen (1960) First natural occurrence of coesite: Science, vol. 132, p. 220.

6. Fahey, J. J. (1964) Recovery of coesite and stishovite from Coconino sandstone of Meteor Crater, Ariz.: Am. Mineralogist, vol. 49, p. 1643-1647.

7. Birch, F. (1952) Elasticity and constitution of the Earth's interior: Jour. Geophys. Res., vol. 57, p. 234.

8. Stishov, S. M., and S. V. Popova (1961) A new dense modification of silica: Geokhimya, No. 10, pp. 837-839 (translated into English in Geochemistry, no. 10, pp. 923-926).

9. El Goresy, A. (1965) Baddeleyite and its significance in impact glass: Jour. Geophys. Res., vol. 70, p. 3453-3456.

10. See Kirby, D. (1944) Pure oxide refractories: Metallurgica, vol. 30, p. 65; also Phase Diagrams for Ceramists (1964) compiled by E. M. Levin, C. R. Robbins, and H. F. McMurdie, The American Ceramic Society, Figs. 361 and 362, p. 141.

11. The term "impactite" was first suggested by H. B. Stenzel (see V. E. Barnes, 1939, North American tektites: Univ. of Texas, Bur. Eco. Geo. Pub. 3945, pt. 2, p. 558) to designate glassy rocks that originated as "splashes" from meteorite impact craters. The term now generally is used to designate all rocks that have been fused partially or totally by any type of hypervelocity impact.

12. For a thorough and well-illustrated treatment of shocked opaque minerals see A. El Goresy (1968) The opaque minerals in impactite glasses: in Shock Metamorphism of Natural Materials, B. M. French and N. M. Short, eds., Mono Book Corp., Baltimore, Md., p. 531-553.

13. Dachille, F., P. Gigly, and P. Y. Simons (1968) Experimental and analytical studies of crystalline damage useful for the recognition of impact structures: in Shock Metamorphism of Natural Materials, B. M. French and N. M. Short, eds., Mono Book Corp., Baltimore, Md., p. 555-569; see also M. E. Lipschutz, and R. R. Jaeger (1966) X-ray diffraction study of minerals from shocked iron meteorites: Science, vol. 152, p. 1055-1057.

14. Milton, D. J., and P. S. De Carli (1963) Maskelynite: formation by explosive shock: Science, vol. 140, p. 670; see also J. Wackerlie (1962) Shock wave compression of quartz: Jour. Appl. Phys., vol. 33, p. 922; and P. S. De Carli, and J. C. Jamieson (1959) Formation of an amorphous form of quartz under shock conditions: Jour. Chem. Phys., vol. 31, p. 1675.

15. Engelhardt, W. v. and D. Stöffler (1968) Stages of shock metamorphism in crystalline rocks of the Ries Basin, Germany: in Shock Metamorphism of Natural

Materials, B. M. French and N. M. Short, eds., Mono Book Corp., Baltimore, Md., p. 159-168.

16. For an example of the behavior of glass under shock wave conditions see J. Arndt, U. Hornemann, and W. F. Müller (1971) Shock wave densification of silica glass: Phys. Chem. Glasses, vol. 12, p. 1-7.

17. Chao, E. C. T. (1967) Impact metamorphism: in Researches in Geochem., P. H. Abelson, ed., Wiley, New York, vol. 2, p. 204-233.

18. McIntyre, D. B. (1962) Impact metamorphism at Clearwater Lakes, Quebec: Jour. Geophys. Res., vol. 67, p. 1647, *abstract*.

19. Short, N. M. (1966) Effects of shock pressures from a nuclear explosion on mechanical and optical properties of granodiorite: Jour. Geophys. Res., vol. 71, p. 1195-1215.

20. Short, N. M., and T. E. Bunch (1968) A worldwide inventory of features characteristic of rocks associated with presumed meteorite impact structures: in Shock Metamorphism of Natural Materials, B. M. French and N. M. Short, eds., Mono Book Corp., Baltimore, Md., p. 255-266.

21. Stöffler, D. (1966) Zones of impact metamorphism in the crystalline rocks of the Nördlinger Ries crater: Contr. Mineral. and Petrol., vol. 12, p. 15-24; see also E. C. T. Chao (1967) Shock effects of certain rock-forming minerals: Science, vol. 156, p. 192-202; and T. E. Bunch (1968) Some characteristics of selected minerals from craters: in Shock Metamorphism of Natural Materials, B. M. French and N. M. Short, eds., Mono Book Corp., Baltimore, Md., p. 413-432.

22. Hörz, F. (1970) Static and dynamic origin of kink bands in micas: Jour. Geophys. Res., vol. 75, p. 965-977. For a more comprehensive treatment see H. Schneider (1971) Deformation und umwandlung von biotiten aus gesteinin des Nördlinger Rieskraters durch stosswellenmetamorphose: unpubl. doctoral disser., Mineral. and Petrol. Inst., Univ. Tübingen, Germany.

23. Dietz, R. S. (1947) Meteorite impact suggested by orientation of shatter cones at the Kentland, Indiana disturbance: Science, vol. 105, p. 42-43; also Dietz, R. S. (1960) Meteorite impact suggested by shatter cones in rock; Science, vol. 131, p. 1781-1784; and numerous other papers by the same author.

24. Chao, E. C. T. (1968) Pressure and temperature histories of impact metamorphosed rocks — based on petrographic observations: in Shock Metamorphism of Natural Materials, B. M. French and N. M. Short, eds., Mono Book Corp., Baltimore, Md., p. 135-158.

SUGGESTED READING AND GENERAL REFERENCES

Shock Metamorphism of Natural Materials: B. M. French and N. M. Short, eds., Mono Book Corp., Baltimore, Md., 1968, 644 p.

Stöffler, D. (1972) Deformation and transformation of rock-forming minerals by natural and experimental shock processes, I. Behavior of minerals under shock compressions: Fortschr. Mineral., vol. 49, p. 50-113.

Meteorite Impact and Volcanism: F. Hörz, ed., proceedings of a conference at the Lunar Science Institute, Oct., 1970, Jour. Geophys. Res., vol. 76, p. 5381-5798.

We are indeed fortunate that one of the most interesting bodies in the Solar System is our nearest neighbor in space.   Harold C. Urey, 1967

6. The Moon

INTRODUCTION

Many peoples of the Earth, through thousands of years of recorded history, have attached religious or poetic significance to the Moon. The Moon has been observed by billions of untrained observers, generally in pairs, but scientific and systematic observations of the Moon are rather recent. The relative motion of the Moon played a role in the development of models of the Solar System and planetary motion by Heraclides, Aristarchus, Ptolemy, Copernicus, Brahe, Kepler, and many lesser known investigators (1). Leonardo Da Vinci speculated correctly that the dimly lighted dark side of the lunar disk seen just at the new Moon was caused by sunlight reflected from the Earth (2). However, the evolution of scientific thought concerning the features of the lunar surface was slow prior to 1609. In that year, Galileo constructed an astronomical telescope and began serious observations of the Moon and its surface features. His first observations and drawings (Fig. 1) were published in a brief work in Padua in 1610.[1] He clearly observed the light and dark surface areas of the Moon and stated that they were analogous to the continents and oceans of the Earth. In addition, he accurately located and represented a number of major circular depressions (craters), and recognized the dominance of this topographic form on the lunar surface. The use and number of astronomical telescopes grew rapidly, as did the sizes, complexity, and quality of the instruments. Within the next 100 years, *many* maps and drawings of the Moon were published by a number of astronomers.

[1]The title of Galileo's work, ''Nuncius Sidereus'' is translated as ''The Starry Messenger.''

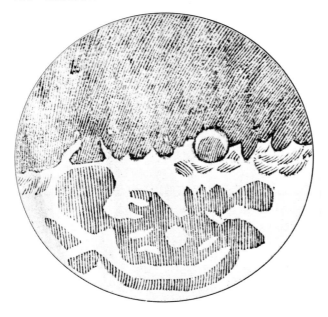

Figure 1 Two of Galileo's drawings of the Moon, published in 1610. Although he identified craters as the dominant topographic form on the lunar surface, several major and easily visible features on the lunar surface are not recorded on his drawings, for example, Copernicus and Tycho. Compare with the full Moon Earth-based telescopic photograph in Fig. 15. (Courtesy of the Anthony Michaelis Collection.)

Sometime in the early part of this century it became fashionable for astronomers to work with more distant objects, and lunar observations were carried on mostly by amateurs and relatively untrained workers. We are indeed fortunate that professional astronomers such as G. P. Kuiper, A. Dollfus, E. J. Öpik, and Z. Kopal continued their interests in lunar and planetary work. Otherwise, our fund of information on which to begin the exploration and observation of the Moon from unmanned spacecraft and by the astronauts in the Apollo Program would have been meager indeed.

The distance from the Earth to the Moon ranges between 356,410 and 406,697 km, the diameter of the Moon is approximately 3474 km, and the period of the Moon's rotation is equal to the period of revolution. The latter fact has the result that the Moon always maintains the same side toward the Earth. The Moon has some minor motions (librations) that result in a slight variation of the area of the visible disk. The libration in lunar longitude is more than 7° but only approximately 1° in lunar latitude. The Moon's atmosphere is virtually nonexistent; in fact, the total atmospheric pressure near the lunar surface is that of a very "hard" vacuum, approximately 10^{-13} torr or less. The small mass of the Moon, compared with the Earth's, results in a lunar gravitational constant that is one sixth that of the Earth. Lunar surface temperatures range from roughly $-158°C$ to more than $130°C$.

TELESCOPIC AND SPACECRAFT IMAGERY

HISTORY

The craters of the Moon had been observed for hundreds of years, yet fierce arguments raged as to their origin. G. K. Gilbert, in 1893 (3), developed a comprehensive theory of impact to explain the craters of the Moon, including the concept that a huge impact formed Mare Imbrium.[2] In 1924, A. C. Gifford (4) compared the craters on the Moon with Meteor Crater, Arizona and concluded that the lunar craters were explosive craters caused by the impacts of large meteorites. He explained the existence of central peaks in some lunar craters by this mechanism, and included calculations of the energy budgets of large craters. Although some of his calculations are incorrect in the light of later data, his was a pioneering work that arrived mostly at the correct conclusions. A far different conclusion was drawn by S. Mohorovicic (5) in 1928, who believed that the lunar craters were explosion features, but who thought that they were caused by immense volcanic explosions. He even conducted experiments to show that the rays and streaks that extend from many lunar craters were similar to those features produced by small

[2]It is interesting to note this incisive work of Gilbert at this early date and to contrast it with his later *denial* that Meteor Crater had been formed by meteoritic impact (page 98).

explosions in cement powder. Alfred Wegener conducted similar experiments by throwing spoonfuls of cement powder onto a planar surface of the same material (6). He concluded that the craters of the Moon were a simple result of the impacts of meteorites. Wegener is better known for his work and speculations on continental drift, many of which are only now gaining general acceptance by the scientific community.

An early statement of the stratigraphic principles that could be applied to the geologic mapping of the Moon was made by Barrell (7); however, apparently he did not apply his insight to the problem. L. J. Spencer (8) concluded, in a remarkably farsighted paper, that lunar craters were the result of meteorite impacts, and he compared the mare materials to the great basalt flows of the Columbia River Plateau in the United States and to the Deccan Traps of India. There was no doubt in his mind that the lunar maria were flooded with mafic rocks from lava flows. In more recent years, just prior to the Apollo landings, the argument continued. E. M. Shoemaker championed the impact origin of the majority of lunar craters, and the opposing view (volcanic origin) was ably argued by Jack Green, who relied heavily on Earth analogs to lunar structures. The arguments were summarized by Baldwin (9), who added his own interpretations and conclusions, but chiefly supported the impact origin. Similar summary works, also containing original contributions and interpretations, were published by Fielder (10) and others.

The fundamental differences between the relatively low, smooth, lightly cratered and low albedo lunar maria and the higher elevation, rougher, brighter and more densely cratered lunar highlands (terrae) were recognized by virtually all observers. However, the chemical or physical reasons or both for these differences were the subject of much speculation (9).

GEOLOGIC MAPPING

A major step in lunar exploration, prior to the direct investigation of the Moon by men, was the systematic geologic mapping of the Moon's visible face, which was accomplished mostly by workers with the U. S. Geological Survey. R. J. Hackman, a photogeologist, began recent attempts to unravel the stratigraphy of the Moon and to map its surface (11). He later collaborated with Shoemaker (12), who considerably extended and developed the ideas and concepts used in the original work, as well as developed new interpretations. NASA funding led to a rapid acceleration in lunar geologic mapping, which was organized and managed by Shoemaker and involved a large number of investigators. Along with the progress in the systematic mapping came increased understanding of lunar stratigraphy. The state of lunar stratigraphic knowledge that resulted from telescopic observations of the lunar surface has been summarized by D. E. Wilhelms (13), who detailed the rationale and methodology of lunar geologic mapping. In addition to Earth-based telescopic observations, later lunar mappers had the benefit of high resolution imagery of the lunar surface that was obtained by the Ranger, Surveyor, and Orbiter spacecrafts. (For examples, see Figs. 2 to 6.) These missions played an important role in the identification of prospective landing sites for the Apollo lunar missions, as well as gathered much more detailed surface information than could be obtained from Earth-based instruments. The Lunar Orbiter Program was especially effective in obtaining high resolution imagery of a large portion of the lunar surface.[3] Little work currently exists that interprets

[3]It should be emphasized that much of this imagery still has not been exhaustively studied or used in lunar geologic work. The number of images and the areas covered are simply far too large.

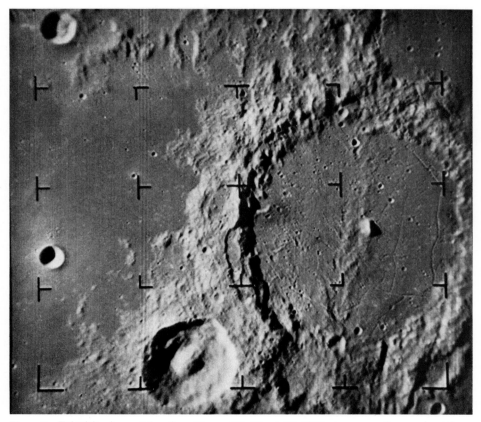

Figure 2   Television image taken from Ranger IX, March 24, 1964, at a range of approximately 400 km above the lunar surface, 2 min and 50 sec before impact. Field of view is 174 by 194 km. The rilles, dark haloed craters, and central ridge and peak in the floor of the Alphonsus are clearly visible. Note the smooth margin of Mare Nubium at left. (Courtesy of NASA, Jet Propulsion Laboratory.)

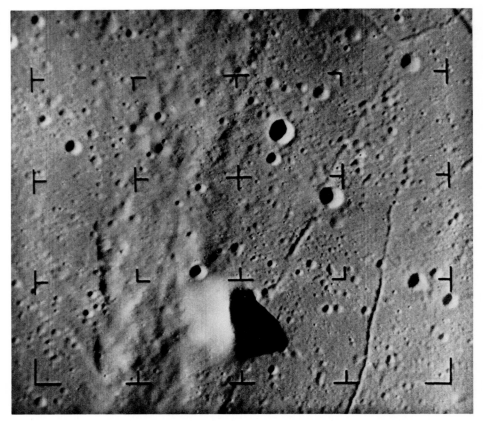

Figure 3 Close-up television image from Ranger IX of the central peak, central ridge, and northeastern floor of the crater Alphonsus taken 39 sec before spacecraft impact. The final images transmitted by Ranger IX permitted the recognition of craters less than 1 m diameter. The field of view is 42 by 45 km. (Courtesy of NASA, Jet Propulsion Laboratory.)

the stratigraphy of the averted face of the Moon in any but the most general terms; however, many images of the backside exist and this effort could be undertaken by interested scientists or students. The most recent lunar mapping and geologic work has had the use of higher resolution photography obtained by several different cameras that were flown in lunar orbit on the Apollo missions (14, Figs. 7 and 8).

The basic observation that makes large-scale lunar photogeology possible is that each of the circular mare basins on the near side has associated with it a number of concentric scarps, radiating fractures, and extensive ejecta

deposits. From simple transection and overlap relations, the relative ages
of the mare basins can be deduced (Fig. 9). These structural features and the
deposition of the ejecta are assumed to be approximately contemporaneous
with the formations of the basins by large impacts that have fractured the
lunar surface and have thrown out extremely large volumes of lunar rock in
widespread ejecta and fallback deposits. On a smaller scale, the same types of
observations can be applied to virtually all lunar craters and terrains. Relative
ages can be established by simple observations of the interruption of one
structure or feature by another. However, the mare basins and their

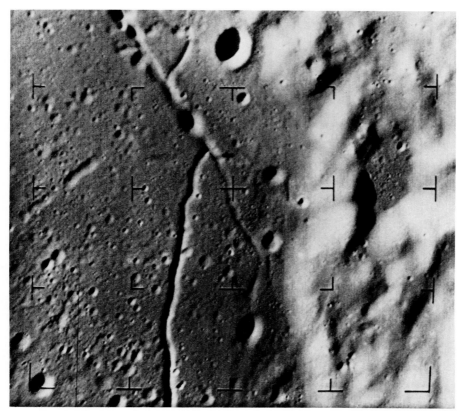

Figure 4   Television image taken from Ranger IX, 1 min and 12 sec before its impact in the
floor of the crater Alphonsus. The rilles and dark haloed craters (probably small pyroclastic
cones) on the northeastern floor of the crater are shown as well as a portion of the eroded
eastern wall. The field of view is 28.3 by 33.3 km. (Courtesy of NASA, Jet Propulsion
Laboratory.)

Figure 5 Spherical photomosaic of television images transmitted from Surveyor VII, which landed approximately 29 km north of the crater Tycho in the southern lunar highlands. The landing site is on the ejecta from Tycho, a prominent rayed lunar crater. The small crater in the foreground is approximately 2 m in diameter and appears to have been formed by the impact of a block of secondary ejecta from some other nearby crater. Broken fragments of the block can be seen scattered about the crater and the nearby surface. The horizon is about 13 km away in the central portion of the mosaic. (Courtesy of NASA, Jet Propulsion Laboratory.)

associated ejecta establish the large-scale markers in lunar time-stratigraphic units.

The geologic units mapped on the Moon were defined by surface texture, albedo, color, relative age, and stratigraphic position (wherever possible). In addition, in some definitions of stratigraphic units there are implicit, if not explicit, assumptions or interpretations as to how the units were formed or deposited. Clearly maps constructed on these bases are not geologic maps in the terrestrial sense, on which units of clearly different lithology are delineated. However, it is reasonable to assume that a stratigraphic unit with

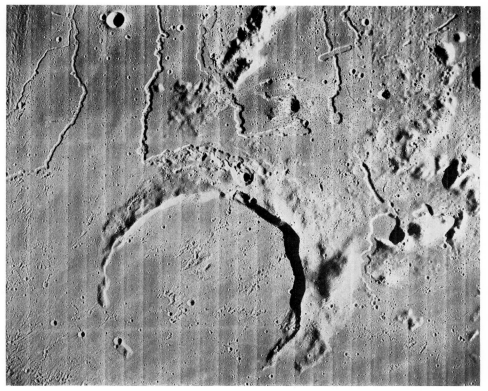

Figure 6   An example of a Lunar Orbiter image, the partially flooded crater Prinz with a striking array of sinuous rilles. Notice also the abundant secondary craters and crater clusters on the mare surface. (Lunar Orbiter V image, courtesy of the NASA, Langley Research Center.)

homogeneous appearance probably has some common lithologic or other features throughout its extent. These features may be texture, composition, mineralogy, time of deposition, structural history, or a combination of these. In any event, it is difficult to conceive how one could have improved on this method for lunar geologic mapping with the meager data available at the time, and many maps were constructed based on data from telescopic observation and spacecraft imagery (Fig. 10). Whatever the mapped stratigraphic units may be, it is hoped that samples collected from any individual map unit will be representative of the entire map unit in some way. A composite geologic map of the near side of the Moon has been compiled by Wilhelms and

Figure 7   Hasselblad photograph from lunar orbit by the crew of Apollo 8, showing details of crater morphologies, rilles, and grabens in and around the crater Goplenius. The diameter of the crater is approximately 65 km, and it is located on the eastern half of the visible face at about 45° E, 10° S. North is at the bottom of the photograph. (Courtesy of NASA, Johnson Space Center.)

McCauley as well as a geological province map (15, Figure 11). Even though most lunar geologic maps have been made without the aid of surface information, the interpretations and assumptions mostly have been borne out by closer observation. However, the Apollo 16 landing site was selected based at least partly on geologic interpretations that later proved to be false.

NATURE OF THE LUNAR SURFACE

Although the large-scale geological features seemed to be interpretable, prior to close-up Ranger and Surveyor imagery and prior to the actual return of the

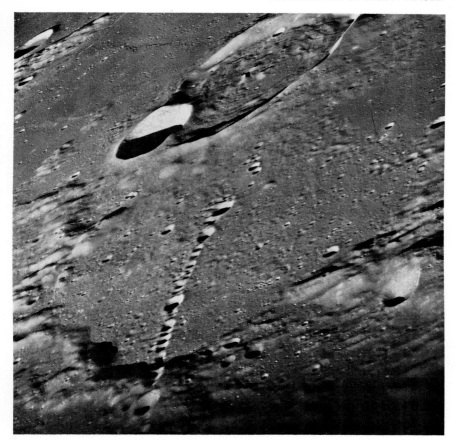

Figure 8   Hand-held Hasselblad view of the lunar surface near the center of the visible face by the Apollo 12 crew. The chain of secondary craters in the foreground is Davy Rille. The two craters in the background are of strikingly different ages. Crater at left (smaller) is relatively fresh and recent, although ejecta and ray patterns from the crater are not readily visible. The older (larger) crater has a much smaller depth to diameter ratio and a much more softened outline and degraded rim. The view is southwest. (Courtesy NASA, Johnson Space Center.)

first lunar samples, there was intense interest in the exact nature of the lunar surface at small scales. It is the physical state, composition, and texture of the immediate surface and the shallow subsurface that must account for most of the optical, electrical, thermal, and mechanical properties. Many types of remote measurements had been made of the lunar surface, for example, albedo, polarization, radar characteristics, and variations of different properties with phase angle, and numerous attempts were made to interpret these data in terms of lunar surface structure and chemistry (16). However,

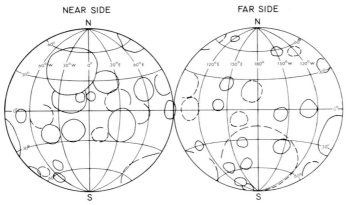

NEAR SIDE
N

FAR SIDE
N

Figure 9 Large lunar basins and multiringed basins. Circles indicate the main basin rings. The ejecta blankets and structures associated with these basins provide the foundation for lunar stratigraphy. (Compiled by D. E. Wilhelms, K. A. Howard, D. E. Stuart-Alexander, B. K. Lucchitta, and M. N. West, U. S. Geological Survey.)

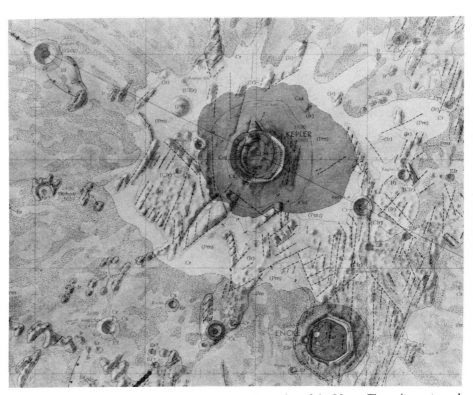

Figure 10 A portion of Hackman's map of the Kepler region of the Moon. The units portrayed indicate relative ages and modes of emplacement: old terrain (Ir) and old crater materials (Iar and Iaf); mare material (Pm); post-mare crater materials (Er, Ef, Elr, Elf; Cr, Cf); and recent slope materials (Cs). (From Hackman, 1962, 11.)

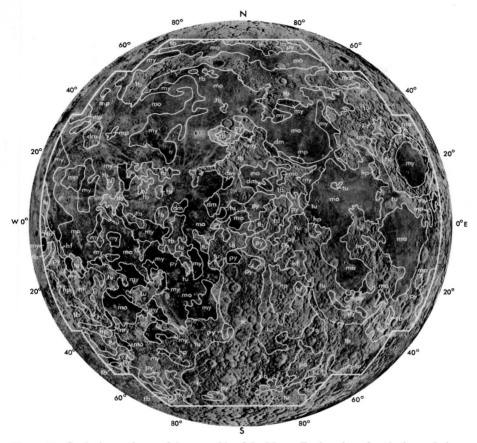

Figure 11 Geologic provinces of the near side of the Moon. Explanation of geologic symbols: dm — dark mantles; mp — mare plateaus; my — mare, young; mo — mare, old; py — plains, young; po — plains, old; hf — hilly and furrowed; hp — hilly and pitted; tb — terra of younger basins; tu — terra, undivided; te — terra, cratered. Solid line indicates province boundary, dashed where obscured by younger (unmapped) deposits. (From McCauley and Wilhelms, 15; Courtesy of Academic Press.)

these results were the subjects of hot debates by many lunar scientists, and no coherent model of the lunar surface gained wide acceptance except for a rather general one.

It was generally proposed by most workers that the lunar surface was covered by a layer of loose particulate material (the lunar regolith) whose particles resulted mostly from the erosive force of meteorites, secondary ejecta, and mass wasting. Also volcanic ash or tephra were thought to be possible components. The composition, thickness, mechanical properties, grain size distribution, and the like of the regolith were very difficult to deduce from information prior to the first Surveyor landing, and they were the object of much argument. Salisbury and Smalley (17) stated the prelanding model well and pictured the lunar surface as composed chiefly of overlapping ejecta blankets from lunar impact craters (Fig. 12). Shoemaker (18), as well as many other geologists and astronomers studying the Moon, concurred in this model. The currently available data from the Apollo explorations leave little room to doubt the reality of this basic structure for the surface layer over a large portion of the Moon.

Unfortunately for some aspects of lunar exploration, the thickness of the regolith is such that actual outcrops of lunar rocks in situ are very scarce on the lunar surface in areas that are readily accessible to men and relatively simple machines. Thus sampling the lunar surface became, in part, a sophisticated statistical problem associated with the collection of samples of loose debris. The regolith was estimated to be 3 to 6 m thick at the Apollo 11 landing site (19), thought to be only approximately 2 m at the Apollo 12 landing site, 8.5 m at the Apollo 14 site (20), and even thicker at the other highlands localities.

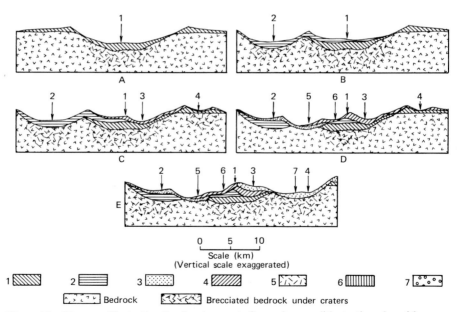

Figure 12   Diagrams illustrating the development of complex regolith stratigraphy with only seven impacts of different sizes. This was the model proposed for the evolution of the regolith in the lunar highlands by Salisbury and Smalley (17) in 1964. (Courtesy of Academic Press.)

ESTIMATION OF SURFACE AGES

Relative ages of surface features, especially craters, can be estimated by their sharpness of detail and outline, as well as total relief, if compared with other features of similar size. The longer a surface feature is exposed to the bombardment of meteoroids at the lunar surface, the more eroded and "softened" it appears. Thus craters with obvious ray patterns of ejecta and very sharp detail in their morphologies, such as Copernicus, Tycho, Aristarchus, and Kepler, are comparatively young and recent (Fig. 13). Craters with battered outlines, softened morphology, and no visible ray pattern, such as Gassendi, Catherina, Cavendish, and Maurolycus, have been present on the lunar surface for much longer and are considerably older (Fig. 2). This type of relation as a guide to the ages of craters and the ages of terrains on which craters can be seen was developed extensively by Pohn and Offield (21). They arrived at first-order conclusions about the relative ages of various areas of the visible face by identifying the oldest crater on the terrain in question.

If it is assumed that the predominant number of craters on the lunar surface have resulted from the impacts of meteoroids, asteroids, and comets, and that the surface distribution of impacts is random, it should be possible to determine the relative ages of surfaces on the Moon by counting the density of craters (craters per unit area). The longer a surface is exposed on the Moon, the more craters it should have because of the greater opportunity for impacts. The earliest counts of lunar craters were made on Earth-based telescopic photography (22, 23) and only included craters greater than 10 km in diameter. This work was extended down to craters approximately 1 km in diameter by Öpik (24), but these data were subject to large uncertainties because of the resolution of the images, variations in lighting conditions of the surface, and other problems. Various other authors have developed the statistics and the rationale for accuracies of crater counts to a high degree (25). The statistics of lunar crater counts were discussed at length by Shoemaker (18) who estimated the meteoroid bombardment exposure ages of various portions of the Moon from crater density distributions, based on Ranger and other spacecraft imaging. Numerous other authors have used similar procedures to attempt to establish the ages of various parts of the lunar surface or to estimate the size frequency distribution of objects in the inner part of the Solar System during early geologic time (26). However, there are fairly large uncertainties in this method for determining *absolute* ages of the lunar surface, including: (a) the size frequency distribution of meteoroids in near lunar space at present and through geologic time is largely unknown; (b) many of the lunar craters may be volcanic, secondaries, collapse craters, or of other origins that cannot be recognized and disregarded in such counts; and (c) both the velocity and mass of the object that caused any specific crater are unknown. The absolute errors inherent in this method are

Figure 13   Low sun angle Hasselblad photograph of the crater Eratosthenes showing the typical morphology of a large impact crater. Notice the numerous secondary craters from both Copernicus and Eratosthenes; in particular, note the "herring bone" structure associated with the ejecta from secondary craters in the foreground. Such structure is typical of fresh chains of secondary craters. At center right is the southwestern end of the Apennine Mountains. View is north-northwest across a wide expanse of Mare Imbrium (distance), showing the very smooth topography of the mare surface. The rim to rim diameter of Eratosthenes is approximately 66 km. Photograph is from Apollo 12. (Courtesy NASA, Johnson Space Center.)

pointed out by the estimates of mare surface ages from crater counts that were made immediately prior to the return of samples from the Moon, most of the estimates were an order of magnitude too young. A post-Apollo view and reinterpretation of lunar cratering has been published by Hartmann (27). After radiometric dating of surfaces on the Moon from returned samples, it is apparent that the flux of large objects in the vicinity of the Moon decreased very rapidly in early Solar System history.

SURFACE TRANSIENT PHENOMENA

There is a long history of reported changes and transient conditions on the lunar surface. Many of them have been reported by amateur or unknown observers and their authenticity cannot be established. However, a number of observations have been made by qualified and experienced observers. Sir William Herschel reported what he believed to be ''an erupting volcano'' near the lunar crater Aristarchus in 1783. Again, he observed three bright spots on the portion of the Moon that was illuminated by earthshine. A number of other observers have contributed to this literature, including W. Goodacre, W. H. Pickering, F. H. Thornton, W. H. Wilkins, D. Alter, and many others. Moore (28) has reviewed the previous literature of reported lunar changes and has even tested the resolving power and optical characteristics of older instruments to determine whether or not some observers *could* have seen what they reported.

One of the most interesting and well documented recent observations of a transient lunar event was recorded by N. A. Kozyrev, who obtained a spectrogram of a transient phenomenon on the central peak of the crater Alphonsus (29). Kozyrev also visually observed that the central peak appeared brighter than usual at the time that the spectrum was obtained (30).

Another important observation of lunar bright spots and color phenomena was made in 1963, by Greenacre and Barr (31) through the 61 cm refracting telescope at the Lowell Observatory. They observed pink to red spots on and around the crater Aristarchus on two different occasions. These observations were confirmed by several observers. They also observed a violet to bluish haze in the crater immediately following the disappearance of the red spots (32).

There is a tendency for the phenomena to be observed on crater rims, central peaks, and around the edges of the mare basins, possibly where steep slopes occur most commonly. Although there have been a large number of ideas developed to explain these events, it now seems probable that these phenomena are the result of surface and near surface luminescence of silicates and gases, probably under the bombardment of solar radiation (33). It is highly doubtful that these phenomena are linked directly to lunar volcanic activity.

LUNAR SAMPLE ANALYSES

INTRODUCTION

Until the direct analysis of the lunar surface by the alpha backscatter experiment, our only clue to the bulk chemistry of the Moon was its calculated density, 3.34 gm/cm^3. The alpha backscatter experiment (34) carried a source emitter of alpha particles, which were collimated such that they impinged on the lunar surface. A large number of these alpha particles were backscattered to a detector with energies proportional to the weights of the atomic nuclei of the elements that composed the lunar material. Also, a few protons were obtained from nuclear reactions of the alpha particles with the nuclei of lighter elements. The alpha backscatter analysis technique was chosen for the Surveyor Program because of its simplicity and high probability that it would function. The technique can discriminate between lighter elements with little difficulty; but the spectrum for heavier elements tends to smear together, and these components are lumped in a single or few categories.

Preliminary backscatter results from Surveyor V (Fig. 14), which landed in Mare Tranquillitatis (23°E, 1°N), indicated that the surface was basaltic (35). Later data reductions of the same spectra were reported immediately prior to the Apollo 11 landing in Mare Tranquillitatis, which not only indicated that the surface was basaltic but that it contained an abnormally high amount of titanium (36). This report was largely ignored and disbelieved by the scientific community, but later analyses of samples returned by Apollo 11 showed that the analysis was accurate. The experiment also was flown on Surveyor VI and VII with good results (37). However, the remote analyses were capable of

 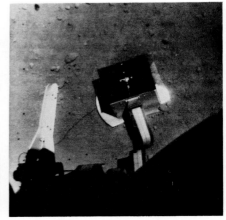

Figure 14   Two television views of the alpha backscatter experiment (Dr. Anthony Turkevich, University of Chicago, Principal Investigator) deployed from Surveyor V onto the surface of Mare Tranquillitatis. The image at left shows the initial deployment of the instrument, and the image at right shows the position of the instrument for another analysis after a ½-sec firing of the rocket engine of the spacecraft. At right, the flange of the instrument has dug into the surface slightly as the instrument moved downhill (slope 19°) and to the right. During the movement, the instrument rotated slightly counterclockwise. It is obvious that many small fragments and clumps of particles on the lunar surface moved also. (Courtesy NASA, Jet Propulsion Laboratory.)

Table I   Weights of Lunar Sample Returned by Each of the Apollo Missions

Mission	Weight of Sample (kg)
Apollo 11	20.7
Apollo 12	34.1
Apollo 14	42.8
Apollo 15	76.6
Apollo 16	95.4
Apollo 17	110.4
Approximate total	380.0 kg

distinguishing only the gross characteristics of lunar material, and the detailed work remained for terrestrial laboratories to perform on returned samples.

Apollo Missions 11, 12, 14, 15, 16, and 17 landed successfully on the Moon and returned samples for detailed examination (Table I).

In addition, unmanned spacecraft flown by the USSR (Luna 16 and Luna 20) returned several hundred grams from additional sites on the eastern limb of the Moon (Fig. 15). The results of lunar sample analyses now comprise an extensive literature (38), and these papers (as well as the discussion that follows) contain many broad generalizations. However, it should be remembered that we are attempting to deduce the history and genesis of a rather large planetary body from only eight sample return sites and remote instrument emplacements whose locations are restricted to only one side of the body.

The materials returned from the Moon can be classified broadly into three categories:

1. Crystalline rocks with igneous textures.
2. Uncohesive particulate material (also called "soil" and "fines").
3. Breccias and microbreccias.

CRYSTALLINE ROCKS WITH IGNEOUS TEXTURES

The somewhat awkward name for this class of lunar samples is necessary for accuracy. Although most of these samples appear to be simple igneous rocks (in the conventional sense) that have crystallized and solidified from magma at or near the surface of the Moon, some samples with virtually identical textures are crystallized impact melts. For example, sample 14310[4] (Fig. 16) may be such a rock (39).

[4]The system by which lunar samples are numbered may appear mystic at first. The *first two digits* of samples returned by Apollo 11 are "10"; for Apollo 12 they are "12"; for Apollo 14 "14"; and for Apollo 15 "15". Samples from Apollo 16 and 17 have "6" and "7", respectively, for their first digits. It is also common to find numbers such as 14310,24, which indicates the 24th split, fraction or subdivision of sample 14310.

Figure 15 Earth-based full Moon photograph showing the Apollo and Luna landing and sample return sites. Also, notice the difference in albedo between the dark, relatively uncratered maria and the bright, much cratered highland surfaces. The rays from relatively recent lunar craters such as Tycho (south, center) are prominent surface features at high angles of illumination but are barely visible at low illumination angles. (Courtesy of the Lick Observatory.)

Figure 16   Section of rock 14310,214. Although this rock appears to have ordinary igneous texture, there has been considerable debate concerning its possible origin as impact melt. Crossed polarizers, length of field of view is approximately 3 mm.

In the maria landing sites, that is, Apollo 11, 12, 15 (plains samples) and 17 (plains samples) the igneous rocks are mostly basalts or microgabbros composed chiefly of three major minerals: calcic plagioclase, clinopyroxene, and ilmenite (Table II). Many of these rocks, especially the finer grained ones, also contain olivine as a major or minor component. These rocks include a wide range of textures — ophitic, subophitic, granular, intersertal, harrisitic, diktytaxitic, variolitic, and intrafasciculate. The grain sizes of these rocks range from 0.1 to more than 1 mm, with a few rocks containing individual crystals of plagioclase and/or pyroxene more than 1 cm long. A number of the rocks are poryphyritic, but most seem to have rather even textures with crystals of comparable size. Several rocks have textures that indicate rapid cooling during a portion of their history (Fig. 17). It should be anticipated that many lunar rocks cooled very quickly from magma erupted onto the lunar surface as lava flows. Figures 17 through 20 illustrate the textures and major mineralogies of several mare rocks with igneous textures.

Lunar highlands rocks tend to have considerably more plagioclase than mare rocks, and generally are anorthosite, troctolite, norite, or

Table II   Mineralogical Modal Analyses of Some Mare Rocks with Igneous Textures[a]

Mineral Component	Rock No. 10017	Rock No. 10045	Rock No. 10058
Clinopyroxene[b]	59.4	53.2	45.7
Plagioclase	25.1	26.9	37.1
Olivine	—	3.1	—
Cristobalite	tr.	1.8	5.1
Fe-Ti oxides (mostly ilmenite)	14.5	11.3	10.5
Troilite	0.36	1.3	0.27
Iron	0.04	0.18	0.03

[a] From Brown et al. (1970) Mineralogical, chemical and petrological features of Apollo 11 rocks and their relationship to igneous processes: Proc. Apollo 11 Lunar Sci. Conf., Geochim. et Cosmochim. Acta, Suppl. 1, vol. 1, A. A. Levinson, ed., p. 197, Pergamon Press. All values in volume percent.
[b] Includes pyroxferroite (9.3% in 10045 and present in all). Counts made in transmitted light and corrected after counting in reflected light (approx. 2300 counts in each sample from approx. 3 cm²).

Figure 17   Olivine vitrophyre 12009 showing rapid crystallization textures of olivine crystals. There are two distinct sizes and morphologies of olivine. The ascicular olivines include melt and some grew in parallel with the larger crystals. Plane polarized light, length of field of view is approximately 3 mm.

Figure 18
Figures 18-20 Photomicrographs of rocks from the lunar maria illustrating the wide variety of grain size and texture. **Figure 18** is a section of rock 12002,157 showing radiating and intrafasciculate plagioclase and clinopyroxene. **Figure 19** is a section from rock 12052,89 showing olivine phenocrysts, which apparently are reacting with the groundmass to form clinopyroxene, and a trachytic plagioclase and clinopyroxene matrix. **Figure 20** shows a section of rock 12040,5, which is a relatively coarse grained basaltic rock. Lengths of fields of view of all three sections are approximately 3.2 mm. **Figures 18 and 20** are in crossed polarizers. **Figure 19** is in plane polarized light.

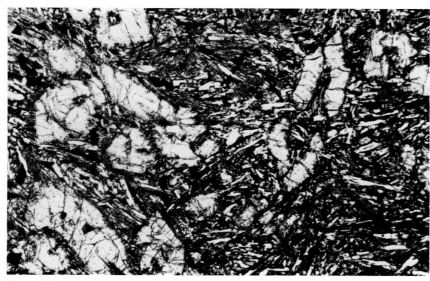

Figure 19

Figure 20

plagioclase-rich gabbro. A large number of other rock names have been used in the literature, but those generally fall within the suite listed above. Figures 21 through 24 illustrate some typical highlands rock textures and mineralogies.

The individual mineral phase compositions are virtually identical to those of minerals from similar terrestrial rocks: plagioclase — mostly close to $An_{90} \pm 5$; olivine — $Fo_{85} \pm 10$; ilmenite — close to ideal $FeTiO_3$; pyroxene — a wide range of composition. Lunar pyroxenes range in composition virtually all over the pyroxene quadrilateral, including augite, subcalcic augite, ferroaugite, pigeonite, and a few orthopyroxenes. In addition, one new iron-rich pyroxenoid mineral was discovered in the Apollo 11 samples (40), which was named pyroxferroite (Fig. 25). This phase was discovered by the Apollo 11 Preliminary Examination Team, because of its abundance in some samples and its bright yellow color. Many pyroxferroite crystals analyzed are close to $(Fe_{0.85}Ca_{0.15})SiO_3$, but it has a wide composition range. Pyroxferroite commonly contains small amounts of magnesium and manganese. Minor minerals in the lunar igneous rocks include: potassium feldspar, cristobalite, tridymite, quartz, baddeleyite, native iron, troilite, spinel, perovskite, rutile, fayalite, chromite, ulvöspinel, apatite, zircon, and a few other very rare

Figure 21

Figures 21-24 Photomicrographs of examples of lunar highlands rocks with igneous textures in thin section. **Figures 21 and 22** are gabbroic anorthosites (68415 and 68416) collected in the southern lunar highlands by Apollo 16. **Figure 23** is an interstitial to variolitic and intrafasciculate norite (62295,69) with complex intergrowths of plagioclase (~57%) and orthopyroxene (~24%). The rock contains abundant microcrystalline and glassy mesostasis (~16%). **Figure 24** is a coarse grained norite (76535) with approximately 60 percent calcic plagioclase and 40 percent orthopyroxene. **Figure 23** is in plane polarized light; all others are in crossed polarizers. Lengths of fields of view are approximately 3.2 mm.

Figure 22

177

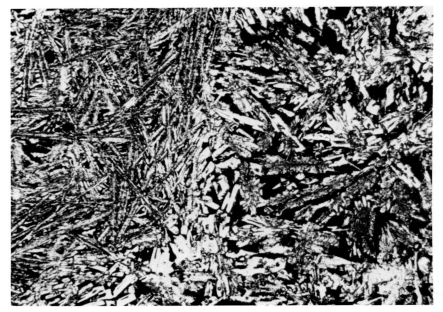

Figure 23

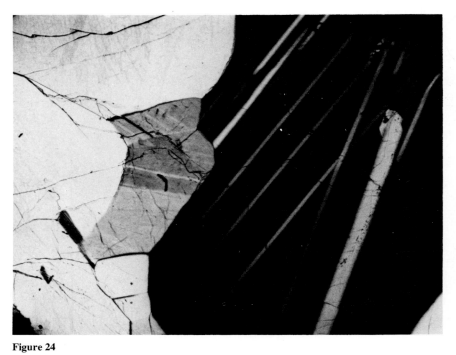

Figure 24

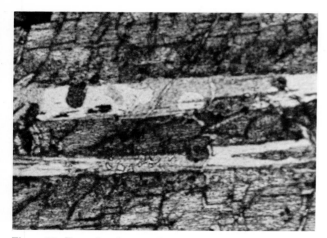

Figure 25 Pyroxferroite associated with plagioclase in rock 12021. The mineral is triclinic, yellow, and the iron-rich analog of pyroxmangite. Pyroxferroite was recognized during the preliminary examination of the Apollo 11 samples in the Lunar Receiving Laboratory and is an abundant constituent of some mare basalts. Plane polarized light, length of field of view is approximately 0.8 mm.

phases. Another new mineral that occurs in lunar samples was named armalcolite[5] (Fig. 26), which commonly occurs with ilmenite as an associated phase (41). The composition of armalcolite has a small range, but mostly fits very well the generalized formula $(Fe_{0.5}Mg_{0.5})Ti_2O_5$. Armalcolite is isostructural with pseudobrookite. Tranquillityite (Fig. 27) also is a new mineral described from the Apollo 11 and Apollo 12 samples (42). This mineral was recognized by several research groups that published a joint description of the phase. Tranquillityite has the approximate formula $Fe_8^{2+}(Zr+Y)_2Ti_3Si_3O_{24}$, but also contains minor amounts of Ca, Al, Mn, Cr, Nb, Hf, U and rare earth elements. It is "foxy" red in color, semiopaque, nonpleochroic, and is isotropic or weakly anisotropic. The average refractive index is near 2.12, and it has a theoretical density of approximately 4.7 gm/cm³. It occurs as thin laths and sheaves of laths with maximum dimensions commonly less than 100 μm. It seems to be a late stage phase associated with the crystallization of mesostasis material. Associated phases are glass, alkali feldspar, troilite, iron, pyroxferroite, and tridymite/cristobalite. Preliminary X-ray structural data indicate it is a new mineral structure that is completely unrelated to any known terrestrial group. A number of other phases have

[5]The name is derived from the names of the flight crew of Apollo 11, Neil Armstrong, "Buzz" Aldrin, and Mike Collins. *After* the discovery of this mineral in lunar samples, it also was found in terrestrial rocks.

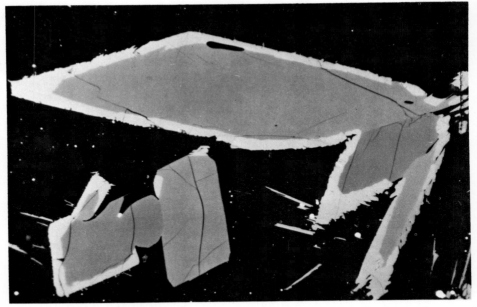

Figure 26   Armalcolite (gray) mantled by ilmenite (white) in a polished section from Apollo 17 sample 74242. Reflected light, length of field of view is approximately 0.2 mm. (Courtesy of A. El Goresy.)

been recognized that almost certainly are new minerals; however, crystallographic data are lacking and new mineral names have not yet been approved.

   Rock-forming Minerals.   Descriptions of each of the four lunar rock-forming minerals follow. For a more detailed discussion of other lunar minerals see the various Proceedings of the Lunar Science Conferences and related volumes (38).

   Lunar plagioclase is virtually indistinguishable from the calcic plagioclase in many terrestrial plutonic mafic rocks. Some lunar highlands rocks are composed of more than 85 modal percent plagioclase. Albite twinning is extremely common, but Carlsbad, albite-Carlsbad, and pericline twins also are common. One example of a Baveno-r twin has been recorded (43). The widths of individual crystals in polysynthetic twins ranges widely even though the range of An content is small. The composition of most lunar plagioclase ranges from approximately An_{97} to An_{85}. Optical characteristics of lunar plagioclase and comparisons with major element compositions and structural states of the same grains also are similar to terrestrial samples (43, 44). Analyses for minor elements by Smith (45) indicate that there are several populations of plagioclase crystals that can be distinguished by their iron, potassium, and magnesium contents. Zoning in lunar plagioclase rarely can be found optically; however, some lunar plagioclase grains do tend to have potassium-rich rims. Lunar plagioclases display a wide range of crystal habits and textures. Platy, lathlike, and equant grains are common, probably reflecting differences in the environments of origin and, in some instances, subsequent history. Some lunar plagioclase is poikilitic, enclosing pyroxene, ilmenite, and other minerals. In a few lunar rocks, plagioclase occurs mantling pyroxene (46). This texture has been termed "intrafasciculate," and apparently results from the skeletal and rapid crystallization of plagioclase followed by the crystallization of the included (or partially included) melt (47).

Pyroxenes in lunar samples have a wide range of composition (Fig. 28) and include orthopyroxene, augite, ferroaugite, subcalcic augite, and pyroxferroite (actually a pyroxenoid). Some lunar pyroxenes contain substantial amounts of titanium, as much as 7 weight percent in the Apollo 11 and Apollo 17 samples. It appears that some of the titanium in lunar

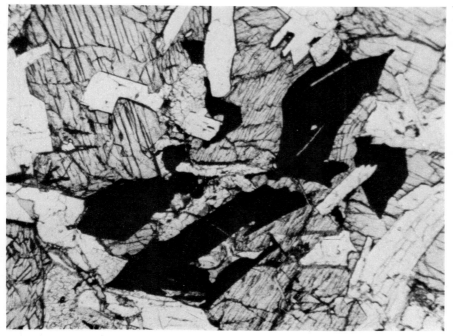

Figure 27   Photomicrograph of tranquillityite (opaque) in transmitted light in sample 10047. In very thin crystals it has a characteristic "foxy" red color, but appears opaque at normal thin section thicknesses. The mineral commonly occurs in laths and sheaves of laths with overall dimensions less than 100 μm associated with late crystallizing interstitial phases. Plane polarized light, length of field of view is approximately 250 μm.

pyroxenes occurs as Ti^{3+}, as opposed to Ti^{4+}, which is the occurrence in terrestrial rocks. This conclusion is based on the interpretation of crystal field spectra (48) as well as other techniques of investigation which include formula computations from major element analyses. Many lunar rocks contain augite crystals with exsolved pigeonite lamellae and bronzite crystals with exsolved augite lamellae. Studies of the orientation and size of exsolution lamellae and other microstructures in lunar pyroxenes have led to interpretation of the subsolidus cooling histories of a number of lunar rocks (49). Based on these data, it appears that some lunar rocks have been reheated as many as three times, possibly by mild shock metamorphism.

A few lunar pyroxene crystals are more than 2 cm long, and poikilitic pyroxene is common. Some of the large clinopyroxenes exhibit "hourglass" zoning. Exhaustive studies of the zoning of lunar pyroxenes have been performed by Bence and Papike (50), who have attempted to show crystallization trends in lunar basaltic rocks as indicated by the zoning of pyroxenes (Fig. 29). Lunar orthopyroxenes have similar M_1-M_2 site Fe/Mg distributions, and some are enriched in Al. Overgrowths of pigeonite on lunar orthopyroxene are common in some specimens. For example, many pyroxene crystals in rock 14310 have twinned magnesian pigeonite overgrown on bronzite with common (100). This occurrence is distinct from mare basalts and may be a useful indicator for characterization of the crystallization trends of some lunar highlands rocks (51). In general, X-ray studies of cation

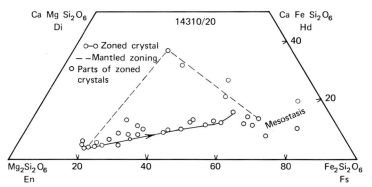

Figure 28   The extreme range of composition of lunar pyroxenes is illustrated by this plot of pyroxenes in a *single* lunar sample, 14310,20. The plot is on the "pyroxene quadrilateral," a portion of the ternary system $Mg_2Si_2O_6$-$Fe_2Si_2O_6$-$Ca_2Si_2O_6$. (Reprinted from Mineral-chemical variations in Apollo 14 and Apollo 15 basalts and granitic fractions by G. M. Brown et al. Proc. Third Lunar Sci. Conf., Geochim. et Cosmochim. Acta, Suppl. 3, vol. 1, p. 143, by permission of the MIT Press, Cambridge, Mass.)

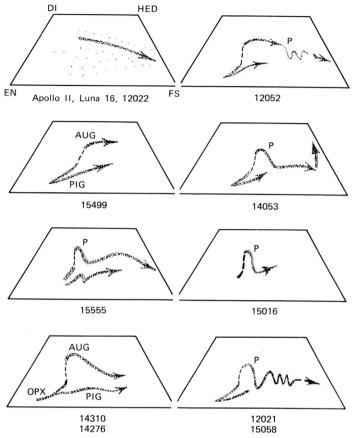

Figure 29   Crystallization trends of lunar pyroxenes plotted on the pyroxene quadrilateral (50; see Fig. 28) showing the wide range of composition and variability of crystallization paths. (Reprinted from Pyroxenes as recorders of lunar basalt petrogenesis: Chemical trends due to crystal-liquid interaction by A. E. Bence and J. J. Papike by permission of the MIT Press, Cambridge, Mass.)

distribution and crystal structure of orthopyroxenes at the Fra Mauro (Apollo 14) landing site indicate that these orthopyroxenes are quite similar to terrestrial orthopyroxenes from volcanic rocks (51). Other lunar pyroxenes from clasts in breccias at the Apollo 14 and Apollo 15 landing sites are unzoned larger crystals that probably originated in a plutonic-metamorphic environment (52).

Ilmenite is a major component of virtually all mare samples, and composes as much as 48 modal percent of some rock specimens. However, most lunar samples contain only a few percent of ilmenite, with the greatest amount occurring in the Apollo 11 and Apollo 17 (mare) samples. It appears that the composition of most lunar ilmenite is close to the ideal formula ($FeTiO_3$) and that the titanium is tetravalent, but that ilmenite in some lunar samples, such as 14053, appears to contain some trivalent titanium (53). This difference from terrestrial ilmenites probably is caused by the extremely reducing conditions under which this sample crystallized. Ilmenite crystals commonly are tabular, in long blades, and both skeletal and dendritic crystals are abundant in some rocks, particularly from Apollo 11. Ilmenites rarely are zoned, and most appear to be homogeneous throughout. However, some ilmenites do contain exsolution lamellae of TiO_2, presumably rutile (54).

Olivine from the Moon is mostly forsteritic, but there is some fayalite in mesostasis material in a number of lunar samples. The composition of most lunar olivine analyzed to date ranges about $Fo_{85}\pm$ approximately 10 mole percent. The contents of Ca and Cr in lunar olivines have permitted the recognition of different olivine populations. The high and low Ca olivines may represent volcanic and plutonic conditions of crystallization by analogy with terrestrial olivines, and the high and low Cr groups may represent different oxidation conditions (45). It appears that much of the Cr in lunar olivines is divalent, probably reflecting the extremely reducing conditions under which lunar rocks crystallized. There do not appear to be different populations of lunar olivines that can be recognized on the basis of Ti or Mn contents; however, the total number of lunar olivine analyses is still rather limited.

Lunar olivines show a wide variety of textures and crystal habits. Rock 12009 (Fig. 17) is particularly striking in its two distinct habits of olivine crystals, skeletal and ascicular with groundmass inclusions. Studies of the optic orientations of olivine microlites in rock 12009 indicate that there are only weakly preferred orientations, not sufficient to suggest significant motion of the melt relative to the crystals (55). The habits of the olivine in rock 12009 have been seen in a number of other lunar rock fragments, and these habits probably indicate supersaturation of olivine in the melt and very rapid cooling. Some lunar olivines, particularly in the Apollo 11 and 12 samples show apparent resorption and replacement by pyroxene (Fig. 30); however, this texture also has been interpreted as crystallization of pyroxene that has nucleated on the interior surface of a skeletal olivine crystal. The abundance of olivine in lunar rocks ranges from zero to as much as 91 modal percent (e.g., 72415) in a brecciated dunite from the Apollo 17 samples. Olivine-rich clasts have been found in the small rock fragments (1 to 4 mm fraction) and in the breccias from all of the landing sites. Small, euhedral olivine crystals occur in glass in some partially crystallized lunar glass spherules.

Chemical Compositions.   Several analytical studies have attempted to identify the most prevalent lunar rock compositions. However, the number of individual lunar rocks with igneous texture that are available for analysis is

Figure 30   Fibrous clinopyroxene forming at the expense of olivine phenocrysts in rock 12052. Plane polarized light, length of field of view is approximately 3.2 mm.

relatively small, and many of them may be crystallized impact melts and thus not represent magmas from the interior of the Moon. Sample analyses of basaltic lunar igneous rocks are presented in Table III (56). Reid et al. (57) have used clusters of lunar glass particle analyses to deduce the preferred compositions of lunar rocks. They, and other investigators, find that there are, at least, two major nonmare rock types: (a) anorthositic gabbro (also termed highland basalt, feldspathic basalt, gabbroic anorthosite, and other names); and (b) Fra Mauro basalt (also termed KREEP[6] and other names). The nonmare rocks are characterized by high contents of Al_2O_3, low FeO, low Cr_2O_3, and low calcium/aluminum ratios (Fig. 31) compared with the mare materials. The maria glass compositions from all maria sites make a broad cluster around the same area on various geochemical plots. Other rock types in both the highlands and maria are less abundant, but other significant compositions do occur.

Several geochemical trends and properties of lunar rocks in general were recognized in the early lunar sample returns. Lunar rocks with igneous

[6] The term KREEP derives from the fact that this rock type is enriched in potassium (K), rare earth elements (REE), and phosphorus (P), as well as other minor and trace elements.

Table III   Examples of Major Element Analyses and Averages of Basaltic Lunar Rocks with Igneous Textures[a]

Oxide	Average Compositions[b]			15085,34[c]	15555,157[c]
	Apollo 11	Apollo 12	Apollo 14		
SiO_2	40.10	47.10	47.70	46.39	44.75
Al_2O_3	8.60	12.80	21.44	5.79	9.85
Fe_2O_3	0.00	0.00	0.00	—	—
FeO	18.90	17.40	7.78	26.75	23.40
MgO	7.74	6.80	7.29	8.20	8.03
CaO	10.70	11.40	13.05	9.12	10.72
Na_2O	0.46	0.64	0.70	0.21	0.30
K_2O	0.30	0.07	0.48	0.07	0.09
TiO_2	12.20	3.17	1.16	3.07	2.64
P_2O_5	0.2	0.17	0.42	0.09	0.07
MnO	0.25	0.24	0.11	0.37	0.32
Cr_2O_3	0.37	0.31	0.25	0.67	0.77

[a] All values in weight percent.
[b] From Rose et al. (56).
[c] From Mason et al. (56).

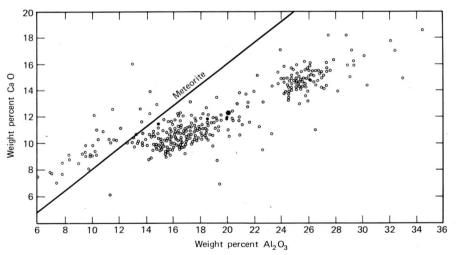

Figure 31   Plot of weight percent CaO versus weight percent Al_2O_3 in glasses from Apollo 14 landing site regolith (57). The two groups may represent two distinctly different parent rock types. The composition of lunar rock 14310 is shown by the filled hexagon. The typical calcium oxide/aluminum oxide values in meteorites are indicated by the solid diagonal line. Similar plots for other elements, based on electron microprobe analyses, and for samples from other landing sites tend to group around the compositions of major lunar rock types. (Reprinted from The major element compositions of lunar rocks as inferred from glass compositions in the lunar soils by A. M. Reid et al., by permission of the MIT Press, Cambridge, Mass.)

texture, in general, tend to be depleted in alkalies and some volatile elements (bismuth, mercury, zinc, cadmium, thallium, lead, germanium, chlorine, and bromine) relative to presumed solar abundances and to terrestrial rocks of comparable composition. The same rocks tend to be enriched in refractory elements, such as titanium, scandium, zirconium, hafnium, yttrium and trivalent rare earths. These compositions are significantly different from any meteorite material that is available for analysis at present. Much attention has been drawn to the fact that lunar samples from Apollo 11 and Apollo 12 have large depletions of europium, as compared with the other rare earth elements (58). However, many feldspar-rich rocks from the lunar highlands have positive europium anomalies, indicating that much europium probably occurs as a divalent cation in the lunar environment as opposed to its normal terrestrial trivalent state and substitutes for Ca^{2+} in lunar plagioclase. Comparisons of elemental abundances in lunar rocks with presumed Solar System composition and other materials are shown in Figs. 32 through 34.

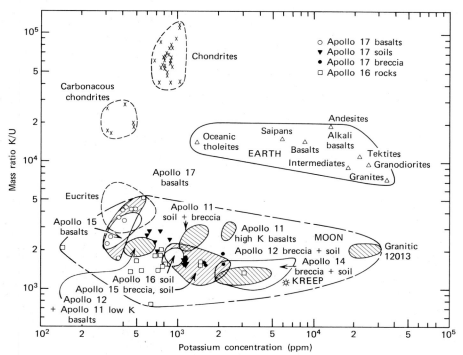

Figure 32   Comparison of the mass ratio K/U versus potassium concentration in lunar samples with other Solar System samples. The value indicated for carbonaceous chondrites probably best represents primordial Solar System composition. The values for lunar samples overlap only with values from some eucritic meteorites and with no other materials. (From Eldridge, J. S., G. D. O'Kelley, and K. J. Northcutt, 1974, Primordial radioelement concentrations in rocks and soils from Taurus-Littrow Lunar Sci. Conf., Geochim. et Cosmochim. Acta, Suppl. 5, vol. 2, p. 1029, Courtesy of Pergamon Press.)

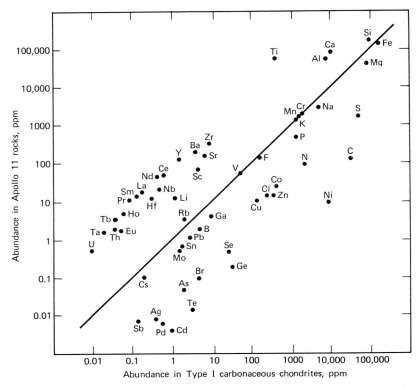

Figure 33 Comparison of elemental abundances in Apollo 11 crystalline rocks with Type I carbonaceous chondrites. Note that the scales for both axes are logarithmic. Points falling on the solid diagonal line are of equal abundance in both materials; points above the line are relatively enriched in lunar rocks, but points plotted below the line are relatively depleted in lunar rocks. An assumption in the significance of this plot is that primordial Solar System abundances of the elements are approximated by the abundance of a given element in carbonaceous chondrites. (From Mason, B., and W. G. Melson, 1970, The Lunar Rocks, John Wiley and Sons, New York, p. 151; Courtesy of John Wiley and Sons.)

The isotopic compositions of lunar samples also are somewhat distinct from terrestrial rocks and meteorites. The rocks at the lunar surface are especially distinct because of their exposure to cosmic and solar radiation. Exposure to cosmic rays and high energy solar radiation allows many high energy atomic interactions and causes concentrations of Al^{26}, Na^{22}, Mn^{54}, Co^{56}, V^{48}, Co^{60}, and other nuclides. Analyses of lunar samples for gamma emitters have been carried out by O'Kelley et al. (59) and other investigators (Table IV). On the basis of the amounts and relative concentrations of various cosmic-ray-induced nuclides, it is concluded that lunar rocks have surface exposure ages that mostly range from tens of millions to hundreds of millions of years. The He³ exposure ages (60) and other mass spectrometric determinations generally agree with these results. During the same analysis for gamma emitters, the concentrations of K, U, and Th were determined. Values for the K/U mass ratio in lunar crystalline rocks range from approximately 2400 to 3100, and the Th/U mass ratio typically ranges between

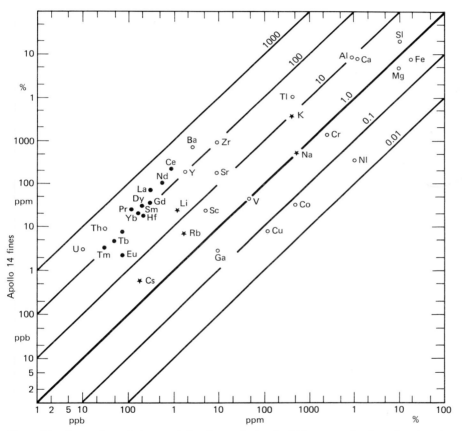

Figure 34   Comparison of elemental abundances in Apollo 14 fines samples (less than 1 mm size fraction) with carbonaceous chondrites. Note that the scales of both axes are logarithmic. Points falling on the heavy solid diagonal line are of equal abundance in both materials; points plotted above the line are enriched in Apollo 14 fines and those plotted below the line are depleted, relative to the abundance in Type I carbonaceous chondrites. Other diagonal lines indicate various enrichment and depletion factors as labeled. An assumption in the significance of this plot is that primordial Solar System elemental abundances are approximated by Type I carbonaceous chondrites. (Reprinted from Composition of the lunar uplands: Chemistry of Apollo 14 samples from Fra Mauro by S. R. Taylor et al., Proc. Third Lunar Sci. Conf., Geochim. et Cosmochim. Acta, Suppl. 3, vol. 2, p. 1244, 1972; by permission of the MIT Press, Cambridge, Mass.)

Table IV Gamma-ray Analyses of Whole Rocks and Fines Samples from Apollo 11[a]

	Sample Number							
	10057-1	10072-1	10003-0	10017-0	10018-1	10019-1	10021-1	10002-6
Weight (g)	897	399	213	971	211.5	234	157	301.5
Bulk density (g/cc)	2.73 ± 0.14	2.37 ± 0.24	2.88 ± 0.35	3.00 ± 0.15	2.0 ± 0.2	2.02 ± 0.15		1.55 ± 0.05
K (ppm)[b]	2550 ± 130	2300 ± 120	480 ± 25	2430 ± 120	1420 ± 70	1200 ± 60	1600 ± 80	1100 ± 60
Th (ppm)[b]	3.30 ± 0.20	2.80 ± 0.17	1.01 ± 0.06	3.25 ± 0.18	2.30 ± 0.20	1.90 ± 0.19	2.50 ± 0.25	1.92 ± 0.10
U (ppm)[b]	0.79 ± 0.06	0.76 ± 0.06	0.26 ± 0.03	0.83 ± 0.07	0.60 ± 0.09	0.43 ± 0.06	0.54 ± 0.08	0.49 ± 0.04
^{7}Be (dpm/kg)[c]	<70		<100	<60				<80
^{22}Na (dpm/kg)[c]	41 ± 4	46 ± 5	41 ± 4	39 ± 4	55 ± 8	47 ± 7	55 ± 8	51 ± 5
^{26}Al (dpm/kg)[c]	75 ± 8	73 ± 8	74 ± 8	73 ± 8	108 ± 16	101 ± 15	110 ± 15	120 ± 12
^{44}Ti (dpm/kg)[c]	<2.5	<2.5		2.1 ± 1.3				<2.5
^{46}Sc (dpm/kg)[c]	10 ± 2	8 ± 2	13 ± 3	13 ± 3	13 ± 4	10 ± 3	13 ± 4	8 ± 2
^{48}V (dpm/kg)[c]			12 ± 9	11 ± 7				
^{52}Mn (dpm/kg)[c]			35 ± 20				33 ± 21	
^{54}Mn (dpm/kg)[c]	32 ± 6	20 ± 4	35 ± 7	33 ± 7	38 ± 10	28 ± 9	21 ± 6	28 ± 7
^{58}Co (dpm/kg)[c]	31 ± 8	40 ± 10	43 ± 10	26 ± 7	33 ± 10	35 ± 10	50 ± 15	40 ± 7
^{60}Co (dpm/kg)[c]				1.1 ± 0.8				

[a] From O'Kelley et al. (59). Values for short-lived nuclides have been corrected for decay to 0000 hr, CDT, 21 July, 1969.

[b] Standardization for assay of K, Th, and U with reference to terrestrial isotopic abundances. Equilibrium of Th and U decay series also assumed.

[c] Disintegrations per minute per kilogram.

3.3 and 3.8. In addition, these analyses have detected the nuclide concentrations resulting from solar flare events.

Other isotopic measurements on lunar samples have detected and characterized solar wind implanted isotopic species in the surfaces and near surface portions of lunar rocks (61), but some of the most interesting data have come from the various crystallization age measurements. Prior to the return of samples for laboratory dating, the estimates of the ages of the youngest lunar mare surfaces were 200 to 300 million years by most research groups (page 167), but the actual crystallization ages that have been measured are *much* older (Table V).

Ages. Interestingly, most of the average lunar rock ages range through a relatively short span of time, approximately 600 my. Also, we have not dated any very young lunar igneous rocks. It appears that the massive eruptions of mare basalt flows began at some unknown time ($\sim$500 my) after the formation of the Moon and ended abruptly approximately 3.3×10^9 years ago. Tera et al. (62) have interpreted the distribution of ages of lunar rocks to indicate that there was a cataclysm in lunar history at approximately 3.95×10^9 yr, possibly the Imbrian impact event. The age of formation of the Moon is interpreted to be approximately 4.6×10^9 yr because many Rb-Sr model ages for soils and rocks (including mare basalts) are near that value (62). It must be remembered in the interpretation of lunar rock ages that the sample is extremely small. Furthermore, the samples of mare rocks probably are biased toward the younger rocks. There should be many more samples from the uppermost (youngest) flows at any mare site than from the first ones that

Table V   Average Crystallization Ages of Lunar Rocks from Sample Return Sites[a]

Mission and Site[b]	Approximate Average Rock Crystallization Age
Apollo 11, Mare Tranquillitatis	3.6×10^9 yr
Apollo 12, Oceanus Procellarum	3.3×10^9 yr
Apollo 14, Fra Mauro highlands	3.9×10^9 yr
Apollo 15, Imbrium Basin mare	3.3×10^9 yr
Apennine Front	3.9×10^9 yr
Apollo 16, Descartes highlands	3.9×10^9 yr
Apollo 17, Mare Serenitatis	3.7×10^9 yr
NE lunar highlands	4.0×10^9 yr
Luna 16, Mare Fecunditatis	3.4×10^9 yr
Luna 20, Eastern highlands	3.9×10^9 yr

[a] Averaged from numerous sources, but primarily from Sr-Rb, Ar^{40}-Ar^{39} and U-Pb work. See numerous papers in vols. 2 of each Lunar Science Conference Proceedings.

[b] General site names are indicated, but the average ages may only be applicable to the immediate area at the landing site.

began to fill the mare basins. Similarly, samples from the lunar highlands sites seem to be dominated by ejecta from the most recent large basins, that is, Mare Imbrium and possibly Mare Orientale. However, sample 72417, which is a brecciated dunite, has been dated at 4.6×10^9 yr on an internal Sr/Rb isochron.

Crystallization History.   The mineral assemblages present in lunar rocks, such as ilmenite-troilite-native iron, indicate that they crystallized from extremely dry and oxygen-poor melts. The paragenetic sequence of opaque minerals indicates that the melts from which many lunar rocks crystallized became increasingly reduced at lower temperatures (63). The best estimates of the oxygen fugacity during the crystallization of some lunar rocks is as low as 10^{-13} to 10^{-14} atm, and abundant evidence has been found for subsolidus reduction (64). Some lunar rocks from Apollo 14 (14053, 14072) show subsolidus reduction of chromian ulvöspinel to ilmenite + aluminian chromite + native iron. The same mineral assemblages, as well as the general absence of hydrous phases[7] also indicate that the melts and magmas were extremely dry.

The liquidus temperatures of many lunar rocks appear to have been extremely high, $\sim 1300°C$, and experimental work and calculations have shown that lunar basalts are very fluid, low viscosity melts compared with terrestrial analogs (65). Considerable discussion has occurred relative to the identification of liquidus phases in lunar rocks. Plagioclase, ilmenite, olivine, clinopyroxene, and spinel are the most commonly identified phases on the liquidus in various lunar rocks (66). The identity of the liquidus phase or phases is, of course, dependent on the temperature, pressure, and bulk composition of the melt.

The fine grain size of many of the lunar rocks, together with the skeletal textures of some minerals in some lunar rocks (Fig. 17), indicate that these rocks cooled and crystallized rapidly. Hence, it seems likely that most of the rocks, particularly from the mare sample collection sites, are extrusives or very shallow intrusives. This conclusion is supported by the presence of abundant vesicles in some samples (e.g., 10022, Fig. 35). However, the identity of the gas or gases that were dissolved in the melt and came out of solution to form the vesicles is still in question.

Some of the coarser grained rocks may be simple intrusives, but some of the very feldspar-rich rocks from the lunar highlands (e.g., anorthosite), may be low density cumulates. Similarly, it seems possible that some of the dunites and very olivine-rich rocks (e.g., 72415) are high density cumulates. However, many lunar rocks that appear to have had relatively simple cooling

---

[7]Only a few occurrences of hydrous minerals have been reported in lunar samples under very special petrologic conditions, for example impact melted rocks, or in lunar regolith such that the minerals *might* be contamination from terrestrial laboratories. The hydrogen for some of the hydrous phases may be solar wind gas that has been remobilized by impact.

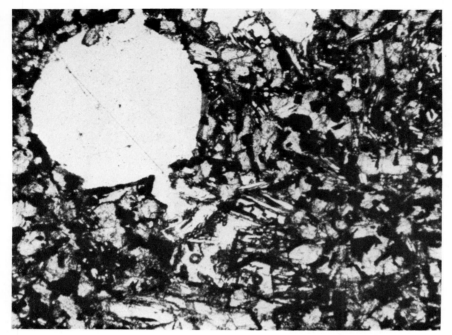

Figure 35 Photomicrograph of a portion of a thin section of rock 10022 showing a section through a roughly spherical vesicle. Such vesicles are abundant in this and some other mare basalts. The composition of the gas responsible for the vesicles is still a speculative matter. The abundant opaque mineral is ilmenite. The remainder of the sample is composed of plagioclase and clinopyroxene with rare olivine. Plane polarized light, length of field of view is approximately 3.2 mm.

or crystallization histories at first examination are found to have had one or more episodes of reheating (67), probably as a result of shock metamorphism.

The liquid line of descent for lunar magmas has been the subject of several investigations. Three major magma types have been suggested or implied by different workers: mare basalts, KREEP basalts, and feldspar-rich magmas related to the formation of the highlands anorthosites and other feldspar-rich rocks (68, 69). Although several different types of mare basalts have been recognized (i.e., high vs. low K, Apollo 11, Apollo 12, etc.), the genetic relations between these groups are not at all clear. Melt inclusions in a large number of lunar rocks have been analyzed by Roedder and Weiblen (70) in an attempt to trace the liquid line of descent. The inclusions (Fig. 36) occur in virtually all lunar rock forming minerals. Roedder and Weiblen find that the compositions of the melt inclusions in all minerals from all of the Apollo landing sites show remarkably similar liquid lines of descent (Fig. 37), and that liquid immiscibility occurs late in the crystallization of all samples. The

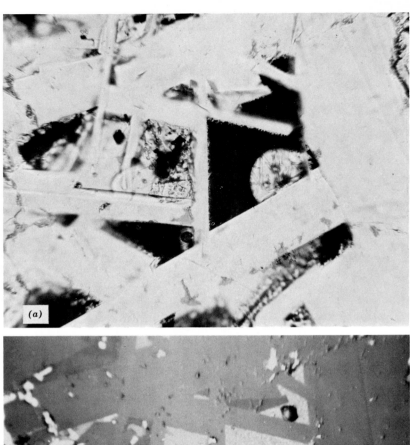

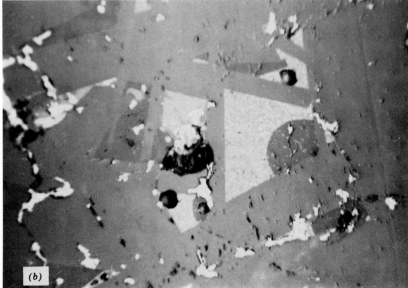

Figure 36 Immiscible melt trapped interstitially between plagioclase laths in lunar rock 14310. *(a)* Transmitted light; *(b)* reflected light. The globule of high silica melt contains a few acicular crystals as well as glass, and the adjacent high iron melt has mostly crystallized into several different phases (70). Maximum dimension of the high silica melt inclusion is approximately 30 μm. (Reprinted from Petrographic features and petrologic significance of melt inclusions in Apollo 14 and 15 rocks by E. Roedder and P. Weiblen by permission of the MIT Press, Cambridge, Mass.)

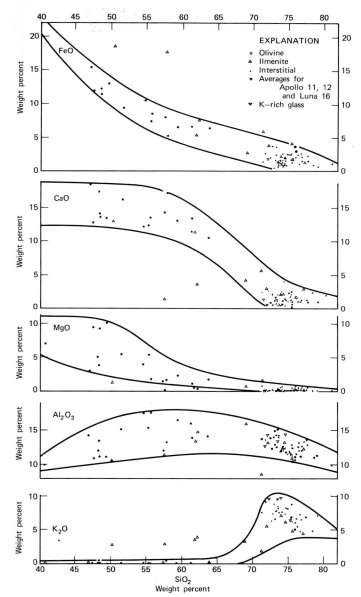

Figure 37 Melt inclusion data presented in the form of silica variation diagrams (70). These compositions may delineate a portion of the liquid line of descent of lunar magmas. *Explanation*: inclusions in olivine are indicated by open circles, ilmenite is shown by triangles with point up, interstitial is shown by dots, K-rich glass is shown by triangles with point down, and filled circles indicate averages of Apollo 11, Apollo 12, and Luna 16 data. Roedder and Weiblen regard the data from olivine host crystals as best for tracing the liquid line of descent for parent magmas. (Reprinted from Petrographic features and petrologic significance of melt inclusions in Apollo 14 and 15 rocks by E. Roedder and P. Weiblen by permission of the MIT Press, Cambridge, Mass.)

late crystallizing immiscible liquid is rich in silica and potash, with an overall "granitic" composition.

   Origin of Lunar Magmas.   Partial melting of parent materials within the upper 200 km of the Moon seems to be the favorite model of most researchers. There are, however, some disagreements as to the amount of partial melting and the composition of the parent material. The source of nonmare basalts is approximately three times richer in refractory elements than the source of mare basalts according to Hubbard et al. (68). Furthermore, they suggest that, to a first approximation, all nonmare basalts can be derived by varying degrees of partial melting from a single source material that is enriched 7 to 10 times in refractory elements. They also speculate that the depth of the source material for the nonmare basalts must be greater than 100 km, in order to satisfy the conditions of their favored thermal model (71).

   Lunar rock compositions in the system forsterite-silica-fayalite-anorthite have been projected onto the forsterite-silica-anorthite face of this tetrahedron by Walker et al. (72). The result is a pseudoternary liquidus diagram (Fig. 38), which they have used to plot lunar rock compositions and trace crystallization paths. From their representation, it appears that a good source for many lunar magmas would be partial melting of a spinel-anorthite-olivine rock or some later derivative from this material.

   Again, one of the great unknowns in trying to determine the origin of lunar magmas is the role that large impacts may play in their generation. Also, many of the rocks collected on the surface may actually be impact melted and crystallized regolith or mixtures of rock and mineral fragments and may not represent primary lunar magmas.

   Shock Metamorphism.   The crystalline rocks returned from the Moon show moderately abundant evidence of shock metamorphism (Fig. 39) including: deformation structures in plagioclase, pyroxene and olivine, diaplectic plagioclase glass; abundant microfractures; abundant mechanically induced twins in lunar minerals and abundant impactite glass (see Chapter 5). However, evidence of shock metamorphism is much more abundant in the small rock fragments from the lunar breccias and regolith than in the larger rocks (68, 69). This observation is not surprising because only a small percentage of the large ejecta from known terrestrial impact craters shows obvious evidence of shock metamorphism. The greater abundance of shock-induced features in rocks from the lunar regolith probably is due to the multiple impact events that most lunar surface material has experienced.

   One ubiquitous shock metamorphism feature of rocks collected from the lunar surface is microcraters, caused by the hypervelocity impacts of small meteoroids (Fig. 40). These craters occur on all lunar surface materials, not only the igneous rocks (see Fig. 43, page 202), and range in size down to the very limit of both optical microscopes and scanning electron microscopes (75).

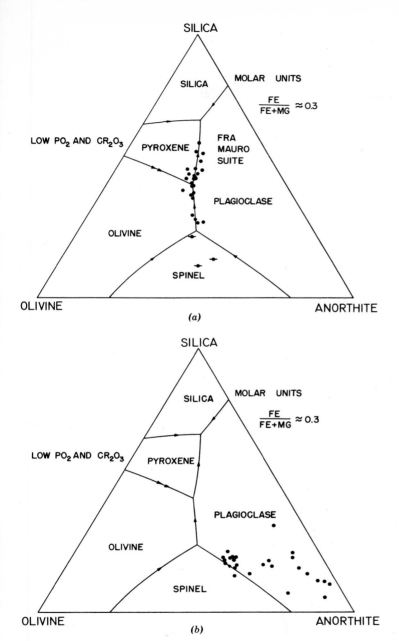

Figure 38 *(a)* Pseudoternary liquidus diagram for the system SiO_2 - $CaAl_2Si_2O_8$ - Fe_2SiO_4 - Mg_2SiO_4. Filled circles represent individual Fra Mauro site rock compositions. Open circles are the averages of spinel-troctolites from Luna 20 and Apollo 16. *(b)* A compilation of average glass and lithic fragment compositions reported for lunar anorthosites and anorthositic norites. (After Walker et al., 72.)

Figure 39   Badly shock damaged plagioclase grain in Apollo 16 anorthosite breccia 60215,13. Crossed polarizers, length of field of view is approximately 3 mm.

UNCOHESIVE PARTICULATE MATERIAL

Much of the lunar regolith is composed of individual grains of many different origins (76). Abundant rock and mineral fragments are derived from larger lunar rocks by erosion and comminution, caused by both large and small meteoroids. Glasses of many different types are an abundant component of most lunar "soil."[8] Impactite glasses are present that represent the entire range of known lunar rock compositions, although melt glasses from individual minerals are relatively rare. Diaplectic and thetomorphic glass, especially plagioclase glass, is abundant.

Regularly Shaped Glass.   One of the most striking glass components of the lunar regolith is the regularly shaped glass particles: spheres, flattened spheres, dumbbells, tear drops, and other fluid forms (Fig. 41). These regularly shaped colored glasses also have a wide range of lunar rock and soil compositions, and range in color from light gray-green to wine red, orange,

[8]The term *soil* has been applied widely to the finer rock and mineral material of the lunar regolith. The term does not imply any similarity to terrestrial soil (i.e., organic content, weathering, etc).

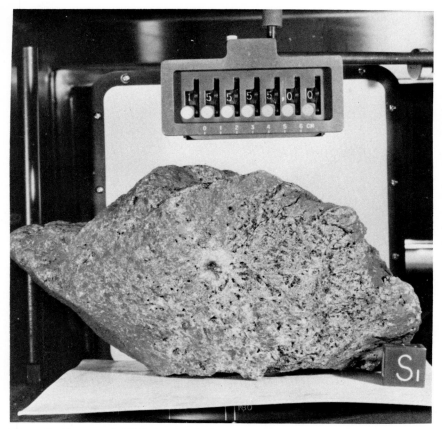

Figure 40 Lunar rock 15555 showing the large crater approximately 1 cm in diameter that resulted from the hypervelocity impact of a micrometeoroid. It appears that the impact almost split the rock. The various letters, numerals, scales, and gray disks are to record the size, orientation, and identity of the sample. Several such ''mug shots'' are part of the permanent record for each lunar sample of more than 50 gm. (NASA photograph S-71-43393.)

bright green, brown, and yellow. These glass spheres commonly contain a few shocked and/or partially melted mineral grains, and almost certainly result from rock fusion by the impacts of small meteoroids. The energy of the impact fuses some of the target rock which is ejected into near lunar space; a spherical shape is formed because of the surface tension of the melt; and the sphere cools and solidifies before it falls back to the surface. Tear drops and dumbbells are fluid ejecta that have significant angular momentum, and the flattened spheres were still plastic when they fell back to the surface. The

Figure 41 Glass spherules in Apollo 11 sample 10010 as photographed by the Apollo 11 Preliminary Examination Team against a brushed aluminum background. The spherules have a wide range of color and composition as do similar dumbbell and tear-drop shaped particles in the lunar regolith. Spherules approximately 1 cm in diameter have been observed, but most are much smaller; and they tend to be most abundant in the fine size fractions. The largest spherule above is approximately 0.4 mm in diameter. (NASA photograph S-69-45181.)

regularly shaped glass particles range in size from less than 1 μm to more than a centimeter; however, abundant spheres commonly are no larger than a few millimeters.

The surfaces of glass spheres have been studied extensively both optically and by scanning electron microscopy (77). It has been suggested by some observers (78) that distinctive green glass spheres (Apollo 15, Apennine Front, Fig. 42) and orange glass spheres (Apollo 17) are volcanic in origin. However, other investigators have suggested that impacts could have formed these spheres also. These glass surfaces, as well as well formed crystal faces or surfaces of glass fragments, are excellent locations on which to view extremely small hypervelocity impact craters (Fig. 43). These craters are

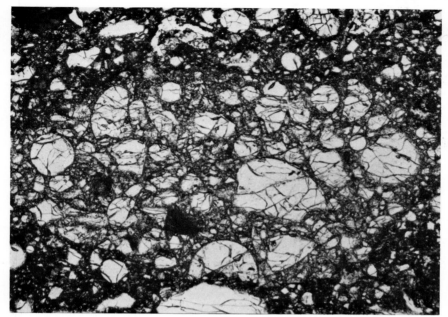

Figure 42   Lithic fragment in Apollo 15 breccia sample 15426 that is composed almost entirely of green glass spherules and spherule fragments. Length of field of view is approximately 1.2 mm, plane polarized light.

ubiquitous on particles in the lunar regolith and generally show the following morphology: central glass-lined pit surrounded by radiating fractures and a large spall zone in turn surrounded by redeposited ejecta in streaks and rays (79).

   Glassy Agglutinates. [9]   These are another type of glassy particle found in the lunar regolith that are distinctive to lunar samples. They are fragile, glass-welded aggregates of mineral grains and small rock fragments. Much of the molten ejecta from meteoroid impacts will intersect the surface of the regolith while the ejecta is still molten and plastic. This molten ejecta cements regolith grains together, generally forming delicate networks of grains that are held together by brittle filaments and adhesions of glass (Fig. 44). Thus, not only do micrometeoroid impacts tend to comminute and break down previously existing lunar material, but they also tend to construct agglutinates of larger grain sizes than were previously available near the impact site. The percentage of agglutinates in regolith samples can be used as a crude index of the length of exposure of the regolith to meteoroid bombardment; hence, the percentage of agglutinates in regolith samples has been used as an index of

[9]These particles also have been called glass-welded aggregates, agglomerates, glazed agglutinates, glassy spatter, and other terms.

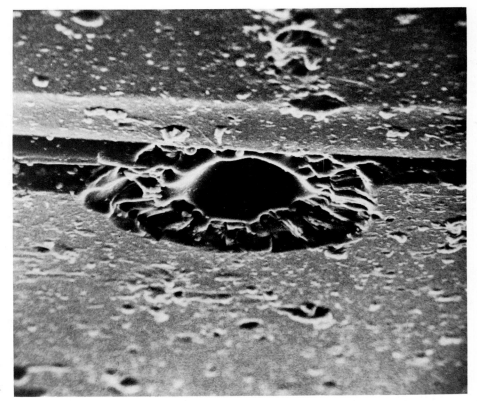

Figure 43   Hypervelocity impact crater on the surface of a lunar glass fragment. Diameter of the central pit is approximately 10 μm, and it is surrounded by a spall zone approximately 5 μm wide. This is a scanning electron microscope image. (Courtesy of NASA, Jet Propulsion Laboratory.)

"maturity" by some workers (80). Another abundant component of most regolith particulate samples is breccia and microbreccia fragments, which are described in a following section (page 207).

Modal Analyses.   This kind of analysis of particle types in regolith samples can be used to interpret origins and histories of regolith deposits at different lunar surface sites. Regolith samples from near large, fresh impact craters (e.g., Cone Crater, Apollo 14) tend to have many rock fragments and mineral grains and a very low percentage of glassy agglutinates and glass fragments. Surface samples away from recent impact craters, especially those with very long meteoroid bombardment exposure ages, tend to have a high percentage of glassy agglutinates, few rock fragments, and many glass fragments. Distinctive rock fragment and glass types can be identified in modal analyses

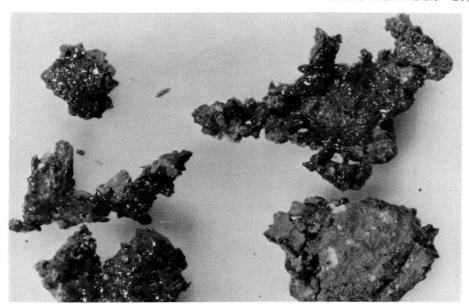

Figure 44 Typical lunar regolith agglutinate particles. Rock fragments, mineral grains, glass, and fine dust are bonded together by brownish vesicular glass. The glass probably originates from regolith materials melted in micrometeorite impacts and spattered onto the nearby surface. Length of field of view is approximately 15 mm. (Courtesy of NASA, Johnson Space Center, photograph S-73-25448.)

of some samples, indicating a different source area or history for the particles. Sample modal analyses of particle types in regolith samples from Apollo 16 are shown in Fig. 45.

One particularly interesting result of regolith particle modal analyses occurred in investigations of the first samples returned by Apollo 11. Four different groups of investigators (81) recognized a small percentage of anorthosite and other feldspar-rich rock fragments in the Apollo 11 regolith samples and suggested that the feldspar-rich rocks might have come from the lunar highlands as ejecta from distant craters. This was the first suggestion from returned sample data that the lunar highlands might be composed largely of feldspar-rich rocks. Wood et al. (82) constructed a geophysical model of the Moon based on this observation.

Grain Size Analyses.   These analyses of lunar regolith also can be used to characterize various surface and subsurface units. Most of the grain size analyses have been performed on the less than 1 mm size fraction of the regolith. The coarser particles are separated and treated with greater care in distribution for study because these size fractions may contain unique or rare

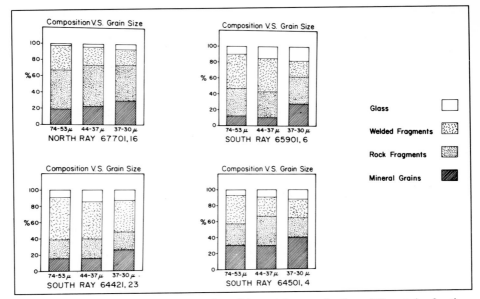

Figure 45   Examples of modal analyses of regolith particle types for three different size fractions each of four Apollo 16 samples. Many additional categories of particle types can be counted. By similar analyses of many regolith samples from one site, a view of the lateral and vertical variations in the regolith can be gained and correlated with surface features. (From Butler et al., 1973, Grain size frequency distributions and modal analyses of Apollo 16 fines: Proc. Fourth Lunar Sci. Conf., Geochim, et Cosmochim. Acta, Suppl. 4, vol. 1, p. 267-278; Courtesy of Pergamon Press.)

rock fragments. The size frequency distribution of the fine fractions of lunar soil are especially interesting to lunar scientists who have speculated concerning the possible electrostatic transport of lunar dust; however, no sample evidence has been found to support this hypothesis. Grain size analyses of lunar samples have been performed by a number of different investigators (83), but most of the results are similar to those shown in Fig. 46. The size frequency distributions of the 1 mm to 10 μm size fractions are approximately log-normal on a weight fraction basis. However, King et al. (76, 83) noted a tendency for lunar samples to be slightly bimodal, with a broad mode in the 1ϕ to 4ϕ size interval due to the formation of glassy agglutinates and a more narrowly defined mode centered at about 5ϕ, which probably results from the influx of very fine ejecta from distant craters in old terrains. In general, lunar samples in this size range are characterized by poor to very poor sorting, platykurtosis, and low values of skewness (nearly symmetrical). Various investigators have had considerable difficulty in achieving reproducible and accurate grain size analysis of lunar samples, primarily because they contain so many very fine grains and because they are so dry. There is great difficulty with ordinary dry sieving of lunar samples because of clumping of the finer size fractions unless the relative humidity is rather high, apparently because of electrostatic attraction between the very fine particles. Also, violent sieving action destroys numerous delicate glassy agglutinates. Analyses of the less than 10 μm fraction of some lunar samples have been accomplished using scanning electron microscopy (84) and by optical photomicroscopy (83). However, the significance and interpretation of these measurements is still in doubt.

A general inverse correlation of the grain size of lunar fines (less than 1 mm) samples with site meteoroid bombardment exposure age has been suggested by King et al. (83) and is supported by other laboratories. The longer the regolith is exposed to meteoroid bombardment, the finer the mean size of the regolith material. This relation is especially easy to recognize if site geology is taken into account (i.e., samples of ejecta from large fresh impact craters and other sources of coarse particles are avoided).

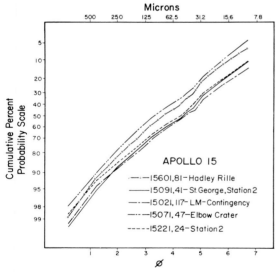

Figure 46 Examples of grain-size analyses of the less than 1 mm size fraction of the lunar regolith. Notice that the samples are slightly bimodal, poorly sorted, and platykurtic. (From Butler et al., 1972, Size frequency distributions and petrographic observations of Apollo 15 lunar samples: *in* The Apollo 15 lunar samples, J. W. Chamberlain and C. Watkins, eds., p. 46; Copyright © Lunar Science Institute 1972.)

However, some sites are so complex that it is difficult to recognize any samples that do not contain recent coarse ejecta.

Although there are numerous reversals in the general trend, it appears that the lunar regolith tends to be coarser grained with depth, based on observations from lunar surface cores and trenches. Thus the regolith may grade in grain size from coarse blocks, barely fractured loose from their crystallization sites, at the base of the regolith to finer and finer size material with increasing evidence of shock metamorphism at the surface. However, it should be anticipated that the general trend will be complicated by overlapping ejecta deposits from craters of different sizes, much as was suggested by Salisbury and Smalley (17).

The suite of rock fragments that is available in the coarser fractions of the regolith (1 to 5 mm) is much more diverse than that represented by the larger hand specimens that have been returned by the Apollo missions (85). Much of the effort concerning these size fractions has been to identify possibly important lunar rock types that are not represented in the large samples (81).

Physical Properties.   The physical properties of the lunar regolith have been investigated in some detail, with the investigations generally falling into three different categories: (a) engineering studies — with few or no scientific objectives; (b) calibration studies — to attempt to interpret older remote measurements or provide base data for new ones; and (c) measurements with no problem orientation — mostly for the joy of precise physical measurements with few practical applications and little fundamental scientific value. These studies have been made in situ and on returned samples in vacuum and in air.

The in situ density of fine regolith at the Apollo 12 site has been estimated at 1.6 to 2.0 gm/cm^3 (86). The density of the regolith increases rapidly with depth, as has been noted by astronauts trying to implant flagstaffs or other equipment into the lunar surface layer. The increase in bulk density of the regolith with depth (Fig. 47) has been calculated by Carrier et al. (87), based on vacuum oedometer tests made on 200 grams of returned sample. The amount of densification calculated from these data is not great, and it has been speculated that the impact of meteoroids might significantly affect (increase) the rate of densification with depth (87). Various other estimates and measurements of lunar surface soil density range from 1.4 to 2.0 gm/cm^3 (88).

Lunar soil properties vary with location on the lunar surface and as a result of local geological setting. Houston et al. (89) made this conclusion based on observations of astronaut footprint depths, and they have also computed porosity of the lunar soil based on friction angle measurements of other workers (88). They obtain a figure of approximately 43 percent mean porosity, with a standard deviation of ± approximately 3 percent for intercrater areas.

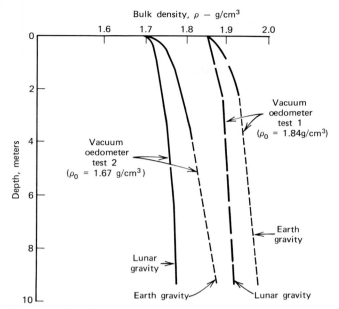

Figure 47   Density versus depth for lunar soil 12001. Vacuum oedometer data (87) have been used to calculate densification due to the self-weight of the soil in lunar and earth gravity. (Reprinted from Strength and compressibility of returned lunar soil by W. D. Carrier, III, et al., by permission of the MIT Press, Cambridge, Mass.)

Porosities of crater rim material and crater and rille slopes were found to be greater, mostly about 46 to 47 percent, but some localities are considerably greater.

The most probable values of cohesion for lunar soils lie between 0.1 and 1.0 kN/m² and the most probable lunar soil friction angles are 30 to 50 deg, with the greater values associated with lesser porosities, according to Mitchell et al. (88).

BRECCIAS AND MICROBRECCIAS

All of the particle types that occur in the uncohesive particulate material (soil) also can occur as components of the breccias and microbreccias.[10] Glass spheres, rock fragments, mineral and glass fragments are abundant in the

[10]The term ''microbreccia'' has been used by many investigators and also by some of the Lunar Sample Preliminary Examination Teams. However, in this book the term breccia will be used to include all scales.

breccias (Fig. 48). The delicate glassy agglutinates become difficult to distinguish in many of the breccias, but undoubtedly they are incorporated into many specimens also. It now appears that there are two major breccia types: soil breccias[11] and base surge emplaced autometamorphosed breccias.

Soil Breccias.   These are abundant in the Apollo 11 samples (90) and are somewhat rarer in samples from other missions. One example that has been well described is rock 14313, which has been interpreted as shock-lithified particulate regolith (91). This breccia type contains abundant recognizable clasts of breccia and rocks with igneous textures, undevitrified regularly shaped glass particles and other undevitrified glass fragments. The clasts within this polymict breccia type range from unshocked to shock-melted, but no evidence is found of pervasive thermal metamorphism. The margins of clasts are distinctly visible and the matrix, which is mostly brownish glass, shows no evidence of recrystallization. In hand specimen, these breccias tend to be loosely cohesive or friable compared with the strongly metamorphosed breccia.

[11]The term "soil breccia" is now used by most workers, but is essentially synonymous with "shock-compressed microbreccia," "low-grade" and "low facies, group 1" breccias.

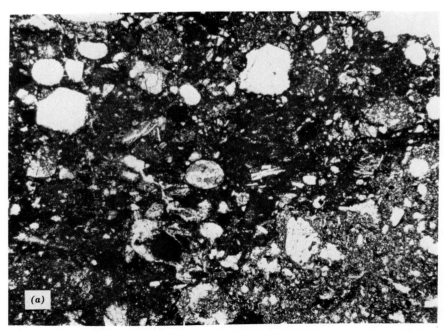

Figure 48 Photomicrographs of thin sections of three types of lunar breccia: *(a)* rock 14318,43 showing the complex texture and diversity of clasts typical of most lunar breccias returned by the Apollo missions;

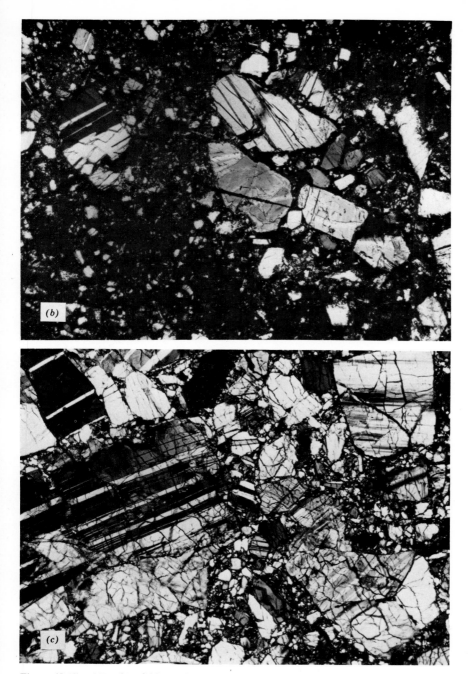

Figure 48 (Cont'd) *(b)* a feldspar-rich breccia (65095,54) collected by the Apollo 16 mission from the lunar highlands; and *(c)* anorthosite breccia (60215,13). Figure 48*a* in plane polarized light; Figs. 48*b*, *c* in crossed polarizers. Lengths of fields of view approximately 3.2 mm.

Autometamorphosed Breccias.   The stratigraphy of the Fra Mauro Formation at the Apollo 14 landing site has been interpreted as an autometamorphosed single cooling unit of hot, impact-emplaced ejecta from the Imbrian Basin by Warner (92) and others. This interpretation was based on the petrographic series of breccia types, ranging from unrecrystallized breccias through fully recrystallized breccias and the structures of the Apollo 14 site that apparently are related to the Imbrian event. The unrecrystallized breccias are identical to the soil breccias described above, except that agglutinates are rare, and solar wind related particle tracks and chemistry may not be present. The thermally metamorphosed and recrystallized breccias are mostly easily characterized by their matrix textures (Fig. 49), and the visible amount of pervasive recrystallization and metamorphism of clasts, spherules, and matrix. The welding and recrystallization cross section through an impact-emplaced base surge and fallback unit should resemble the cross section through a terrestrial ash flow tuff deposit (see Meteorites, Fig. 17, p. 35). The initial emplacement temperatures of parts of the Fra Mauro Formation may have been within the range 700°C to 1050°C, if the estimates of Anderson et al. (93) and Grieve et al. (94) are correct.

Thus it is likely that areas of the lunar surface that are relatively young and away from large impact craters (e.g., mare surfaces) will contain mostly soil breccias, but old areas of the lunar surface and those areas close to large impact craters (e.g., highland surfaces) will contain both types of breccia. The Apollo 16 landing site provided the most complex suite of breccias, and these have been described by Wilshire et al. (95) and others.

Lunar Chondrules.   These were discovered in both types of breccia at the Apollo 14 site (96), which was the first convincing occurrence of chondrules in rocks other than chondritic meteorites.[12] King et al. have discussed the implication of this observation for the origin of chondritic meteorites (page 30, Origin of Chondrules). They conclude that lunar chondrules are formed by at least three mechanisms related to large lunar impact events:

1. Crystallization of shock-melted silicate spherules and droplets (Figs. 50 and 51). This mechanism will form chondrules by the free silicate melt that has been thrown into near lunar space forming a roughly spherical shape, because of surface tension, followed by supercooling and rapid crystallization as has been shown by Nelson et al. (97). Even if the glass does not supercool and crystallize by this method, it may crystallize and devitrify more slowly if incorporated into the hot base surge and ejecta deposits.

2. Rounding of rock clasts and mineral grains by abrasion in impact-generated base surge deposits (Fig. 52). The Fra Mauro Formation at the Apollo 14 landing site is thought to be the base surge

---

[12]Chondrulelike bodies were later discovered in the ejecta from the Lonar Lake, India impact crater.

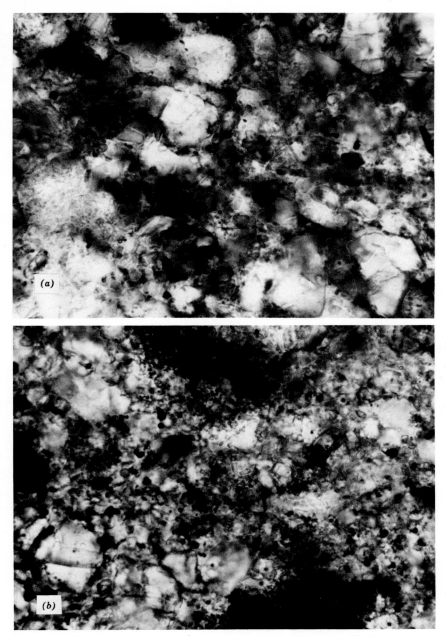

Figure 49 Photomicrographs of matrix textures in Apollo 14 breccias (92) illustrating progressive thermal metamorphism and recrystallization; *(a)* is unmetamorphosed, and *(f)* is most metamorphosed and recrystallized. Longer dimension of each photomicrograph is 220 μm, convergent transmitted light. (Reprinted from Metamorphism of Apollo 14 breccias by Jeffrey L. Warner by permission of the MIT Press, Cambridge, Mass.)

211

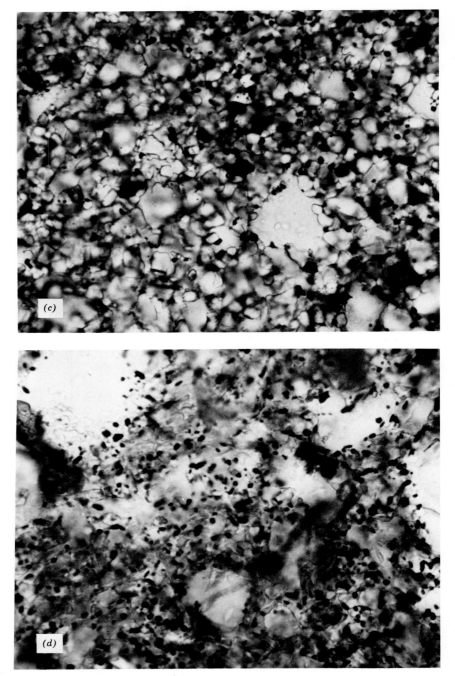

Figure 49 (Cont'd)

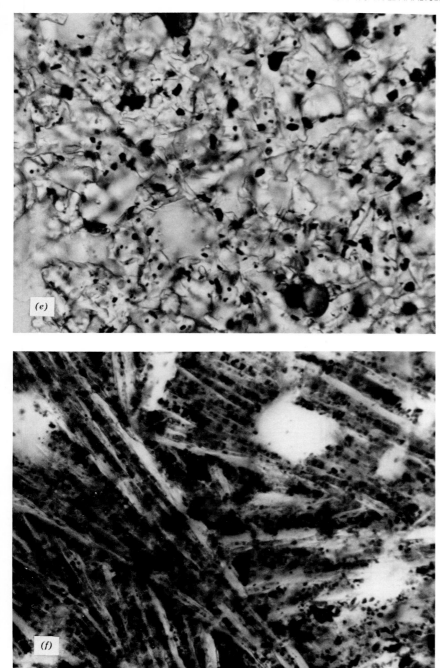

Figure 49 (Cont'd)

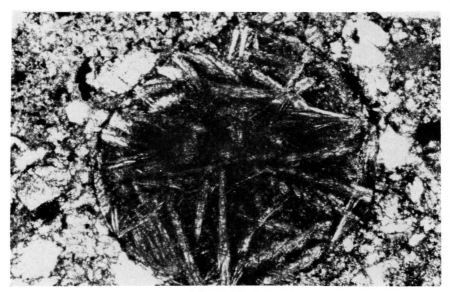

Figure 50   Lunar chondrule in Apollo 14 sample 14318. The chondrule apparently was a fluid drop that assumed a spherical shape due to surface tension, then crystals of plagioclase and pyroxene nucleated at the surface of the sphere, and crystallization proceeded into the interior. The crystals are surrounded by dark brown turbid glass. Diameter of chondrule is 0.5 mm, plane polarized light. (After King et al., 96; Copyright 1972 by the American Association for the Advancement of Science.)

deposit and other ejecta from the crater that formed the Imbrian Basin. There would be ample opportunity for particle interactions and abrasion in the more than 500 km of transport to the Apollo 14 landing site of most of the particles that make up the base surge deposit.

3. Diffusion around rock clasts and mineral grains that are of markedly different composition from the surrounding detritus in hot base surge and fallback deposits (Fig. 53).

Relative Abundance.   Breccias are the most abundant rock type collected at the Apollo 14 and Apollo 16 landing sites and are common in samples returned from the other landing sites, except for Apollo 12 (Table VI). The highland sites and terrains seem to be composed largely of brecciated ejecta from the large mare basins and from large nearby craters. Some of the rocks classified as rock fragments with igneous textures in Table VI may actually be large clasts that have been eroded out of or have been collected from weakly coherent breccias.

Geochemistry, Ages, Physical Properties.   Meteorite fragments have been recognized in lunar breccias and soil by several investigators. These fragments have included stony meteorites, meteoritic nickel-iron, and one

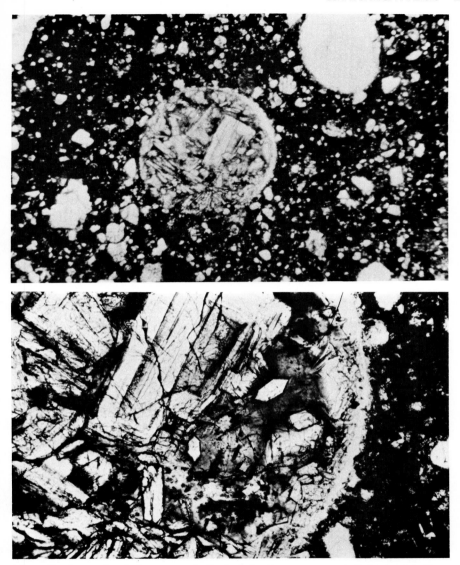

Figure 51   Lunar chondrule composed of euhedral orthopyroxene and olivine crystals in brown transparent glass in Apollo 14 sample 14313. Note that the glass spherule in the upper right of the photomicrograph at top shows no signs of devitrification or crystallization. Diameter of chondrule is approximately 0.8 mm. The bottom photograph is a higher magnification view of the lunar chondrule at top. Note the skeletal, euhedral orthopyroxene crystals and the small euhedral olivines in the light brown glass matrix. There is some devitrification of the glass, but most of the glass is transparent between the crystals. Both views in plane polarized light. (96; Reprinted from Chondrules in Apollo 14 samples and size analyses of Apollo 14 and 15 fines by E. A. King, Jr., et al., by permission of the MIT Press, Cambridge, Mass.)

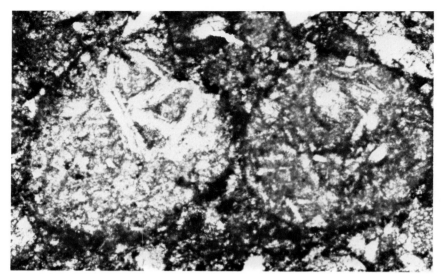

Figure 52 Lunar chondrules apparently formed by rounding of rock clasts by abrasion. This type of chondrule is abundant in Apollo 14 samples, and the abrasion apparently has taken place in the base surge from the Imbrian event as it traveled more than 500 km to the Fra Mauro landing site. Length of field of view is 0.8 mm, plane polarized light. (96; Reprinted from Chondrules in Apollo 14 samples and size analyses of Apollo 14 and 15 fines by E. A. King, Jr., et al., by permission of the MIT Press, Cambridge, Mass.)

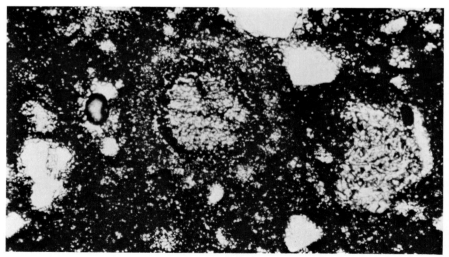

Figure 53 Dunite clast surrounded by diffusion halos (aureoles) of pyroxene and opaque minerals in an Apollo 14 sample. Such clasts surrounded by a recrystallized zone are abundant in the Apollo 14 breccias, and similar recrystallization is common around chondrules in slightly recrystallized chondritic meteorites. Length of field of view is 0.8 mm, plane polarized light. (After King et al., 96; Copyright 1972 by the American Association for the Advancement of Science.)

Table VI   Numbers of Breccias and Rocks with Igneous Textures in the Apollo Sample Returns[a]

	Breccias	*Rocks with Igneous Textures*
Apollo 11	8	5
Apollo 12	4	41
Apollo 14	30	2
Apollo 15	29	27
Apollo 16	60	21
Apollo 17	46	47
Totals	177	143

[a]For the larger than 50 gm samples only. The reader should be aware that crew selection bias may have seriously affected the proportions for any given site. (*Sources.* Apollo lunar sample catalogs and lunar sample inventories. Does not include samples collected as chips from surface boulders.)

possible piece of a mesosiderite. Extensive trace element analytical studies by Ganapathy, Anders, and co-workers (98) have estimated the amount of meteoritic contamination in Apollo rocks and soils. The average value for Apollo soils and breccias is approximately 2 percent of meteoritic CI (carbonaceous chondrite) composition.

Numerous analyses of soils and breccias have been performed (Table VII) that generally show a slight concentration of refractory elements (Fig. 34). Otherwise, the analyses mostly are similar to the rocks with igneous textures.

Model ages for lunar soils and breccias determined by radiometric dating techniques generally are older than those obtained for the rocks with igneous texture and cluster around 4.6×10^9 yr (99). The interpretation of these model ages is not clear, but it appears that there has been a significant amount of lead mobility in the soils, possibly as a result of impact heating. However, rubidium-strontium ages agree with lead ages, and it has been suggested that the regolith, which is an accumulation of diverse particles with different ages, represents some sort of aggregate sample of the lunar crust (100). The 4.6×10^9 yr age may represent the preponderant age of the rocks from which the regolith is derived.

The magnetic properties of lunar breccias and other rocks have been the subject of considerable interest. Gose et al. (101) have correlated the magnetic properties of breccias from Apollo 14, which have pronounced viscous remanent magnetization, with the metamorphic grade of the breccias. It appears that the type of magnetization can be correlated with the size of the interstitial iron grains present in the breccia sample, which in turn can be correlated with the metamorphic grade. Thus they conclude that the remanent magnetization in the lunar breccias is of thermal origin and is contemporaneous with the time of breccia formation. However, lunar

Table VII   Examples of Average Major Element Compositions of Lunar Breccias and Soils[a]

Oxide	Breccias			Soils (Less than 1 mm Fraction)		
	Apollo 11	Apollo 12	Apollo 14	Apollo 11	Apollo 12	Apollo 14
SiO_2	41.80	46.52	47.78	42.04	46.40	47.93
Al_2O_3	13.10	14.64	16.76	13.92	13.50	17.60
Fe_2O_3	0.00	0.00	0.00	0.00	0.00	0.00
FeO	15.90	13.85	10.24	15.74	15.50	10.37
MgO	7.70	9.06	10.57	7.90	9.73	9.24
CaO	11.80	11.15	10.48	12.01	10.50	11.19
Na_2O	0.46	0.61	0.83	0.44	0.59	0.68
K_2O	0.16	0.40	0.56	0.14	0.32	0.55
TiO_2	8.49	2.17	1.68	7.48	2.66	1.74
P_2O_5	<0.2	—	0.54	0.12	0.40	0.53
MnO	0.22	0.19	0.13	0.21	0.21	0.14
Cr_2O_3	0.32	0.33	0.21	0.30	0.40	0.25

[a]From Rose, H. J., F. Cuttitta, C. S. Annell, M. K. Carron, R. P. Christian, E. J. Dwornik, L. P. Greenland, and D. T. Ligon, Jr. (1972) Compositional data for twenty-one Fra Mauro lunar materials: Proc., Third Lunar Sci. Conf., vol. 2, D. Heymann, ed., Geochim. et Cosmochim. Acta, Suppl. 3, vol. 2, p. 1215-1229. All values in weight percent.

breccias also commonly have a component of magnetization that is as stable as the magnetization which is typical of the lunar igneous rocks, but this component is generally very small. Typical values of the natural remanent magnetization of lunar rocks range from approximately 2×10^{-6} to 2×10^{-3} emu/gm (102). On a statistical basis, the breccias have higher natural remanent magnetizations than the rocks with igneous textures.

Most investigators agree that the "hard" component of lunar rock magnetization is thermoremanent magnetization acquired in a magnetic field of approximately $10^3 \gamma$ (102). It should be noted that the remanent magnetic fields measured on the Moon to date range from approximately 6 to slightly more than 100γ. Subsatellites launched by Apollo 15 and 16 have determined that lunar magnetic anomalies are on the scale of tens of kilometers. Also, it is observed that the field intensities over highland sites are stronger than over mare sites. Some large anomalies are associated with craters such as Van de Graff. Runcorn and co-workers (103) strongly support the idea that much of the magnetization of lunar rocks was caused by an internal lunar magnetic field as a result of molten core motion early in lunar history. However, there are other possibilities, and the origins of the sample magnetic characteristics and regional lunar magnetic fields are still in dispute.

Organic and Biologic Analyses.   Lunar samples, particularly breccia and soil samples, have been investigated extensively for their content of organic compounds and possible life forms. For the early lunar sample returns, it was required that the samples be isolated and treated as if they contained ''high hazard pathogens,'' and extensive facilities were constructed to contain any biological agents in the samples or crew while tests were made to ascertain whether or not any organisms present might present a threat to the terrestrial biosphere. A special laboratory, the Lunar Receiving Laboratory, was constructed and equipped to contain the samples and crew and to perform these tests, as well as preliminary geological examinations and other time-critical experiments (104). The results of quarantine tests and life detection tests have, of course, been completely negative (105) as have searches for fossil life forms (106). Organic carbon and compounds have been detected in lunar samples, but only at extremely low levels of concentration which correlate with the contamination history of the sample (107). Small amounts of carbon, methane, and ethane have been found in virtually all lunar surface samples, but the source of these materials is almost certainly the solar wind. The indigenous carbon and carbon compound contents of many meteorites are much greater than lunar samples.

LUNAR GEOPHYSICS

Measurements obtained from instruments placed on the surface of the Moon and from lunar orbiting spacecraft have given us a large-scale geophysical view of the Moon, especially if the data are combined with observations from the lunar samples.

STRUCTURE OF THE LUNAR CRUST

Both the passive and active lunar seismic experiments (108), together with velocity data derived from returned lunar samples, have elucidated the structure of the near surface portion of the Moon. In addition to moonquakes and small explosive charge detonations, the impacts of the Lunar Module ascent stage and Saturn IVB stage have provided excellent energy sources with which to explore the lunar interior (109). The velocity structure of the near-surface and interior is deduced from direct observations and the computation of models of the velocity structure that seem to be best fits of the observed data. It is reasonably certain that the structure of the shallow lunar interior is different from site to site; however, the following generalizations seem to be defensible with the data at hand.

The material at the immediate surface and very shallow subsurface consists of an accumulation of broken and weakly bonded particulate rock fragments and mineral grains with velocities ranging from approximately 180 to 220

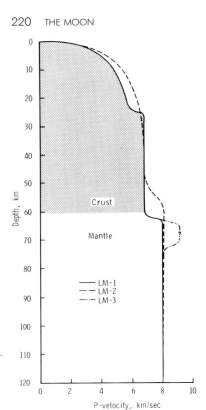

Figure 54   Depth versus P-wave
velocity curve for the outer 120 km of
the Moon. Note the "crust"/"mantle"
interface at approximately 60 km and
the shallower discontinuity at 25 km.
The dashed lines are other lunar
models (LM) that basically satisfy the
same observations. (Courtesy of D. R.
Lammlein, 116; Copyrighted by the
American Geophysical Union.)

m/sec. Immediately underlying this low velocity surface layer at depths of
from 3 to 15 m is a high velocity layer, v = approximately 250 to 300 m/sec.
The velocity gradually increases to a depth of approximately 5 km where the
velocity is approximately 4 km/sec (110). Below this depth the velocity
gradually increases to approximately 6 km/sec at 25-km depth, where there is
a sharp discontinuity in the slope of the depth-velocity curve (Fig. 54) and the
velocity remains virtually constant at 7 km/sec to a depth of 60 km. At 60 km
the velocity again sharply increases until it is approximately 8 km/sec at a
depth of 63 km within the Moon (111).

From these data it may be interpreted that the Moon has a fragmented and fractured outer layer to a depth of a few kilometers. Below this layer, the Moon seems to be composed of basalts (or other rocks with similar velocities) to a depth of 25 km. From 25 to 65 km, the best fit velocities, together with lunar sample comparisons, seem to indicate gabbroic anorthosite as the prime candidate. Below 65 km, the boundary between the lunar crust and mantle, it appears that the rocks are more mafic, probably rich in olivine and pyroxene that would account for the high velocities (111). With several lunar seismic stations still in operation, it may be possible to refine and extend this model as additional data become available.

STATE OF THE LUNAR INTERIOR

There has been a great deal of argument as to the temperature and state of the lunar interior. Various researchers have argued that the lunar interior is cold, hot, or lukewarm. These arguments variously derive from assumed rates (hence, heats) of accretion, assumed radioactive element content, and postulated modes of lunar origin.

Several Apollo Program lunar surface experiments have contributed to this discussion rather recently — the passive seismic experiment, heat flow measurements, and the lunar surface magnetometer. The lunar heat flow experiment (112) has measured lunar heat flow and subsurface temperatures directly at two lunar landing sites, those of Apollo 15 and Apollo 17. The results indicate that the heat flow at the lunar surface is approximately one half that at the surface of the Earth, a surprisingly large figure for a body only one fourth the diameter of the Earth. The measured value of heat flow at the Apollo 15 site is slightly less than 3.3×10^{-6} watts/cm². Similar values are measured for the Apollo 17 site. If these values are typical of a large part of the lunar surface, then the Moon is too hot at present for the "chondritic" compositional models of the Moon. Higher average values of K, Th, or U must be present in the Moon than in chondritic meteorites to provide the necessary heating from the exothermic radioactive decay of these elements.

The data obtained by Sonett et al. (113) seemed to fit a cold to lukewarm Moon based on telemetry from the lunar surface magnetometer. Their model temperatures for the Moon indicated 800°C for the deep core and only 450°C for the lunar interior electrical conductivity peak value. However, Dyal et al. (114) interpreted higher internal temperatures for the Moon.

Data derived very recently from the lunar passive seismic experiment (115) indicate that the Moon may be partially molten below approximately 1050 km, with temperatures of approximately 1000°C. These data derive from the observation that the shear waves from meteoroid impacts and moonquakes that originate on the farside of the Moon are not observable on some of the seismographs that are operating on the front side. The loss of the shear waves indicates a lunar interior of high attenuation (low Q), similar to the Earth, possibly caused by partially molten silicates. Further confirmation of a hot shallow lunar interior would be the occurrence of recent lunar volcanic rocks; however, these have not been found. The youngest lunar rocks known at present apparently are more than three billion years old.

The shallow lunar interior is characterized by very low attenuation (high Q) of seismic energy. This was demonstrated by the exceedingly long seismic records generated by spacecraft and rocket-casing impacts on the Moon, signals generally could be recorded for as long as 2 hr. The reason for this low attenuation is not known. It may be that the outer shell of the Moon is quite rigid and has the strength to support large lithostatic loads such as the mascons (page 224) for long periods of time.

MOONQUAKES AND LUNAR TECTONISM

The passive seismic experiment has demonstrated that the Moon is virtually aseismic compared with the Earth. If it were not for the extremely low seismic noise of the Moon and good sensitivity of the landed seismometers, the Moon would appear to be very dead indeed. However, large numbers of very low energy quakes (approximately 1800 per year) have been observed, but these do not have random times of occurrence. There is a strong tendency for seismic events to occur at times of maximum tidal stress (Fig. 55, 116). Furthermore, many of these events are observed with exactly the same epicenter and focus (Fig. 56), within the accuracy of the seismic net. The identical locations of epicenters and foci suggest that many of the lunar moonquakes result from movement across the same structural discontinuity or fault. This interpretation is supported by the striking similaries of many of the wave trains (Fig. 57). This observation requires that the focal zones must be very small, proably 10 km or less, otherwise it would not be possible to generate such similar wave trains in an inhomogeneous medium (117).

The foci of moonquakes tend to be rather deep within the Moon, approximately 800 km (117), and the Richter magnitudes of the largest events are between two and three. Latham et al. estimate that the total energy release from the Moon is of the order of 10^{15} ergs/yr, nine orders of magnitude less than that of the Earth.

The majority of moonquake signals do not have similar wave trains and commonly occur in "swarms" (117). These events occur at a rate of 8 to 12 per day during a swarm as compared to 1 to 2 per day between swarms. The swarms do not contain any large single or conspicuous event and they begin and end abruptly. Latham et al. speculate that these swarms of moonquakes may correlate with periods of lunar volcanic activity, which may be continuing at a very minor rate at the present time.

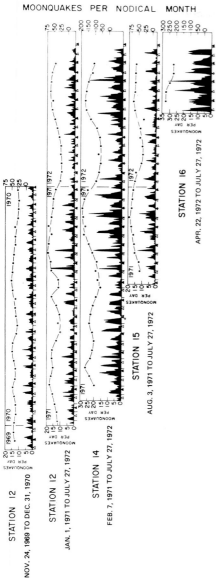

Figure 55 Lunar seismic activity (excluding meteoroid impacts) for the period November 24, 1969 to July 27, 1972. The number of moonquakes per day is plotted as a function of time. Vertical lines on the abscissa indicate the times of maximum latitudinal librations. Solid triangles indicate times of perigee and open triangles indicate times of apogee. Dashed lines indicate the number of events detected per nodical month. A long period variation, approximately 206 days, is evident at stations 12, 14, and 15 during 1971 and 1972 but not at station 12 in 1969 and 1970. For station locations see Fig. 56. (Courtesy of D. R. Lammlein, 116; Copyrighted by the American Geophysical Union.)

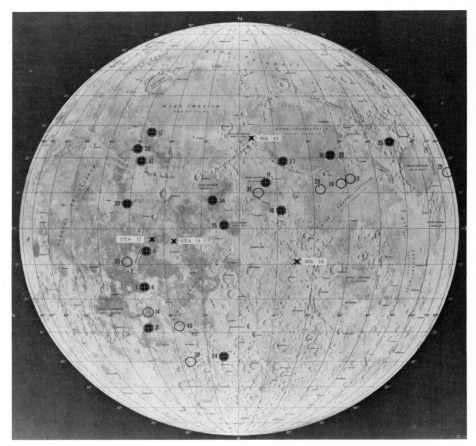

Figure 56   Locations of the epicenters of lunar seismic events of internal origin and the Apollo 12 through 16 seismic stations. Open circles have foci ranging from 730 to 1200 km depth. Solid circles indicate foci for which the depth cannot be determined. Note that epicenters 1 and 6, and 18 and 32 are so close that they cannot be distinguished at this scale. One focus occurs on the farside of the Moon. (Courtesy of D. R. Lammlein, 116; Copyrighted by the American Geophysical Union.)

MASS CONCENTRATIONS

From the detailed analyses of the motions of spacecraft in orbit about the Moon, Muller and Sjogren (118) reported that there were substantial positive gravity anomalies associated with the circular mare basins. They termed these anomalies "mascons," and this term gained wide acceptance. Data from subsequent observations confirmed the existence of the mascons, but there

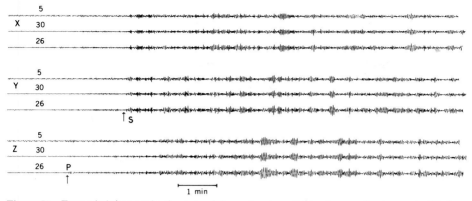

Figure 57 Expanded time scale playouts of three closely matching lunar seismic events. Notice the extreme similarities of the wave trains. *X* and *Y* are the records from the long period horizontal component seismometers and *Z* the records from the long period vertical seismometer. The two most prominent phases are the direct compressional wave (P-wave) and the shear wave (S-wave) arrivals. Such similar events have the same focus and are thought to result from movement induced by tidal strain along the same fault. (Courtesy of D. R. Lammlein, 116; Copyrighted by the American Geophysical Union.)

was not general agreement as to their origin. Urey (119) supposed that the mass concentrations were the result of the capture by the Moon, at low relative velocities, of higher density meteoritic bodies. Other investigators began to reduce lunar data in attempts to find additional mascons, and Campbell et al. (120) reported two new mascon basins, one of which is exceptionally wide and on the lunar farside.

The origin of the mascons seems to be simply explained to a first approximation. From the lunar samples analysis results, we now know that the mare are composed of basaltic rocks with densities ranging from approximately 3.2 to 3.4 gm/cc, as compared with the feldspar-rich rocks and thick deposits of glassy breccias from the lunar highlands, which have densities mostly ranging from less than 2.9 to 3.1 gm/cc. The association of the mascons with the circular mare basins almost certainly reflects the great thickness of higher density rocks filling the crater and associated fractures caused by the large impacts. The absence of mascons associated with mare materials outside the circular mare basins probably results from the shallow thickness of the mare material. It does not seem necessary to involve any higher density meteorite residue to account for the magnitudes of the observed anomalies. However, the problem is complicated by assumptions concerning the amount of isostatic compensation. It may be that the thick, rigid outer shell of the moon is capable of supporting the mascons out of complete isostatic equilibrium for very long times.

LASER ALTIMETRY

An experiment carried for the first time on the Apollo 15 Command and
Service Module, the laser altimeter, confirmed some important facts about the
Moon. In approximately 4½ revolutions of data, the altimeter was able to
determine that there is a 2-km displacement of the center of mass from the
center of figure toward the earthside (121). It also was possible to construct
precise profiles of the surface beneath the flight path of the spacecraft (Fig.
58). These profiles illustrate that the maria are much lower and smoother than
the lunar highlands. Some features were found to be surprisingly deep. For
example, Gagarin has a rim to deep floor depth of approximately 6 km, and
the floor of Mare Smythii is more than 4 km below the elevations of the

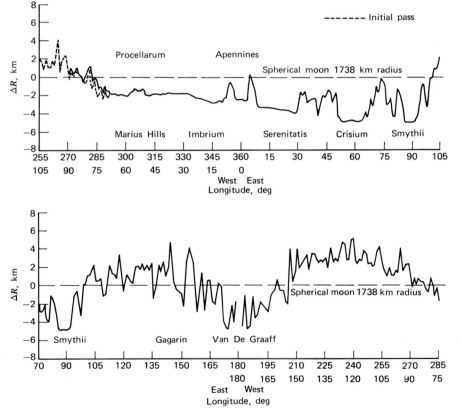

Figure 58 Laser altimetry profiles of the near (top) and farside (bottom) of the Moon as
determined on Apollo 15. (Reprinted from Analysis and interpretation of lunar laser
altimetry by W. M. Kaula et al., by permission of the MIT Press, Cambridge, Mass.)

neighboring highlands. From elevations ranging from 74 to 148 km, the beam illuminated an area approximately 30 m in diameter, and the data have altitude accuracy of ± 2 m. This instrument has demonstrated that some of the maria are extremely level, varying by not more than 150 m about the mean value across 200 to 600 km paths (121).

The moment of inertia ratio (I/MR^2, where I is the mean moment of inertia, M is the mass, and R is the radius) has been calculated by Kaula et al. (121) using constraints from altimetry, seismic, and compositional data. The preferred value is 0.395, compared with 0.4 which would be the ratio for a sphere with a homogeneous density distribution. This places a serious limit on the size of a possible nickel-iron core for the Moon, no more than 3 percent of lunar volume.

ORIBITAL GAMMA-RAY AND X-RAY ANALYSES

The Lunar Orbital Gamma-ray Experiment was flown on both Apollo 15 and Apollo 16 (123). Although the spatial resolution of the instrument is not especially good, it is sufficient to show some important major features. The differences in counting rates over the 0.55 to 2.75 MeV range are due chiefly to differences in Th, U, and K contents. The highest concentrations of radioactivity found in the area covered by the experiment are in the region of Mare Imbrium and Oceanus Procellarum (including the Aristarchus Highlands), 84 to more than 91 counts per second. The levels of radioactivity elsewhere on the Moon are considerably lower, 74 to 84 cps at all other localities. Also, it appears that the K/U ratio everywhere on the lunar surface is lower than on the surface of the Earth.

The Apollo 15 X-ray fluorescence experiment (124) demonstrated that the lunar highlands are systematically higher in Al and lower in Mg than the lunar maria; thus, they probably are rich in plagioclase compared to the maria. The dominant rock type of the highlands has been interpreted as anorthositic gabbro or feldspathic basalt, with possible minor amounts of anorthosite and other rocks by Adler et al. (124). They also postulate that the feldspar-rich highlands rocks may represent the first major differentiation of the Moon.

ORIGIN OF THE MOON

Prior to the Apollo sample returns, the most popular theories for the origin of the Moon were: (a) that it was originally part of the Earth that had broken away, either because of a high rotation rate or through the gravitational influence of a third body, and (b) that the Moon was a large captured asteroid or other body, possibly from a different portion of the Solar System.

G. H. Darwin suggested the "fission hypothesis" for the origin of the Moon in 1908, and included drawings of the way in which he thought that the separation could have occurred (125). This hypothesis and variations of it were later discussed by Wise (126) and O'Keefe (127). It appears that there would be a large mass loss, due to heating by tidal friction, as well as loss of angular momentum immediately after fission. The idea of gravitational interaction with another (unspecified) body has been abandoned by most recent investigators because of the relative improbability and dynamic arguments. However, the means by which the rotational instability of the more massive original Earth would be obtained is not clear either. Darwin (125) and other early workers in this field supposed that thermal contraction could bring the protoplanet to rotational instability. O'Keefe has shown that this mechanism will only make a minor contribution to the rotational instability of the mass. That rotational instability might occur as a consequence of planetary differentiation, that is, core formation, has been postulated by Ringwood (128) and others. It is clear that core formation will increase the rotation rate, but O'Keefe (129) has argued convincingly that core formation will improve the rotational stability of the planet. It is highly unlikely that the Moon could have fissioned from the Earth before formation of the core or large-scale differentiation of the planet. Otherwise, the Moon should have the same bulk density (corrected for gravitational compression) as the Earth ($5.5 \, gm/cm^3$) but it does not ($3.34 \, gm/cm^3$). If the Moon fissioned from the Earth after the major differentiation of the Earth, it should be composed of material from the crust and upper mantle of the Earth (volume of the Moon is only approximately 1/80 that of the Earth), but the differences in chemistry between terrestrial rocks from the crust and mantle and samples returned from the Moon are thought to be too great to be accounted for by this method of lunar origin. However, O'Keefe (129) has argued that the Moon has about the same deficiency in siderophile elements as the mantle and crust of the Earth. He further argues that the lack of volatile elements in the Moon could be accounted for by intense tidal heating after fission, in agreement with the idea of Wise (126). Ringwood and Essene (130) generally have supported the modified fission hypotheses of O'Keefe and Wise, but believe that the chemical fractionation between the Earth and Moon is so great that complete vaporization of the material that later condenses to form the Moon probably is required. Ringwood stresses the point that the material he believes later accreted to form the Moon was precipitated from a huge, hot primordial atmosphere.

The capture hypothesis of lunar origin was popular prior to Apollo 11, but has declined in the post-Apollo era. Lunar tidal evolution studies by Gerstenkorn (131) indicated that the Moon was very close to the Earth only $\sim 2 \times 10^9$ yr ago. Because this time is much younger than the age of the Earth, Gerstenkorn proposed that the Moon had been captured, in a retrograde orbit just before the time of the Moon's closest approach to the Earth. Both MacDonald (132) and Öpik (133) pointed out that a close approach of the Moon to the Earth would provide great internal tidal heating for both bodies.

They pointed out that the present surface features of the Moon should have been formed mostly at this time and subsequently. In addition, there should be a tremendous structural record of this event in old crustal rocks of the Earth, but it has not been recognized. Also, the rocks returned by the lunar missions have crystallization ages that are much older than 2 billion years. Urey (134) also suggested that the Moon was captured approximately 4.5 billion years ago, and that it was a primitive object of solar composition, except for gases, and different in its iron to silicate ratio from the terrestrial planets. However, a redetermination of the abundance of iron in the Sun's atmosphere (135) has invalidated this argument. The recent history of the capture hypothesis has been summarized by Ringwood (130), who points out that "capture of the Moon by the Earth is an event of low intrinsic probability."

The "double planet" hypothesis is that the Moon was always a near neighbor of the Earth and accreted in the same part of the Solar System. This hypothesis assumes that metallic iron was concentrated in the Earth relative to the Moon by differentiation in the Solar Nebula; however, the processes are not defined or specified (136). There is a recent suggestion that the accretion of the Earth may have preceded the Moon, and that the Sun became a luminous body before the material that formed the Moon accreted into a sizable body. If this were true, the retention of volatiles in the Earth would be explained as would their depletion in the Moon; but the mechanisms as to how accretion of the Moon occurred are poorly understood.

Whatever the origin of the Moon, the old ages of the surface rocks, current internal state, and the differences between the composition of lunar rocks, terrestrial rocks, and meteorites must be taken into account. The processes that form moons must not be terribly unlikely events because there are too many of them in the Solar System. Capture of asteroids seems to be a plausible explanation for the origin of the outer moons of Jupiter and the moons of Mars, but the origins of the larger moons in the Solar System, particularly our Moon, remain a speculative and fundamental problem.

REFERENCES AND NOTES

1. Koestler, A. (1959) The sleepwalkers — A history of man's changing vision of the Universe, Macmillan, New York, 624 p.

2. Emanuelli, P. (1956) Da Vinci's astronomy: *in* Leonardo Da Vinci, Instituto Geografico De Agostini S. p. A., Novara, Reynal & Co., New York, p. 208.

3. Gilbert, G. K. (1893) The Moon's face: A study of the origin of its features: Bull. Phil. Soc. Wash., vol. 12, p. 241-292.

4. Gifford, A. C. (1924) The mountains of the Moon: New Zealand Jour. Sci. Tech., vol. 7, p. 129-142; also see vol. 11, p. 319-327 (1930).

5. Mohorovicic, S. (1928) Experimentalle Untersuchungen Über die Entstehung der

Mondkrater: ein neuer Beitrag zur Explosionshypothese (Croatian with German summary): Archiv za Hemiju i Farmaciju, Zagreb, vol. 2, p. 66-76.

6. Wegener, A. (1921) Die Entstehung der Mondkrater: Sammlung Vieweg, Heft 55, Braunschweig, 48 p. See also an English translation by A.M.C. Sengör (1975) The Moon, vol. 14, p. 211-236.

7. Barrell, J. (1927) On continued fragmentation, and the geologic bearing of the Moon's surficial features: Am. Jour. Sci., 5th ser., vol. 13, p. 282-314. See also Smithsonian Inst. Ann. Rept., 1928, p. 283-306.

8. Spencer, L. J. (1937) Meteorites and the craters on the Moon: Nature, vol. 139, p. 655-657.

9. Baldwin, R. B. (1963) The measure of the Moon: Univ. of Chicago Press, Chicago, 488 p.

10. Fielder, G. (1961) Structure of the Moon's surface: Pergamon Press, New York, 266 p. See also Fielder, G. (1965) Lunar geology: Lutterworth Press, London, 184 p., and numerous papers with various coauthors.

11. Hackman, R. J. (1961) Photointerpretation of the lunar surface (with fold-in map supplement): Photogramm. Eng., vol. 27, no. 3, p. 377-386; (1961) Geology of the Moon: Space Sci., vol. 10, no. 10, p. 2-6, see also Hackman, R. J., and A. C. Mason (1961) Engineer special study of the surface of the Moon: U. S. Geol. Survey, Misc. Geol. Inv., Map I-351 (4 sheets). See especially Hackman, R. J. (1962) Geologic map of the Kepler region of the Moon: U. S. Geol. Survey, Misc. Geol. Inv., Map I-355.

12. Shoemaker, E. M., and R. J. Hackman (1962) Stratigraphic basis for a lunar time scale: *in* The Moon, Z. Kopal and Z. K. Mikhailov, eds., Symp. No. 14, Intern. Astron. Union, Academic Press, London, p. 289-300; also Shoemaker, E. M., R. J. Hackman, and R. E. Eggleton (1962) Interplanetary correlation of geologic time: Advances in the Astronaut. Sci., vol. 8, p. 70-89.

13. Wilhelms, D. E. (1970) Summary of lunar stratigraphy — Telescopic observations: U. S. Geol. Survey Prof. Pap. 599-F, 47 p.

14. For example, see El Baz, F. (1972) New geological findings in Apollo 15 lunar orbital photography: Proc. Third Lunar Sci. Conf., Geochim. et Cosmochim. Acta, Suppl. 3, vol. 1, E. A. King, Jr., ed., p. 39-61; also El Baz, F. and S. A. Roosa, *ibid.*, p. 63-83.

15. Wilhelms, D. E., and J. F. McCauley (1971) Geologic map of the near side of the Moon: U. S. Geol. Survey, Map I-703; also, McCauley, J. F., and D. E. Wilhelms (1971) Geological province map of the near side of the Moon: Icarus, vol. 15, p. 363-367.

16. This is an extensive literature but for an entry see summaries in Baldwin, R. B. (1963) The measure of the Moon: Univ. of Chicago Press, Chicago, 488 p.; also, Salisbury, J. W., and P. E. Glaser, eds. (1964) The lunar surface layer — Materials and characteristics: Academic Press, New York, 532 p.; and Hess, W. N., D. H. Menzel, and J. A. O'Keefe, eds. (1966) The nature of the lunar surface: Johns Hopkins Press, Baltimore, Md., 320 p.; and many other works.

17. Salisbury, J. W., and V. G. Smalley (1964) The lunar surface layer: *in* The Lunar Surface Layer — Materials and Characteristics, J. W. Salisbury and P. E. Glaser, eds., Academic Press, New York, p. 410-443.

18. Shoemaker, E. M. (1966) Preliminary analysis of the fine structure of the lunar surface in Mare Cognitum: *in* The Nature of the Lunar Surface, W. N. Hess, D. H. Menzel, and J. A. O'Keefe, eds., The Johns Hopkins Press, Baltimore, Md., p. 23-77, also published as a Jet Propulsion Lab., Cal. Inst. Tech., Tech. Rept. No. 32-700.

19. Shoemaker, E. M., M. H. Hait, G. A. Swann, D. L. Schleicher, D. H. Dahlem, G. G. Schaber, and R. L. Sutton (1970) Lunar regolith at Tranquillity Base: Science, vol. 167, p. 452-455; see also Oberbeck, V. R., and W. L. Quaide (1967) Estimated thickness of a fragmental surface layer of Oceanus Procellarum: Jour. Geophys. Res., vol. 72, no. 18, p. 4697-4704.

20. Watkins, J. S., and R. L. Kovach (1972) Apollo 14 active seismic experiment: Science, vol. 175, p. 1244-1245.

21. Pohn, H. A., and T. W. Offield (1969) Lunar crater morphology and relative age determination of lunar geologic units: U. S. Geol. Survey, Interagency Rept., Astrogeol. 13, 35 p.

22. MacDonald, T. L. (1931) The number and area of lunar objects: Jour. Brit. Astron. Assoc., vol. 41, p. 288-290.

23. Young, J. (1940) A statistical investigation of the diameters and distribution of lunar craters: Jour. Brit. Astron. Assoc., vol. 50, p. 309-326.

24. Öpik, E. J. (1960) The lunar surface as an impact counter: Royal Astron. Soc., Monthly Notices, vol. 120, no. 5, p. 404-411. See also, Kreiter, T. J. (1960) Dating lunar features by using crater frequency: Pub. Astron. Soc. Pac. vol. 72, p. 393-398.

25. Gault, D. E. (1970) Saturation and equilibrium conditions for impact cratering on the lunar surface — criteria and implications: Radio Sci., vol. 5, p. 273-291, this paper was not published in the open literature until 1970, but was completed in 1960 and widely circulated among lunar workers. See also, Marcus, A. H. (1964) A stochastic model for the formation and survival of lunar craters: Icarus, vol. 3, p. 460-472; vol. 5, p. 165-177, 178-190, 190-200, 590-605; vol. 6, p. 56-74. An excellent series of related papers. See also, Marcus, A. H. (1967) Statistical theories of lunar and Martian craters: *in* Mantles of the Earth and Terrestrial Planets, S. K. Runcorn, ed., Wiley, Interscience, New York, p. 417-424.

26. For example see Hartmann, W. K. (1965) Terrestrial and lunar flux of large meteorites in the last two billion years: Icarus, vol. 4, p. 157-165; also (1966) Early lunar cratering: Icarus, vol. 5, p. 406-418; (1970) Preliminary note on lunar cratering rates and absolute time-scales: Icarus, vol. 12, p. 131-133; and (1971) Lunar cratering chronology: Icarus, vol. 13, p. 299-301. See also Freyer, R. J., and Titulaer, C. (1970) Counts of small craters in the southern lunar highlands: Earth and Planet, Sci. Letters, vol. 9, p. 6-9. See also Ronca, L. B., and R. R. Green (1970) Statistical geomorphology of the lunar surface: Geol. Soc. Am., Bull., vol. 81, p. 337-352. Also Soderblom, L. A., and L. A. Lebofsky (1971) A technique for rapid determination of relative ages of lunar areas from Orbiter photography: Cal. Inst. Tech., Div. Earth and Planet. Sci., Contr. 1657.; and numerous other papers.

27. Hartmann, W. K. (1972) Paleocratering of the Moon: Review of post-Apollo data: Astrophys. Space Sci., vol. 17, p. 48-64.

28. Moore, P. A. (1965) An evaluation of the reported lunar changes: *in* Geological Problems in Lunar Research, F. E. Whipple, ed., New York Acad. Sci., Annals, vol. 123, art. 2, p. 797-810; also, (1965) Variations on the surface of the Moon, An Evaluation: Astron. Jour., vol. 7, no. 4, p. 106-113.

29. Kozyrev, N. A. (1956) Luninescence of the lunar surface and intensity of solar corpuscular radiation: Krymskaya Astrofiz. Observ., Izv. vol. 16, p. 148-159 (in Russian); see also, Cal. Inst. Tech., Jet Propulsion Lab., Translation No. 18, 18 p. (J. L. Zygielbaum, trans.)

30. Kopal, Z. (1964) The Moon, 2nd ed., Academic Press, New York, Chap. 6, p. 108-118.

31. Greenacre, J. A. (1963) A recent observation of lunar color phenomena: Sky and Teles., vol. 26, no. 6, p. 316-317; also, (1964) Another lunar color phenomenon: Sky and Teles., vol. 27, p. 3.

32. Anonymous (1964) Lunar color phenomena: USAF, ACIC Tech. Paper no. 12, 11 p., May, 1964.

33. Kopal, Z. (1966) Luminescence of the moon and solar activity: *in* The Nature of the Lunar Surface, W. N. Hess, D. H. Menzel, and J. A. O'Keefe, eds., The Johns Hopkins Press, Baltimore, Md., p. 173-183; see also Derham, C. J., and J. E. Geake (1964) Luminescence of meteorites: Nature, vol. 201, p. 62-63.

34. Turkevich, A. L. (1961) Chemical analysis of surfaces by use of large-angle scattering of heavy charge particles: Science, vol. 134, p. 672-674; also, Turkevich, A. L., K. Knolle, R. A. Emmert, W. A. Anderson, J. H. Patterson, and E. Franzgrote (1966) Instrument for lunar surface chemical analysis: Rev. Sci. Instr., vol. 37, no. 12, p. 1681-1686; and other papers.

35. Turkevich, A. L., E. J. Franzgrote, and J. H. Patterson (1967) Chemical analysis of the Moon at the Surveyor V landing site — Preliminary results: Cal. Inst. Tech., Jet Propulsion Lab., Tech. Rept. 32-1246, p. 120-149; also, (1967) same title, Univ. of Chicago, Enrico Fermi Inst. Nuclear Studies, Doc 67-84, 12 p. plus figures. See also, (1967) Chemical analysis of the Moon at the Surveyor V landing site: Science, vol. 158, p. 635-637, and other papers.

36. Turkevich, A. L., E. J. Franzgrote, and J. H. Patterson (1969) Chemical composition of the lunar surface in Mare Tranquillitatis: Science, vol. 165, p. 277-279.

37. Turkevich, A. L., E. J. Franzgrote, and J. H. Patterson (1968) Chemical analysis of the Moon at the Surveyor VI landing site — Preliminary results: Cal. Inst. Tech., Jet Propulsion Lab., Tech. Rept. 32-1262, p. 127-153; also, Turkevich, A. L. (1968) Chemical analysis of the Moon at the Surveyor VII landing site — Preliminary results: Science, vol. 162, p. 117-118.

38. The prime sources of sample information are listed below; however, in addition to these volumes, individual papers have appeared in a wide variety of journals. Science, vol. 167, no. 3918, 30 January, 1970, 792 p. (this is the first publication of the results of the Apollo 11 Lunar Science Conference); Geochimica et Cosmochimica Acta, Suppl. 1, vols. 1-3, Proc. Apollo 11 Lunar Sci. Conf., A. A. Levinson, ed., Pergamon Press; Geochimica et Cosmochimica Acta, Suppl. 2, vols. 1-3, Proc. Second Lunar Sci. Conf., A. A. Levinson, ed., Pergamon Press; Geochimica et Cosmochimica Acta, Supp. 3, vol. 1, E. A. King, Jr., ed., vol. 2,

D. Heymann, ed., vol. 3, D. R. Criswell, ed., Proc. Third Lunar Sci. Conf. MIT Press; Geochimica et Cosmochimica Acta, Suppl. 4, vols. 1-3, Proc. Fourth Lunar Sci. Conf., Pergamon Press; Geochimica et Cosmochimica Acta, Suppl. 5, vols. 1-3, Proc. Fifth Lunar Sci. Conf., Pergamon Press; Geochimica et Cosmochimica Acta, Suppl. 6, vols. 1-3, Proc. Sixth Lunar Sci. Conf., Pergamon Press. After each preliminary examination of lunar samples prior to distribution to investigators, a brief description of the samples has appeared in Science, authored by the Lunar Sample Preliminary Examination Team (LSPET). Also, the lunar sample investigators have received catalogs with sample descriptions and other data after each mission; however, these were distributed only to scientists involved in the program and to some NASA document depositories and are not readily available. For example, the Apollo 11, Lunar Sample Information Catalog, NASA, Manned Spacecraft Center, August 31, 1969, 412 p. plus plates was duplicated in only 400 copies and is now a collector's item. A volume containing studies of a portion of the Luna 16 sample is Earth and Planetary Sci. Letters., vol. 13, no. 2, p. 223-471: a group of papers concerning the Luna 20 sample are in Earth and Planetary Sci. Letters., vol. 17, no. 1, p. 3-63; also, Geochimica et Cosmochimica Acta, vol. 37, no. 4, 719-1110, April, 1973.

39. For example, see Dence, M. R., and A. G. Plant (1972) Analysis of Fra Mauro samples and the origin of the Imbrium Basin: Proc. Third Lunar Sci. Conf., Geochim. et Cosmochim, Acta, Suppl. 3, vol. 1, E. A. King, Jr., ed., p. 379-399; Green, D. H., A. E. Ringwood, N. G. Ware, and W. O. Hibberson (1972) Experimental petrology and petrogensis of Apollo 14 basalts: Proc. Third Lunar Sci. Conf., Geochim. et Cosmochim. Acta, Suppl. 3, vol. 1, E. A. King, Jr., ed., p. 197-206; and other papers in this same volume. Also see Ganapathy, R., J. C. Laul, J. W. Morgan, and E. Anders (1972) Moon: Possible nature of the body that produced the Imbrian Basin, from the composition of Apollo 14 Samples: Science, vol. 175, p. 55-58.

40. Chao, E. C. T., J. A. Minkin, C. Frondel, C. Klein, Jr., J. C. Drake, L. Fuchs, B. Tani, J. V. Smith, A. T. Anderson, P. B. Moore, G. R. Zechman, Jr., R. J. Traill, A. G. Plant, J. A. V. Douglas, and M. R. Dence (1970) Pyroxferroite, a new calcium bearing iron silicate from Tranquillity Base: Proc. Apollo 11 Lunar Sci. Conf., A. A. Levinson, ed., Geochim. et Cosmochim. Acta, Suppl. 1, vol. 1, Pergamon Press, p. 65-79.

41. Anderson, A. T., T. E. Bunch, E. N. Cameron, S. E. Haggerty, F. R. Boyd, L. W. Finger, O. B. James, K. Keil, M. Prinz, P. Ramdohr, and A. El Goresy (1970) Armalcolite: A new mineral from the Apollo 11 samples: Proc. Apollo 11 Lunar Sci. Conf., A. A. Levinson, ed., Geochim. et Cosmochim. Acta, Suppl. 1, vol. 1, Pergamon Press, p. 55-63.

42. Lovering, J. F., D. A. Wark, A. F. Reid, N. G. Ware, K. Keil, M. Prinz, T. E. Bunch, A. El Goresy, P. Ramdohr, G. M. Brown, A. Peckett, R. Phillips, E. N. Cameron, J. A. V. Douglas, and A. G. Plant (1971) Tranquillityite: A new silicate mineral from Apollo 11 and Apollo 12 basaltic rocks: Proc. Second Lunar Sci. Conf., Geochim. et Cosmochim. Acta, Suppl. 2, vol. 1, A. A. Levinson, ed., MIT Press, p. 39-45.

43. Wenk, H. -R., M. Ulbrich, and W. F. Müller (1972) Lunar plagioclase: A

mineralogical study: Proc. Third Lunar Sci. Conf., Geochim. et Cosmochim. Acta, Suppl. 3, vol. 1, E. A. King, Jr., ed., MIT Press, p. 569-579. Also, other papers this volume.

44. Butler, J. C., M. F. Carman, Jr., and E. A. King Jr. (1974) Unit cell parameters, compositions and 2V measurements of selected lunar plagioclases: The Moon, vol. 9, p. 327-334.

45. Smith, J. V. (1971) Minor elements in Apollo 11 and Apollo 12 olivine and plagioclase: Proc. Sec. Lunar Sci. Conf., Geochim. et Cosmochim. Acta, Suppl. 2, vol. 1, A. A. Levinson, ed., MIT Press, p. 143-150.

46. King, E. A., Jr., M. F. Carman, and J. C. Butler (1970) Mineralogy and petrology of coarse particulate material from lunar surface at Tranquillity Base: Science, vol. 167, no. 3918, p. 650-652.

47. Drever, H. I., R. Johnston, P. Butler, Jr., and F. G. F. Gibb (1972) Some textures in Apollo 12 lunar igneous rocks and in terrestrial analogs: Proc. Third Lunar Sci. Conf., Geochim. et Cosmochim. Acta, Suppl. 3, vol. 1, E. A. King, Jr., ed., MIT Press, p. 171-184.

48. Burns, R. G., R. M. Abu-Eid, and F. E. Huggins (1972) Crystal field spectra of lunar pyroxenes: Proc. Third Lunar Sci. Conf., Geochim. et Cosmochim. Acta, Suppl. 3, vol. 1, E. A. King, Jr., ed., MIT Press, p. 533-543.

49. Ghose, S., G. Ng, and L. S. Walter (1972) Clinopyroxenes from Apollo 12 and 14: Exsolution, domain structure, and cation order: Proc. Third Lunar Sci. Conf., Geochim. et Cosmochim. Acta, Suppl. 3, vol. 1, E. A. King, Jr., ed., MIT Press, p. 507-531.

50. Bence, A. E., and J. J. Papike (1972) Pyroxenes as recorders of lunar basalt petrogenesis: Chemical trends due to crystal-liquid interaction: Proc. Third Lunar Sci. Conf., Geochim. et Cosmochim. Acta, Suppl. 3, vol. 1, E. A. King, Jr., ed., MIT Press, p. 431-469.

51. Takeda, H., and W. I. Ridley (1972) Crystallography and chemical trends of orthopyroxene-pigeonite from rock 14310 and coarse fine 12033; Proc. Third Lunar Sci. Conf., Geochim. et Cosmochim. Acta, Suppl. 3, vol. 1, E. A. King, Jr., ed., MIT Press, p. 423-430.

52. Weigand, P. W., and L. S. Hollister (1972) Pyroxenes from breccia 14303: Proc. Third Lunar Sci. Conf., Geochim. et Cosmochim. Acta, Suppl. 3, vol. 1, E. A. King, Jr., ed., MIT Press, p. 471-480.

53. Pavičević, M., P. Ramdohr, and A. El Goresy (1972) Electron microprobe investigations of the oxidation states of Fe and Ti in ilmenite in Apollo 11, Apollo 12, and Apollo 14 crystalline rocks: Proc. Third Lunar Sci. Conf., Geochim. et Cosmochim. Acta, Suppl. 3, vol. 1, E. A. King, Jr., ed., MIT Press, p. 295-303.

54. El Goresy, A., P. Ramdohr, and L. A. Taylor (1971) The opaque minerals in the lunar rocks from Oceanus Procellarum: Proc. Second Lunar Sci. Conf., Geochim. et Cosmochim. Acta, Suppl. 2, vol. 1, A. A. Levinson, ed., MIT Press, p. 219-235. See also Simpson, P. R., and S. H. U. Bowie (1972) Opaque phases in Apollo 12 samples: *Ibid.*, p. 207-218.

55. Drever, H. I., R. Johnston, P. Butler, Jr., and F. G. F. Gibb (1972) Some textures in Apollo 12 lunar igneous rocks and in terrestrial analogs: Proc. Third

Lunar Sci. Conf., Geochim. et Cosmochim. Acta, Suppl. 3, vol. 1, E. A. King, Jr., ed., MIT Press, p. 171-184; also, E. A. King, Jr., unpublished observation.

56. Mason, B., E. Jarosewich, and W. G. Melson (1972) Mineralogy, petrology and chemical composition of lunar samples 15085, 15256, 15271, 15471, 15475, 15476, 15535, 15555, and 15556: Proc. Third Lunar Sci Conf., Geochim. et Cosmochim. Acta, Suppl. 3, vol. 1, E. A. King, Jr., ed., MIT Press, p. 785-796; Rose, H. J., Jr., F. Cuttitta, C. S. Annell, M. K. Carron, R. P. Christian, E. J. Dwornik, L. P. Greenland, and D. T. Lignon, Jr. (1972) Compositional data for twenty-one Fra Mauro lunar materials: Proc. Third Lunar Sci. Conf., Geochim. et Cosmochim. Acta, Suppl. 3, vol. 2, D. Heymann, ed., MIT Press, p. 1215-1229. See also *numerous* other papers from the Proceedings of *each* of the Lunar Science Conferences, vols. 2, for major and minor elements, isotopic and other analytical results from lunar samples.

57. Reid, A. M., J. Warner, W. I. Ridley, D. A. Johnston, R. S. Harmon, Petr Jakeš and R. W. Brown (1972) The major element compositions of lunar rocks as inferred from glass compositions in the lunar soils: Proc. Third Lunar Sci. Conf., Geochim. et Cosmochim. Acta, Suppl. 3, vol. 1, E. A. King, Jr., ed., MIT Press, p. 363-378.

58. Gast, P. W., N. J. Hubbard, and H. Wiesmann (1970) Chemical composition and petrogenesis of basalts from Tranquillity Base: Proc. Apollo 11 Lunar Sci. Conf., Geochim. et Cosmochim. Acta, Suppl. 1, vol. 2, A. A. Levinson, ed., Pergamon Press, p. 1143-1163.

59. O'Kelley, G. D., J. S. Eldridge, E. Schonfeld, and P. R. Bell (1970) Primordial radionuclide abundances, solar proton and cosmic-ray effects and ages of Apollo 11 lunar samples by non-destructive gamma ray spectrometry: Proc. Apollo 11 Lunar Sci. Conf., Geochim. et Cosmochim. Acta, Suppl. 1, vol. 2, A. A. Levinson, ed., Pergamon Press, p. 1407-1423; also, O'Kelley, G. D., J. S. Eldridge, and K. J. Northcutt (1972) Primordial radioelements and cosmogenic radionuclides in lunar samples from Apollo 15: Proc. Third Lunar Sci. Conf., Geochim. et Cosmochim. Acta, Suppl. 3, vol. 2, D. Heymann, ed., MIT Press, p. 1659-1670; and other papers by the Oak Ridge and other research groups.

60. Funkhouser, J. G., O. A. Schaeffer, D. D. Bogard, and J. Zähringer (1970) Gas analysis of the lunar surface: Proc. Apollo 11 Lunar Sci. Conf., Geochim. et Cosmochim. Acta, Suppl. 1, vol. 2, A. A. Levinson, ed., Pergamon Press, p. 1111-1116; see also later papers by this and other research groups.

61. Eberhardt, P., J. Geiss, H. Graf, N. Grögler, U. Krähenbühl, H. Schwaller, J. Schwarzmüller, and A. Stettler (1970) Trapped solar wind noble gases, exposure age and K/Ar-age in Apollo 11 lunar fine material: Proc. Apollo 11 Lunar Sci. Conf., Geochim. et Cosmochim. Acta, Suppl. 1, vol. 2, A. A. Levinson, ed., Pergamon Press, p. 1037-1070; Heymann, D., and A. Yaniv (1970) Inert gases in fines from the Sea of Tranquillity: *Ibid.*, p. 1247-1259; and other later papers.

62. Tera, F., D. A. Papanastassiou, and G. J. Wasserburg (1973) A lunar cataclysm at ~3.95 AE and the structure of the lunar crust: Lunar Science IV, J. W. Chamberlain, and C. Watkins, eds., Lunar Sci. Inst., p. 723-725.

63. Cameron, E. N. (1971) Opaque minerals in certain lunar rocks from Apollo 12: Proc. Sec. Lunar Sci. Conf., Geochim. et Cosmochim. Acta, Suppl. 2, vol. 1,

A. A. Levinson, ed., p. 193-206. Also, see Brett, R., P. Butler, Jr., C. Meyer, Jr., A. M. Reid, H. Takeda and R. Williams (1971) Apollo 12 igneous rocks 12004, 12008, 12009, and 12022: A mineralogical and petrological study: *Ibid.,* p. 301-317.

64. El Goresy, A., L. A. Taylor, and P. Ramdohr (1972) Fra Mauro crystalline rocks: Mineralogy, geochemistry and subsolidus reduction of the opaque minerals: Proc. Third Lunar Sci. Conf., Geochim. et Cosmochim. Acta, Suppl. 3, vol. 1, E. A. King, Jr., ed., MIT Press, p. 333-349.

65. Weill, D. F., I. S. McCallum, Y. Bottinga, M. J. Drake, and G. A. McKay (1971) Mineralogy and petrology of some Apollo 11 igneous rocks: Proc. Apollo 11 Lunar Sci. Conf., Geochim. et Cosmochim. Acta, Suppl. 1, vol. 1, A. A. Levinson, ed., Pergamon Press, p. 937-955.

66. Green, D. H., A. E. Ringwood, N. G. Ware, and W. O. Hibberson (1972) Experimental petrology and petrogenesis of Apollo 14 basalts: Proc. Third Lunar Sci. Conf., Geochim. et Cosmochim. Acta, Suppl. 3, vol. 1, E. A. King, Jr., ed., MIT Press, p. 197-206; also, Ford, C. E., G. M. Biggar, D. J. Humphries, G. Wilson, D. Dixon, and M. J. O'Hara (1972) Role of water in the evolution of the lunar crust; an experimental study of sample 14310; an indication of lunar calc-alkaline volcanism: *Ibid.,* p. 207-229; and other papers.

67. Chao, E. C. T., O. B. James, J. A. Minkin, J. A. Boreman, E. D. Jackson, and C. B. Raleigh (1970) Petrology of unshocked crystalline rocks and evidence of impact metamorphism in Apollo 11 returned lunar sample: Proc. Apollo 11 Lunar Sci. Conf., Geochim. et Cosmochim. Acta, Suppl. 1, vol. 1, A. A. Levinson, ed., Pergamon Press, p. 287-314; also Ghose, S., G. Ng, and L. S. Walter (1972) Clinopyroxenes from Apollo 12 and 14: Exsolution, domain structure, and cation order: Proc. Third Lunar Sci. Conf., Geochim. et Cosmochim. Acta, Suppl. 3, vol. 1, E. A. King, Jr., ed., MIT Press, p. 507-531.

68. Hinners, N. W. (1971) The new moon: A view: Rev. Geophys. Space Phys., vol. 9, p. 447-522; also, Hubbard, N. J., P. W. Gast, J. M. Rhodes, B. M. Bansal, H. Wiesmann, and S. E Church (1972) Nonmare basalts: Part II: Proc. Third Lunar Sci. Conf., Geochim. et Cosmochim. Acta, Suppl. 3, vol. 2, D. Heymann, ed., MIT Press, p. 1161-1179.

69. Wood, J. A., J. S. Dickey, Jr., U. B. Marvin, and B. N. Powell (1970) Lunar anorthosites and a geophysical model of the moon: Proc. Apollo 11 Lunar Sci. Conf., Geochim. et Cosmochim. Acta, Suppl. 1, vol. 1, A. A. Levinson, ed., Pergamon Press, p. 965-988.

70. Roedder, E., and P. W. Weiblen (1972) Petrographic features and petrologic significance of melt inclusions in Apollo 14 and 15 rocks: Proc. Third Lunar Sci. Conf., Geochim. et Cosmochim. Acta, Suppl. 3, vol. 1, E. A. King, Jr., ed., MIT Press, p. 251-279.

71. Gast, P. W., and R. K. McConnell, Jr. (1972) Evidence for initial chemical layering of the moon: Lunar Sci. III, Lunar Sci. Inst., Contr. 88, C. Watkins, ed., p. 289-290.

72. Walker, D., J. Longhi, and J. F. Hayes (1972) Experimental petrology and origin of Fra Mauro rocks and soil: Proc. Third Lunar Sci. Conf., Geochim. et Cosmochim. Acta, Suppl. 3, vol. 1, E. A. King, Jr., ed., MIT Press, p. 797-817.

Also, Walker D., T. L. Grove, J. Longhi, E. M. Stolper, and J. F. Hays (1973) Origin of lunar feldspathic rocks: Earth Planet. Sci. Letters, vol 20, p. 325-336.

73. LSPET (Lunar Sample Preliminary Examination Team) (1969) Preliminary examination of samples from Apollo 11: Science, vol. 165, p. 1211-1227.

74. Abundant evidence of shock metamorphism in lunar samples has been documented by numerous investigators. For examples, see Engelhardt, W. von, J. Arndt, W. F. Müller, and D. Stöffler (1970) Shock Metamorphism in lunar samples: Science, vol. 167, p. 669-670; also, Engelhardt, W. von, J. Arndt, W. F. Müller, and D. Stöffler (1970) Shock metamorphism of lunar rocks and origin of the regolith at the Apollo 11 site: Proc. Apollo 11 Lunar Sci. Conf., Geochim. et Cosmochim. Acta, Suppl. 1, vol. 1, A. A. Levinson, ed., Pergamon Press, p. 363-384; also Chao, E. C. T., O. B. James, J. A. Minkin, J. A. Boreman, E. D. Jackson, and C. B. Raleigh (1970) Petrology of unshocked crystalline rocks and evidence of impact metamorphism in Apollo 11 returned lunar samples: Proc. Apollo 11 Lunar Sci. Conf., Geochim. et Cosmochim. Acta, Suppl. 1, vol. 1, A. A. Levinson, ed., Pergamon Press, p. 287-314, and other papers in all Lunar Science Conference Proceedings.

75. Microcraters from the impacts of micrometeorites were recognized by the Lunar Sample Preliminary Examination Team on the Apollo 11 samples. For quantitative studies see Gault, D. E., F. Hörz, and J. B. Hartung (1972) Effects of microcratering on the lunar surface: Proc. Third Lunar Sci. Conf., Geochim. et Cosmochim. Acta, Suppl. 3, vol. 3, D. R. Criswell, ed., MIT Press, p. 2713-2734· also, Hartung, J. B., F. Hörz, and D. E. Gault (1972) Lunar microcraters and interplanetary dust: Ibid., p. 2735-2753, and Morrison, D. A., D. S. McKay, G. H. Heiken, and H. J. Moore (1972) Microcraters on lunar rocks: Ibid., p. 2767-2791.

76. King, E. A., Jr., J. C. Butler, and M. F. Carman, Jr. (1971) The lunar regolith as sampled by Apollo 11 and Apollo 12: Grain size analyses, modal analyses and origins of particles: Proc. Sec. Lunar Sci. Conf., Geochim. et Cosmochim. Acta, Suppl. 2, vol. 1, A. A. Levinson, ed., MIT Press, p. 737-746; and numerous other papers.

77. Carter, J. L. (1971) Chemistry and surface morphology of fragments from Apollo 12 soil: Proc. Sec. Lunar Sci. Conf., Geochim. et Cosmochim. Acta, Suppl. 2, vol. 1, A. A. Levinson, ed., MIT Press, p. 873-892; also. Fulchignoni, M., R. Funiciello, A. Taddeucci, and R. Trigila (1971) Glassy spheroids in lunar fines from Apollo 12 samples 12070,37; 12001,73; and 12057,60: Ibid. p. 937-948, and numerous other works.

78. Carter, J. L., and E. Padovani (1973) Genetic implications of some unusual particles in Apollo 16 less than 1 mm fines 68841,11 and 69941,13: Proc. Fourth Lunar Sci Conf., Geochim. et Cosmochim. Acta, Suppl. 4, vol. 3, Pergamon Press, p. 323-332; also, McKay, D. S., U. S. Clanton and G. Ladle (1973) Scanning electron microscope study of Apollo 15 green glass: Ibid., p. 225-238.

79. Hörz, F., J. B. Hartung and D. E. Gault (1971) Micrometeorite craters on lunar rock surfaces: Jour. Geophys. Res., vol. 76, p. 5770-5798; also, Hartung, J. B., F. Hörz and D. E. Gault (1972) Lunar microcraters and interplanetary dust: Proc. Third Lunar Sci. Conf., Geochim. et Cosmochim. Acta, Suppl. 3, vol. 3, D. R.

Criswell, ed., MIT Press, p. 2735-2753; also, Schneider, E., D. Storzer, J. B. Hartung, H. Fechtig and W. Gentner (1973) Microcraters on Apollo 15 and 16 samples and corresponding cosmic dust fluxes: Proc. Fourth Lunar Sci. Conf., Geochim. et Cosmochim. Acta, Suppl. 4, vol. 3, Pergamon Press, p. 3277-3290.

80. McKay, D. S., G. H. Heiken, R. M. Taylor, U. S. Clanton, D. A. Morrison, and G. H. Ladle (1972) Apollo 14 soils: Size distribution and particles types: Proc. Third Lunar Sci. Conf., Geochim. et Cosmochim. Acta, Suppl. 3, vol. 1, E. A. King, Jr., ed., MIT Press, p. 983-994.

81. King, E. A., Jr., M. F. Carman, and J. C. Butler (1970) Mineralogy and petrology of coarse particulate material from lunar surface at Tranquillity Base: Science, vol. 167, p. 650-652; Short, N. M. (1970) Evidence and implications of shock metamorphism in lunar samples: Ibid., p. 673-675; Anderson, A. T., Jr., A. V. Crewe, J. R. Goldsmith, P. B. Moore, J. C. Newton, E. J. Olsen, J. V. Smith, and P. J. Wyllie (1970) Petrologic history of the moon suggested by petrography, mineralogy, and crystallography: Ibid., p. 587-590; Wood, J. A., J. S. Dickey, Jr., U. B. Marvin, and B. N. Powell (1970) Lunar anorthosites: Ibid., p. 602-604.

82. Wood, J. A., J. S. Dickey, Jr., U. B. Marvin, and B. N. Powell (1970) Lunar anorthosites and a geophysical model of the moon: Proc. Apollo 11 Lunar Sci. Conf., Geochim. et Cosmochim. Acta, Suppl. 1, vol. 1, A. A. Levinson, ed., Pergamon Press, p. 965-988.

83. King, E. A., Jr., J. C. Butler, and M. F. Carman (1972) Chondrules in Apollo 14 samples and size analyses of Apollo 14 and 15 fines: Proc. Third Lunar Sci. Conf., Geochim. et Cosmochim. Acta, Suppl. 3, vol. 1, E. A. King, Jr., ed., MIT Press, p. 673-686; McKay, D. S., G. H. Heiken, R. M. Taylor, U. S. Clanton, D. A. Morrison, and G. H. Ladle (1972) Apollo 14 soils: Size distribution and particle types: Ibid., p. 983-994; Butler, J. C., G. M. Greene, and E. A. King, Jr. (1973) Grain size frequency distributions and modal analyses of Apollo 16 fines: Proc. Fourth Lunar Sci. Conf., Geochim. et Cosmochim. Acta, Suppl. 4, vol. 1, Pergamon Press, p. 267-278; and numerous other papers.

84. For an example see Görz, H., E. W. White, G. G. Johnson, Jr., and M. W. Pearson (1972) CESEMI studies of Apollo 14 and 15 fines: Proc. Third Lunar Sci. Conf., Geochim. et Cosmochim. Acta, Suppl. 3, vol. 3, D. R. Criswell, ed., MIT Press, p. 3195-3200.

85. King, E. A., Jr., M. F. Carman, and J. C. Butler (1970) Mineralogy and petrology of coarse particulate material from lunar surface at Tranquillity Base: Science, vol. 167, p. 650-652; 1 also (1970) Proc. Apollo 11 Lunar Sci. Conf., Geochim. et Cosmochim. Acta, Suppl. 1, vol. 1, A. A. Levinson, ed., Pergamon Press, p. 599-606.

86. Scott, R. F., W. D. Carrier III, N. C. Costes, and J. K. Mitchell (1971) Apollo 12 soil mechanics investigation: Geotechnique, vol. 21, p. 1-14.

87. Carrier, W. D., III, L. G. Bromwell, and R. T. Martin (1972) Strength and compressibility of returned lunar soil: Proc. Third Lunar Sci. Conf., Geochim. et Cosmochim. Acta, Suppl. 3, vol. 3, D. R. Criswell, ed., MIT Press, p. 3223-3234.

88. For a comprehensive review of lunar soil density estimates and measurements see Mitchell, J. K., W. N. Houston, R. F. Scott, N. C. Costes, W. D. Carrier III, and L. G. Bromwell (1972) Mechanical properties of lunar soil: Density, porosity,

cohesion, and angle of internal friction: Proc. Third Lunar Sci. Conf., Geochim. et Cosmochim. Acta, Suppl. 3, vol. 3, D. R. Criswell, ed., MIT Press, p. 3235-3253.

89. Houston, W. N., H. J. Hovland, J. K. Mitchell, and L. I. Namiq (1972) Lunar soil porosity and its variation as estimated from footprints and boulder tracks: Proc. Third Lunar Sci. Conf., Geochim. et Cosmochim. Acta, Suppl. 3, vol. 3, D. R. Criswell, ed., MIT Press, p. 3255-3263.

90. Chao, E. C. T., J. A. Boreman, and G. A. Desborough (1971) The petrology of unshocked and shocked Apollo 11 and 12 microbreccias: Proc. Sec. Lunar Sci. Conf., Geochim. et Cosmochim. Acta, Suppl. 2, vol. 1, A. A. Levinson, ed., MIT Press, p. 797-816; and numerous other authors.

91. Floran, R. J., K. L. Cameron, A. E. Bence, and J. J. Papike (1972) Apollo 14 breccia 14313: A mineralogic and petrologic report: Proc. Third Lunar Sci. Conf., Geochim. et Cosmochim. Acta, Suppl. 3, vol. 1, E. A. King, Jr., ed., MIT Press, p. 661-671.

92. Warner, J. L. (1972) Metamorphism of Apollo 14 breccias: Proc. Third Lunar Sci. Conf., Geochim. et Cosmochim. Acta, Suppl. 3, vol. 1, E. A. King, Jr., ed., MIT Press, p. 623-643.

93. Anderson, A. T., Jr., T. F. Braziunas, J. Jacoby, and J. V. Smith (1972) Breccia populations and thermal history: Nature of the pre-Imbrian crust and impacting body: Lunar Sci. III., C. Watkins, ed., Lunar Sci. Inst. Contr. 88, p. 24-26; also (1972) Thermal and mechanical history of breccias 14306, 14063, 14270, and 14321: Proc. Third Lunar Sci. Conf., Geochim. et Cosmochim. Acta, Suppl. 3, vol. 1, E. A. King, Jr., ed., MIT Press, p. 819-835.

94. Grieve, R., G. McKay, H. Smith, and D. Weill (1972) Mineralogy and petrology of polymict breccia 14321: Lunar Sci. III, C. Watkins, ed., Lunar Sci. Inst. Contr. 88, p. 338-340.

95. Wilshire, H. G., D. E. Stuart-Alexander, and E. D. Jackson (1973) Apollo 16 rocks: Petrology and classification: Jour. Geophys. Res., vol. 78, p. 2379-2392.

96. King, E. A., Jr., M. F. Carman, and J. C. Butler (1972) Chondrules in Apollo 14 samples: Implications for the origin of chondritic meteorites: Science, vol. 175, 7 Jan., 1972, p. 59-60; King, E. A., Jr., J. C. Butler, and M. F. Carman (1972) Chondrules in Apollo 14 samples and size analyses of Apollo 14 and 15 fines: Proc. Third Lunar Sci. Conf., Geochim. et Cosmochim. Acta, Suppl. 3, vol. 1, E. A. King, Jr., ed., MIT Press, p. 673-686; Kurat, G., K. Keil, M. Prinz, and C. E. Nehru (1972) Chondrules of lunar origin: *Ibid.*, p. 707-721; also, Nelen, J., A. Noonan and K. Fredriksson (1972) Lunar glasses, breccias, and chondrules: *Ibid.*, p. 723-737.

97. Nelson, L. S., M. Blander, S. R. Skaggs, and K. Keil (1972) Use of a CO_2 laser to prepare chondrule-like spherules from supercooled molten oxide and silicate droplets: Earth Planet. Sci. Letters., vol. 14, p. 338-344.

98. Ganapathy, R., R. R. Keys, J. C. Laul, and E. Anders (1970) Trace elements in Apollo 11 lunar rocks: Implications for meteorite influx and origin of moon: Proc. Apollo 11 Lunar Sci. Conf., Geochim. et Cosmochim. Acta, Suppl. 1, vol. 2, A. A. Levinson, ed., Pergamon Press, p. 1117-1142; Ganapathy, R., J. C. Laul, J. W. Morgan, and E. Anders (1972) Moon: Possible nature of the body that

produced the Imbrian basin, from the composition of Apollo 14 samples: Science, vol. 175, p. 55-59; Morgan, J. W., U. Krähenbühl, R. Ganapathy, and E. Anders (1972) Trace elements in Apollo 15 samples: Implications for meteorite influx and volatile depletion on the moon: Proc. Third Lunar Sci. Conf., Geochim. et Cosmochim. Acta, Suppl. 3, vol. 2, D. Heymann, ed., MIT Press, p. 1361-1376, and other papers.

99. For an early review of this subject see Huey, J. M., H. Ihochi, L. P. Black, R. G. Ostic, and T. P. Kohman (1971) Lead isotopes and volatile transfer in the lunar soil: Proc. Sec. Lunar Sci. Conf., Geochim. et Cosmochim. Acta, Suppl. 2, vol. 2, A. A. Levinson, ed., MIT Press, p. 1547-1564.

100. Albee, A. L., D. S. Burnett, A. Chodos, O. J. Eugster, J. C. Hueneke, D. A. Papanastassiou, F. A. Podosek, G. P. Russ II, H. G. Sanz, F. Tera, and G. J. Wasserburg (1970) Ages, irradiation history and chemical composition of lunar rocks from the Sea of Tranquillity: Science, vol. 167, p. 463-466.

101. Gose, W. A., G. W. Pearce, D. W. Strangway, and E. E. Larson (1972) Magnetic properties of Apollo 14 breccias and their correlation with metamorphism: Proc. Third Lunar Sci. Conf., Geochim. et Cosmochim. Acta, Suppl. 3, vol. 3, D. R. Criswell, ed., MIT Press, p. 2387-2395.

102. Nagata, T., R. M. Fisher, F. C. Schwerer, M. D. Fuller, and J. R. Dunn (1972) Rock magnetism of Apollo 14 and 15 materials: Proc. Third Lunar Sci. Conf., Geochim. et Cosmochim. Acta, Suppl. 3, vol. 3, D. R. Criswell, ed., MIT Press, p. 2423-2447. For an excellent review of lunar magnetic problems see Fuller, M. (1974) Lunar magnetism: Rev. Geophys. and Space Phys., vol. 12, p. 23-70.

103. Collinson, D. W., S. K. Runcorn, A. Stephenson, and A. J. Manson (1972) Magnetic properties of Apollo 14 rocks and fines: Proc. Third Lunar Sci. Conf., Geochim. et Cosmochim. Acta, Suppl. 3, vol. 3, D. R. Criswell, ed., MIT Press, p. 2343-2361. See also, Pearce, G. W., D. W. Strangway, and W. A. Gose (1972) Remanent magnetization of the lunar surface: *Ibid.,* p. 2449-2464.

104. McLane, J. C., Jr., E. A. King, Jr., D. A. Flory, K. A. Richardson, J. P. Dawson, W. W. Kemmerer, and B. C. Wooley (1967) Lunar Receiving Laboratory: Science, vol. 155, p. 525-529.

105. For examples see Oyama, V. I., E. L. Merek, and M. P. Silverman (1970) A search for viable organisms in a lunar sample: Proc. Apollo 11 Lunar Sci. Conf., Geochim. et Cosmochim. Acta, Suppl. 1, vol. 2, A. A. Levinson, ed., Pergamon Press, p. 1921-1927.

106. Schopf, J. W. (1970) Micropaleontological studies of Apollo 11 samples: Proc. Apollo 11 Lunar Sci. Conf., Geochim. et Cosmochim. Acta, Suppl. 1, Vol. 2, A. A. Levinson, ed., Pergamon Press, p. 1933-1934.

107. Johnson, R. D., and C. C. Davis (1970) Total organic carbon in the Apollo 11 lunar samples: Proc. Apollo 11 Lunar Sci. Conf., Geochim. et Cosmochim. Acta, Suppl. 1, vol. 2, A. A. Levinson, ed., Pergamon Press, p. 1805-1812.

108. Latham, G., M. Ewing, F. Press, G. Sutton, J. Dorman, Y. Nakamura, N. Toksöz, D. Lammlein, and F. Duennebier (1972) Passive Seismic Experiment: Sec. 8, Apollo 15 Prelim. Sci. Rept., NASA SP-289. See also other Apollo

Mission Prelim. Sci. Repts. Also, Kovach, R. L., and J. S. Watkins (1972) The near-surface velocity structure of the Moon: Lunar Sci. III, C. Watkins, ed., Lunar Sci. Inst., Contr. No. 88, p. 461-462.

109. Latham, G., M. Ewing, F. Press, G. Sutton, J. Dorman, Y. Nakamura, N. Toksöz, R. Meissner, F. Duennebier, and R. Kovach (1970) Seismic data from man-made impacts on the Moon: Science, vol. 170, p. 620-626.

110. Kovach, R. L., J. S. Watkins, A. Nur, and P. Talwani (1973) The properties of the shallow lunar crust: An over-view from Apollo 14, 16 and 17: Lunar Sci. IV, Lunar Sci. Inst. and NASA Manned Spacecraft Center, J. W. Chamberlain and C. Watkins, eds., p. 444-445.

111. Toksöz, M. N., F. Press, A. Dainty, K. Anderson, G. Latham, M. Ewing, J. Dorman, D. Lammlein, G. Sutton, and F. Duennebier (1972) Structure, composition and properties of lunar crust: Proc. Third Lunar Sci. Conf., Geochim. et Cosmochim. Acta, Suppl. 3, vol. 3, D. R. Criswell, ed., MIT Press, p. 2527-2544.

112. Langseth, M. G., Jr., S. P. Clark, Jr., John Chute, Jr., and Stephen Keihm (1972) The Apollo 15 lunar heat flow measurement: Lunar Sci. III, C. Watkins, ed., Lunar Sci. Inst. Contr. No. 88, p. 475-477. See also, Langseth, M. G., J. L. Chute, and Stephen Keihm (1973) Direct measurement of heat flow from the Moon: Lunar Sci. IV, J. W. Chamberlain and C. Watkins, eds., Lunar Sci. Inst., p. 455-456.

113. Sonett, C. P., G. Schubert, B. F. Smith, K. Schwartz, and D. S. Colburn (1971) Lunar electrical conductivity from Apollo 12 magnetometer measurements: Compositional and thermal inferences: Proc. Sec. Lunar Sci. Conf., Geochim. et Cosmochim. Acta, Suppl. 2, vol. 3, A. A. Levinson, ed., MIT Press, p. 2415-2431.

114. Dyal, P., C. W. Parkin, and P. Cassen (1972) Surface magnetometer experiments: Internal lunar properties and lunar field interactions with the solar plasma: Proc. Third Lunar Sci. Conf., Geochim. et Cosmochim. Acta, Suppl. 3, vol. 3, D. R. Criswell, ed., MIT Press, p. 2287-2307.

115. Latham, G., J. Dorman, F. Duennebier, M. Ewing, D. Lammlein, and Y. Nakamura (1973) Moonquakes, meteoroids and the state of the lunar interior: Lunar Sci. IV, J. W. Chamberlain and C. Watkins, eds., Lunar Sci. Inst., p. 457-459.

116. Latham, G., M. Ewing, F. Press, G. Sutton, J. Dorman, Y. Nakamura, N. Toksöz, D. Lammlein, and F. Duennebier (1971) Moonquakes: Science, vol. 174, p. 687-692; see also numerous other papers published by the same authors from 1970 to present. A good recent summary is Lammlein, D. R., G. V. Latham, J. Dorman, Y. Nakamura, and M. Ewing (1974) Lunar seismicity, structure, and tectonics: Rev. Geophys. Space Phys., vol. 12, no. 1, p. 1-21.

117. Latham, G., M. Ewing, J. Dorman, D. Lammlein, F. Press, N. Toksöz, G. Sutton, F. Duennebier, and Y. Nakamura (1972) Moonquakes and lunar tectonism results from the Apollo passive seismic experiment: Proc. Third Lunar Sci. Conf., Geochim. et Cosmochim. Acta, Suppl. 3, vol. 3, D. R. Criswell, ed., MIT Press, p. 2519-2526; see also, Latham, G., J. Dorman, F. Duennebier, M. Ewing,

D. Lammlein, and Y. Nakamura (1973) Moonquakes, meteoroids and the state of the lunar interior: Proc. Fourth Lunar Sci. Conf., Geochim. et Cosmochim. Acta, Suppl. 4, vol. 3, Pergamon Press, p. 2515-2527.

118. Muller, P. M., and W. L. Sjogren (1968) Mascons — lunar mass concentrations: Science, vol. 161, no. 3842, p. 680-684.

119. Urey, H. C. (1968) Mascons and the history of the Moon: Science, vol. 162, no. 3860, p. 1408-1410.

120. Campbell, M. J., B. T. O'Leary, and Carl Sagan (1969) Moon — Two new mascon basins: Science, vol. 164, no. 3885, p. 1273-1275.

121. Kaula, W. M., G. Schubert, R. E. Lingenfelter, W. L. Sjogren, and W. R. Wollenhaupt (1972) Analysis and interpretation of lunar laser altimetry: Proc. Third Lunar Sci. Conf., Geochim. et Cosmochim. Acta, Suppl. 3, vol. 3, D. R. Criswell, ed., MIT Press, p. 2189-2204. See also, Kaula, W. M., G. Schubert, R. E. Lingenfelter, W. L. Sjogren, and W. R. Wollenhaupt (1973) Lunar topography from Apollo 15 and 16 laser altimetry: Proc. Fourth Lunar Sci. Conf., Geochim. et Cosmochim. Acta, Suppl. 4, vol. 3, Pergamon Press, p. 2811-2819.

122. Kaula, W. M., G. Schubert, R. E. Lingenfelter, W. L. Sjogren, and W. R. Wollenhaupt (1974) Apollo laser altimetry and inferences as to lunar structure: Proc. Fifth Lunar Sci. Conf., Geochim. et Cosmochim. Acta, Suppl. 5, vol. 3, Pergamon Press, p. 3049-3058.

123. Metzger, A. E., J. I. Trombka, L. E. Peterson, R. C. Reedy, and J. R. Arnold (1972) A first look at the lunar orbital gamma-ray data: Proc. Third Lunar Sci. Conf., Geochim. et Cosmochim. Acta, Suppl. 3, vol. 3, D. R. Criswell, ed., MIT Press, frontispiece, 4 p.

124. Adler, I., J. Gerard, J. Trombka, R. Schmadebeck, P. Lowman, H. Blodget, L. Yin, E. Eller, R. Lamothe, P. Gorenstein, P. Bjorkholm, B. Harris, and H. Gursky (1972) The Apollo 15 x-ray fluorescence experiment: Proc. Third Lunar Sci. Conf., Geochim. et Cosmochim. Acta, Suppl. 3, vol. 3, D. R. Criswell, ed., MIT Press, p. 2157-2178; see also, Proc. Fourth Lunar Sci. Conf., Geochim. et Cosmochim. Acta, Suppl. 4, vol. 3, Pergamon Press, p. 2783-2791.

125. Darwin, G. H. (1908) On the tidal friction of a planet attended by several satellites, and on the evolution of the solar system: Scientific Papers, vol. 2, Cambridge Univ. Press., London, p. 406-458.

126. Wise, D. U. (1963) An origin of the moon by fission during formation of the earth's core: Jour. Geophys. Res., vol. 68, p. 1547-1554; also, Wise, D. U. (1969) Origin of the moon from the earth: Some new mechanisms and comparisons: Jour. Geophys. Res., vol. 74, p. 6034-6045.

127. O'Keefe, J. A. (1969) Origin of the moon: Jour. Geophys. Res., vol. 64, p. 2758-2767.

128. Ringwood, A. E. (1960) Some aspects of the thermal evolution of the earth: Geochim. et Cosmochim. Acta, vol. 20, p. 241-259.

129. O'Keefe, J. A. (1970) The origin of the Moon: Jour. Geophys. Res., vol. 75, p. 6565-6574.

130. Ringwood, A. E., and E. Essene (1970) Petrogenesis of Apollo 11 basalts, internal constitution and origin of the moon: Proc. Apollo 11 Lunar Sci. Conf., Geochim. et Cosmochim. Acta, Suppl. 1, vol. 1, A. A. Levinson, ed., Pergamon Press, p. 769-799.

131. Gerstenkorn, M. (1955) Uber Gezeitenreibung bei Zweikorpenproblem: Zeit. Astrophys, vol. 26, p. 245-274.

132. MacDonald, G. J. F. (1964) Tidal friction: Rev. Geophys., vol. 2, p. 467-541.

133. Öpik, E. J. (1969) The moon's surface: Ann. Rev. Astron. Astrophys., vol. 7, p. 473-526.

134. Urey, H. C. (1962) Origin and history of the moon: *in* Physics and Astronomy of the Moon, Z. Kopal, ed., Academic Press, Chap. 13, p. 481-523.

135. Garz, T., M. Kock, J. Richter, B. Baschek, H. Holweger, and A. Unsold (1969) Abundances of iron and some other elements in the sun and in meteorities: Nature, vol. 223, p. 1254.

136. For examples see Latimer, W. M. (1950) Astrochemical problems in the formation of the Earth: Science, vol. 112, p. 597-599; also, Orowan, E. (1969) Density of the moon and nucleation of planets: Nature, vol. 222, p. 867.

SUGGESTED READING AND GENERAL REFERENCES

Mutch, T. A. (1972) Geology of the Moon: Princeton Univ. Press, Princeton, N.J., Revised edition, 391 p.

Science, 30 January, 1970, vol. 167, no. 3918, p. 417-792, The Moon Issue. The first reports of the lunar sample principal investigators from the Apollo 11 Lunar Science Conference.

Taylor, S. R. (1975) Lunar Science: A Post-Apollo View: Pergamon Press, 372 p. A comprehensive review of lunar science with a strong chemical flavor.

Freeberg, J. H. (1970) Bibliography of the lunar surface: U. S. Dept. of Commerce, National Tech. Info. Serv., PB 194 206, 344 p.

Levinson, A. A., and S. R. Taylor (1971) Moon rocks and minerals: Pergamon Press, New York, 222 p. A condensed and readable introduction to the Apollo 11 samples.

Frondel, J. W. (1975) Lunar Mineralogy; Wiley-Interscience, Wiley, New York, 323 p. An excellent summary and compilation of lunar mineral data, primarily for mineralogists and petrologists.

Proceedings of the Apollo 11 Lunar Science Conference (1970) Geochim. et Cosmochim. Acta, Suppl. 1, vols. 1-3, A. A. Levinson, ed., Pergamon Press; Proc. of the Second Lunar Sci. Conf. (1971) Geochim. et Cosmochim. Acta, Suppl. 2, Vols. 1-3, MIT Press, A. A. Levinson, ed.; Proc. of the Third Lunar Sci. Conf. (1972) Geochim. et Cosmochim. Acta, Suppl. 3, vols. 1-3, MIT Press, E. A. King, Jr., D. Heymann, and D. R. Criswell, eds.; Proc. of the Fourth Lunar Sci. Conf. (1973) Geochim. et Cosmochim. Acta, Suppl. 4, vols. 1-3, Pergamon Press; Proc. of the Fifth Lunar Sci. Conf. (1974) Geochim. et Cosmochim. Acta, Suppl. 5, vols. 1-3, Pergamon Press, and later conference proceedings.

For perspectives of the Moon prior to lunar sample return see: Baldwin, R. B. (1963) The measure of the Moon: Univ. of Chicago Press, Chicago, 488 p.; also Fielder, G. (1965) Lunar Geology: Lutterworth Press, London, 184 p.; also Markov, A.V., ed (1962) The Moon: A Russian view: Univ. of Chicago Press, Chicago, 391 p.

Soon it became apparent that almost all generalizations about Mars derived from Mariners 4, 6, and 7 would have to be modified or abandoned. The participants in earlier flyby missions had been victims of an unfortunate happenstance of timing. It was almost as if spacecraft from some other civilization had flown by Earth and chanced to return pictures only of its oceans.   J. F. McCauley, H. F. Hipsher, and R. H. Steinbacher, 1974

7. Mars

INTRODUCTION AND HISTORY

Ancient observers noted that among the stars were objects that looked like stars but whose apparent motions were very different from stars. These "wandering" stars were called planets, after the Greek word *planetes* meaning wanderers. Mars was one of the earliest known planets, and its apparent motions played a central role in the development of ideas about the structure and motions of the Solar System (1). Galileo also turned his astronomical telescope to Mars, but found it extremely difficult to observe much of interest with his small instrument. Francisco Fontana made the earliest known drawings of Mars, but these mostly showed his own instrumental imperfections, with the exception of one important drawing made in August of 1638, which shows a gibbous phase of the planet (2). The earliest positive identifications of surface features were made by Huygens about 1659, who identified variations in the surface albedo and clearly observed the southern polar cap. Numerous later observations filled in detail of the planet's surface and atmosphere as larger and better astronomical instruments became available.

A number of early Mars astronomers reported linear features and "canali" (channels). The best known of these early investigators is Giovanni

Schiaparelli,[1] who accurately located many Martian features on maps of the planet and seems to have observed more canals than any other single observer. He apparently regarded the canals and the associated dark areas as Martian waterways. Schiaparelli was careful to point out that only a few of the canals were visible at any one time. Percival Lowell later became very active in observations of Mars, particularly in the observation and interpretation of the canals. He founded the Lowell Observatory in Flagstaff, Arizona to take advantage of the high altitude and clear air for planetary, particularly Mars, observations. Lowell recorded more than 500 different Martian canals and was firmly convinced that the canals were constructed by intelligent beings to distribute water about their rather dry planet. Much of the literature of this period is devoted to the canals and debates about their nature and origin and is of little other than historical interest. Excellent summaries of early Mars observations can be found in Glasstone (1) and Flammarion (2). During this period, especially during Schiaparelli's observations, a complex classical Martian nomenclature developed that is rather foreign to modern scientists. The origins of many of these terms from classical literature have been summarized by MacDonald (3).

Modern geological exploration of Mars began in the mid 1960s with the use of sophisticated unmanned spacecraft and their imaging systems. Mariner IV was a "flyby" mission that obtained imagery of approximately 1 percent of the Martian surface at a maximum resolution of 3 km (4). As chance would have it, this 1 percent was relatively uninteresting cratered terrain that closely resembled some portions of the Moon (Fig. 1). This caused a slump in interest in Mars among some scientists who concluded (prematurely) that Mars was very much like the Moon and not nearly as interesting as had been previously thought. The area of imagery coverage by Mariner IV was insufficient to resolve the matter of the existence and nature of the canals. Only about 20 close encounter images were obtained by Mariner IV, and the quality of the images was far less than that obtained by later missions.

In 1969, Mariners 6 and 7 made close approaches to Mars on flyby missions and obtained more than 200 images of high quality (Figs. 2 and 3), some of which have 0.3 km resolution (5). These missions extended the coverage of the Martian surface to more than 10 percent of the total area. The imagery from these missions had sufficient coverage and detail to cast extreme doubt on the existence of the Martian canals.[2] The imagery of the southern polar cap demonstrated that it was an extensive surface deposit of possibly greater

---

[1]Schiaparelli was neither the first observer to report linear features on Mars nor to use the term "canali." However, he commonly is credited with both of these items. Father Pierre Angelo Secchi used the term "canale" to describe Martian surface features as early as 1869 (1).
[2]Early NASA versions of Martian maps had canals faintly indicated from previous Earth-based observations, but as additional data became available from Mariners 6, 7, and 9, the canals rapidly grew fainter and disappeared completely on later maps.

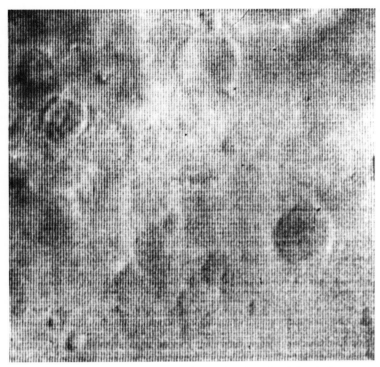

Figure 1   Mariner 4 image of the Martian surface in Atlantis bordering on Mare Sirenum. The image area is approximately 265 × 265 km. The image shows albedo variations on the surface and large Martian craters. One of the craters at upper left apparently has a central peak. (Courtesy of NASA, Jet Propulsion Laboratory.)

thickness than was earlier anticipated (6). Terrains of different crater densities and size frequency distributions were clearly delineated (7). It was concluded that there were surfaces of different ages exposed on the Martian surface, and that windblown surface sediments might be important in Martian surface processes and morphology.

Our current knowledge of Mars and the geology of its surface is based mostly on the spectacularly successful Mariner 9 mission. Mariner 9 was inserted into a Mars orbit on November 14, 1971, during an exceptionally intense planetwide dust storm (8). The dust storm obscured virtually all of the surface (Fig. 4). However, during the next several months the dust storm

Figure 2   Mariner 7 far encounter image of Mars taken from a distance of approximately 471,750 km, north is at the top. The southern polar cap is clearly visible as are numerous light clouds in the central portion of the disk. The ring-shaped structure is Olympus Mons (Nix Olympica) and associated clouds. (Courtesy of NASA, Jet Propulsion Laboratory.)

subsided and the entire surface of the planet was imaged in excellent detail, some frames having resolutions of approximately 100 m.

Basic physical data for the planet are summarized in Table I.

SURFACE FEATURES

Volcanic Features.   Olympus Mons, which was previously suspected to be a large crater, clearly was distinguished on Mariner 9 imagery as an immense

Figure 3   Mariner 7 near encounter image of the Martian surface showing far greater detail than was available from Mariner 4 imagery. (Courtesy of NASA, Jet Propulsion Laboratory.)

volcanic pile with a summit caldera (Fig. 5). Similarly, three other large spots on the Tharsis Ridge were found to be large volcanoes (Fig. 6). Such objects have not been recognized on the Moon; and, in fact, it appears that they do not exist there, at least not at this scale. Olympus Mons rises more than 20 km above the surrounding plains and is more than 500 km across at the base. The multiple summit caldera is approximately 65 km in diameter. Numerous large volcanic constructs now have been identified on the Martian surface from the Mariner 9 imagery. These volcanoes seem to have a wide range in apparent age as judged from crater counts on their slopes and differences in their apparent amounts of erosion. Mars has many smaller volcanic cones, some with associated flows, that have been recognized in virtually every part of the planet (9; Fig. 7).

There is evidence that great areas of the Martian surface are covered with extrusive igneous rocks, somewhat analogous to the lunar maria. Ridged and furrowed, relatively smooth plains with low albedo and associated volcanic

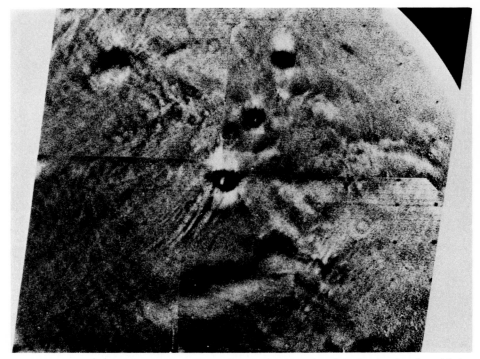

Figure 4 Photomosaic of approximately one fourth of the Martian disk shortly after the orbital insertion of Mariner 9. The surface is almost completely obscured by a planetwide dust storm. The mosaic is centered on the Tharsis Ridge, with Olympus Mons (Nix Olympica) barely visible at upper left. "North," "Middle," and "South spots," which are large shield volcanoes and calderas, are in the upper central part of the mosaic. (Courtesy of NASA, Jet Propulsion Laboratory.)

Table I   General Characteristics of the Planet Mars[a]

Mass	6.419×10^{23} kg
Radius (equatorial)	~3400 km
Radius (polar)	~3360 km
Radius (average)	3393 km
Radius Mars/Radius Earth	0.531
Density	3.9 grams/cm³
Gravitational acceleration	375 cm/(sec)(sec)
Day (sidereal)	24 hr, 37 min, 22.67 sec
Day (solar)	24 hr, 39 min, 35.23 sec
Speed of rotation	0.26 km/sec

[a]Mostly after Glasstone (1), with some later data from Mariner 9.

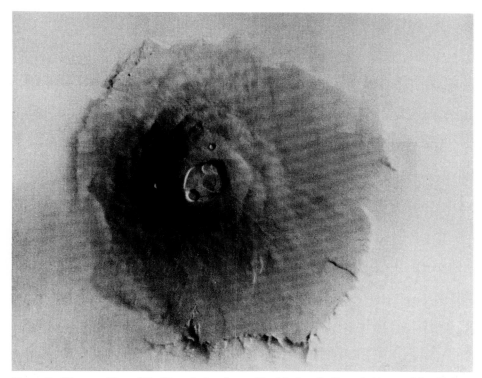

Figure 5   Olympus Mons (Nix Olympica) as imaged by Mariner 9 after subsidence of the dust storm. The feature clearly is an immense pile of volcanic rock 500 km in diameter across the base with a complex summit caldera that is 65 km across. The most recent ultraviolet spectrometer data indicate that the mountain rises more than 20 km above the surrounding plains. There is a steep, 2 km high scarp around the base of the volcano which has not been satisfactorily explained. The slopes of the volcano are very slight, indicating lava of rather low viscosity, possibly basalt. (Courtesy of NASA, Jet Propulsion Laboratory.)

shields cover tens of thousands of square kilometers, for example, Dandelion and environs (lat. 22° S, long. 253° W; Fig. 8).

   Aeolian Features.   The dust observed in the atmosphere of Mars at the time of the Mariner 9 approach, as well as the previous telescopically observed planetwide dust storms. indicate that the atmosphere is quite capable of moving surface detrital particles even though the total atmospheric pressure is only about 1/100 that of the Earth. In fact, some of the particles must have achieved altitudes of more than 20 km above the Martian surface to obscure so completely the character of Olympus Mons during the early portion of the Mariner 9 mission.

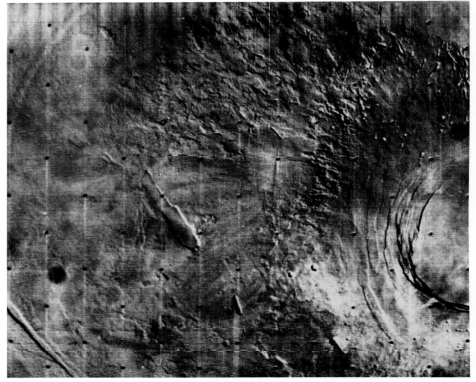

Figure 6 Mariner 9 image of a portion of "South Spot," an extremely large caldera complex, after the dust storm had subsided. The main caldera is approximately 115 km in diameter. (Courtesy of NASA, Jet Propulsion Laboratory.)

Windblown sediment deposits are common on the Martian surface and are visible in many Mariner 9 images, both "A" frames and "B" frames.[3] Many Martian craters have "wind plumes" (Fig. 9); wind "streaks" are common in the equatorial part of the planet, and large fields of dunes have been recognized on the floors of some craters (Fig. 10). The exact sizes of the particles that the Martian atmosphere can move or can move most effectively is speculative, and most writers refer to the detritus as "fine grained." However, it has been suggested that Martian winds exceed 70 m/sec., based on

[3]"A" frames were taken with the 50mm focal length wide angle lens system and have resolution ranging from 1 to 3 km. "B" frames were imaged with a narrow angle 500mm lens system and have resolution ranging from 100 to 300 m.

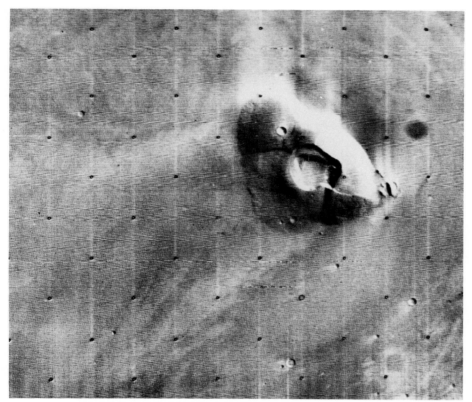

Figure 7   A smaller shield structure and caldera that has been normally faulted, possibly along the fissure through which the magma passed to get to the surface. Note the "wind streaking" of the surface caused by accumulations of surface aeolian sediments (lat. 90° W, long. 15° N; courtesy of NASA, Jet Propulsion Laboratory.)

rates of advance of smaller dust storms across the surface and other atmospheric motions. The common accumulations of detrital particles that are visible on the Martian surface indicate that aeolian transport and erosion are important processes on Mars. Prior to Mariner 9, Sagan and Pollack (10) constructed a model to account for some seasonal albedo and color changes by wind transport of particles. Later work by Woronow and King (7), based on the Mariner 6 and 7 images, suggested that the paucity of small craters in some higher albedo regions of Mars might be the result of filling or partial filling with aeolian detritus. Smith has suggested that many Martian craters are effective aeolian detritus traps under the proper wind regime and has

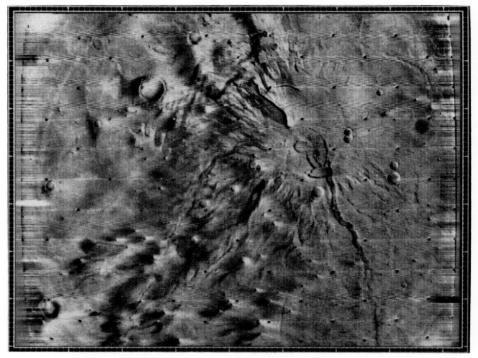

Figure 8 The "Dandelion" shield volcano and caldera complex in Mare Tyrrhenum, Mars (approx. 260° W, 28° S). Numerous individual flows, probably basalt, are visible on the gentle slopes of the shield. The flows merge into the surrounding plains, which probably also are composed mostly of basalt. The irregular summit calderas are easily distinguished from nearby circular craters that probably have resulted from impacts. (Courtesy of NASA, Jet Propulsion Laboratory.)

illustrated several terrestrial analogs (11). These general conclusions seem to be supported by the Mariner 9 imagery.

Channels. Many extremely long and wide channels have been identified in the Mariner 9 imagery. These features are not similar to lunar sinuous rilles and are much larger (Fig. 11). The morphology and possible origins of these features has been discussed by Milton (12), who concludes that many of the channels and other landforms are the result of running water on the Martian surface sometime in the past. Some of the channels show braiding and bars; and it seems certain that surface fluids have played a part in their development, but the composition and properties of the fluid(s) is still in dispute. A number of the channels have large meanders (Fig. 12), and they do seem to trend generally downslope. However, our Martian slope information is only approximate. The large channels tend to get wider toward their

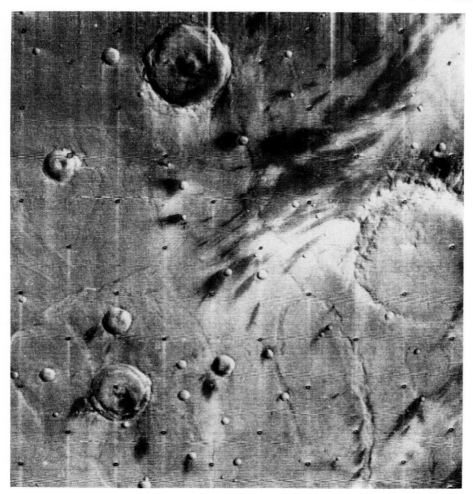

Figure 9   Light and dark albedo markings on the Martian surface are common over a wide area of the planet. These almost certainly originate by aeolian deposition and denudation. (Courtesy of NASA, Jet Propulsion Laboratory.)

mouths, just as in the case of terrestrial rivers, and they appear to have discharged onto low, relatively featureless smooth plains for the most part. Release of volatiles from a thick layer of Martian permafrost (13) has been suggested as a possible source of the fluid.

Extremely numerous smaller channels, dendritic drainage patterns, and arroyo-like features are present throughout the equatorial latitudes, especially on what appear to be steep slopes. Milton (12) has speculated that these

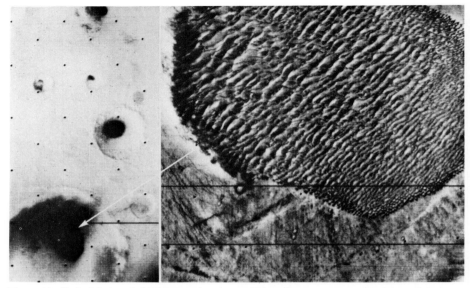

Figure 10   Left: Mariner 9 "A" frame showing a 150 km diameter crater with light and dark markings in the floor. Right: a higher resolution "B" frame showing that the very dark area in the crater floor is really a 128 × 64 km dune field of windblown surface detritus. The dune crests mostly are quite parallel but some bifurcate. The sizes of the dunes near the edge of the field are smaller than those in the central area. This field probably has formed under the influence of strong and nearly constant direction surface winds. (Courtesy of NASA, Jet Propulsion Laboratory.)

features may have resulted from the runoff of precipitation. This appears to be a reasonable hypothesis to many Mars researchers.

Canyons.   Several large canyons are present on the Martian surface. Some of these features are more than twice the size of the Grand Canyon of Arizona (Fig. 13). However, it does not appear that the canyons have formed solely by the action of fluids on the Martian surface. In fact, one very large canyon that is aligned with an even larger canyon is clearly a closed depression (Fig. 14). The origins of the canyons are not known. Their origins may be related to that of the "chaotic terrain," which has been suggested to result from the removal of a thick layer of permafrost or the withdrawal of magma (13). Some investigators have speculated that the canyons are formed along transform faults, plate margins or spreading centers in the Martian crust where continental drift is just beginning to gain topographic expression (14). The answer must await further planetary exploration.

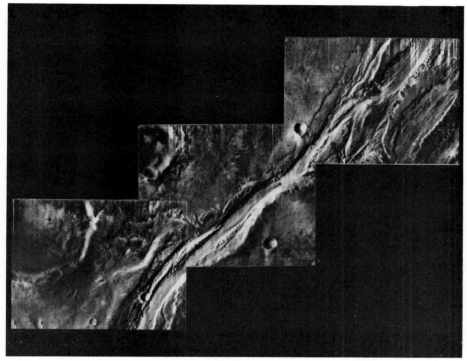

Figure 11   Mosaic of three Mariner "B" frame images showing a 75 km long segment of a large channel in Amazonis. Note the bars and braiding in the upper right frame. The channel apparently has been eroded or at least sculptured by fluid moving on the surface, possibly water. (Courtesy of NASA, Jet Propulsion Laboratory.)

Scarps and Cliffs.   There are numerous high scarps and cliffs on Mars, many of them as high as 2 km, which appear to be actively retreating and forming low knobby, chaotic, and smooth plains units at the expense of higher plateaus (Fig. 15). Some of the cliffs are continuous for more than 1000 km. Sapping by running water and melting ground ice and permafrost may be the prime agents in the erosion of many Martian cliffs (12). The action of permafrost and subsurface ice removal on Mars to account for surface features was first suggested by Sharp et al. (15). Many cliffs also show erosion from aeolian processes, but these seem to be only modifying agents and not the chief cause of cliff retreat.

Layering is prominent in many large Martian cliffs (Fig. 15), indicating stratigraphic units or groups of similar units of substantial thickness. Many of

Figure 12   Mariner 9 image showing several large channels with sinuous courses and apparent meanders. The large channel is several hundred kilometers long. (Courtesy of NASA, Jet Propulsion Laboratory.)

these layers are almost certainly extrusive volcanic rocks; however, it has been speculated that vast deposits of stratified loess may occur on Mars also.

Polar Areas and Caps.   Since the discovery of the southern polar cap of Mars in the latter part of the 17th century, the nature of the polar caps has been in dispute. Lowell supposed that the polar caps of Mars were thick water ice accumulations similar to the polar caps of the Earth. We now believe that the polar caps are composed mostly of CO_2, with possibly minor amounts of CO_2-H_2O clathrate and water ice. During the Mariner 9 mission, the southern Martian polar cap receded rather rapidly at first, but later slowed its retreat. This observation led to speculation that either the residual cap is much thicker than the periphery of the cap, or that the central portion of the cap contains frozen volatiles that are stable at higher temperatures, or both. There was a

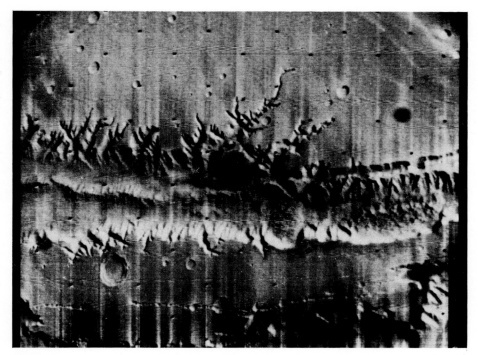

Figure 13   A large canyon on the Martian surface that is approximately twice as deep and several times wider than the Grand Canyon of Arizona. Notice the branched tributaries on the upper side of the canyon. (Courtesy of NASA, Jet Propulsion Laboratory.)

dense haze over the northern polar area until late in the Mariner 9 mission, the northern polar "hood"; however, late mission imagery obtained comprehensive coverage.

The origins of geologic features in the polar regions of Mars have been discussed by Soderblom et al. (16), Cutts (17), Sagan (18), Sharp (19), and others. "Laminated" terrain underlies the transient polar caps at both poles (Fig. 16). This surface unit is composed of relatively thin, flat-lying beds which are estimated to be 10 to 30 m thick. It has been speculated that these deposits may be loess, interbedded loess and frozen volatiles, or possibly even stratified outwash from glaciers at the Martian poles. It appears to be the youngest surface unit that is visible in either of the polar areas (16). Large amounts of water ice may be trapped with dust in laminated deposits beneath the perennial frost caps according to Cutts (17).

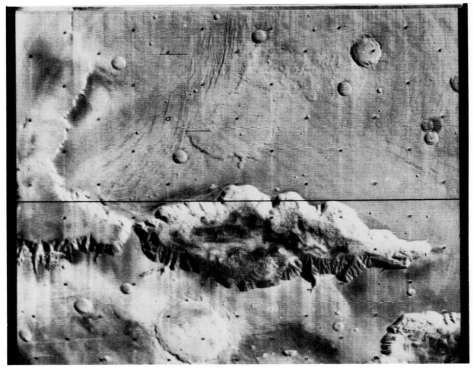

Figure 14 A large closed canyonlike Martian depression as imaged by Mariner 9. The origin of such a depression by surface running fluid erosion seems extremely unlikely. (Courtesy of NASA, Jet Propulsion Laboratory.)

The laminated and thinly layered deposits of the central polar areas are surrounded mostly by smooth, "etch-pitted," and "rippled" plains. The etch-pitted plains are named for their appearance (Fig. 17) and are thought to be formed, at least in part, by wind deflation of smooth plains or other plains morphological units. Soderblom et al. (20) have concluded that there is a general transport of aeolian detritus away from both of the polar areas because of the distribution of surface deposits that are interpreted to be wind deposited. These "mantling" deposits are distributed at high to mid-latitudes about the Martian polar areas. Similar conclusions have been reached by Cutts (21), based on the orientations and abundance of albedo markings and wind eroded surface features. Large ridges and troughs are found at the

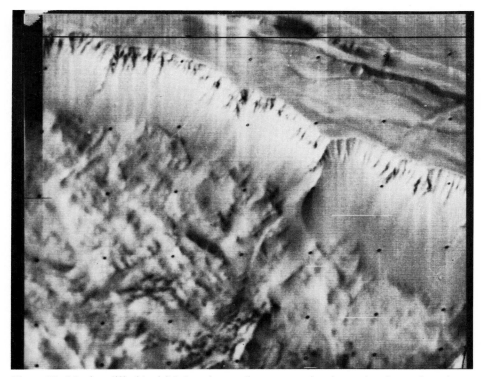

Figure 15   Martian cliff with visible layering eroding away to give rise to chaotic terrain at the base. The cliff is approximately 2 km high. (''B'' frame from Mariner 9; Courtesy of NASA, Jet Propulsion Laboratory.)

margins of the polar caps, particularly in the south. The origin of these features is uncertain, but they appear to have been modified by or to have resulted from the wind regime near the poles. However, to some observers these ridges have the appearance of glacial moraines.

   Craters.   Although the new and unusual features of Mars have received much well deserved attention, the dominant topographic forms on the Martian surface are craters. Öpik (22) first suggested in 1950 that craters might be an abundant landform on Mars. However, it was not until the successful flyby of Mariner IV in 1965 that this hypothesis was demonstrated to be correct. Mariner IV imaged such a small area of the surface that few data were available for cratering statistics; however, Mariners 6 and 7 covered much

Figure 16   Mariner 9 "B" frame image of laminated terrain. This type of layered deposit is present at both Martian polar regions. It appears to consist of layers as thin as 10 m and may be loess interbedded with frozen volatiles. (Courtesy of NASA, Jet Propulsion Laboratory.)

more area with better resolution and provided data for serious discussion. Regional variations in crater density and degradation were investigated by McGill and Wise (23), who concluded that there were real variations in densities for the smallest craters visible, as did Woronow and King (7).

Our current data (Mariner 9 "B" frames) indicate that Martian craters as small as a few hundred meters are common on some parts of the planet. Other parts of Mars, particularly near the edges of the polar caps, have very low densities of small craters, possibly due to crater filling by wind blown detritus (see *Aeolian features*, page 251). Some of the areas of previous crater filling near the polar regions are now probably exhumed (20).

The morphology and density distribution of Martian craters has been reviewed by Hartmann (24). He concludes that much of the terrain around the centers of volcanic activity on the Tharsis Ridge is only approximately 3 $\times$

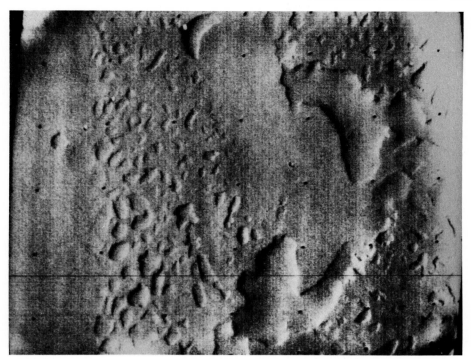

Figure 17 ''Etch-pitted'' plains with apparent wind deflation hollows located approximately 800 km from the Martian south pole. The small pits are 1 to 3 km across. ''B'' camera image taken 3343 km from the surface by Mariner 9. (Courtesy of NASA, Jet Propulsion Laboratory.)

10^8 years old, much younger than any presently known surface on the Moon. Large craters are abundant in what are interpreted to be older terrains. Mars has several excellent examples of huge multiringed basins (25) that are similar to Orientale or Imbrium on the Moon. Hellas is the most striking example, half again as large as the Imbrian Basin, but the floor and many of the outer features of the basin have been obscured by surface detritus and later volcanic flows.

The majority of Martian craters appear to be of impact origin, as on the Moon. Many Martian craters have visible central peaks (Fig. 18) and extensive ejecta deposits surrounding the crater. There are very few Martian craters that show prominent ray systems of ejecta, possibly because none of the close approach images are taken at very high sun angles and also probably because of the Martian winds, which may blow fine ray material away or

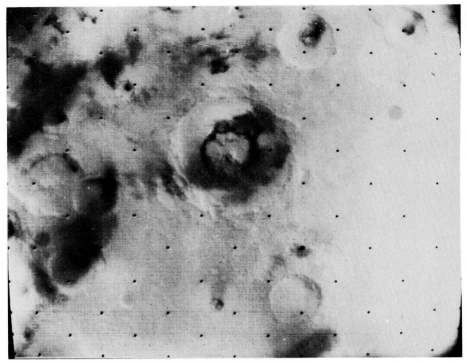

Figure 18 Large Martian crater (diameter approx. 125 km) with prominent central peaks and other central structure. Note also the albedo variations on the crater floor. (Courtesy of NASA, Jet Propulsion Laboratory.)

deposit fine detritus from elsewhere on top of the rays. The dust storm that was in full development at the time of orbital insertion of Mariner 9 may have obscured the ray systems from very recent impact craters. Such storms apparently are frequent events on Mars. Secondary craters are visible around a number of larger craters, particularly on Mariner 9 "B" frames, but even secondary craters are not as abundant as on the Moon. Doublet craters, with the rims nearly tangent, occur on Mars with a much greater frequency than would be expected for a random process. A quantitative treatment of this subject has been presented by Oberbeck and Aoyagi (26), who attributed the doublet craters to the fission of some incoming meteoroids from tidal stresses caused by Mar's gravitational field.

Surfaces and terrains of different probable ages have been identified on Mars by a number of researchers based on the size frequency distributions and density distribution of Martian craters. A composite

geologic-physiographic map of the whole planet has been compiled by
Saunders and co-workers (Fig. 19). On Mars this task is more complicated
than on the Moon because of the nonuniformity of surface erosion and
deposition processes. However, as more detailed work is available on the
surface geology of the planet, it is likely that our estimates of relative and
absolute ages of portions of the Martian surface will improve.

GEOPHYSICAL OBSERVATIONS

Evaluation of inferred topographic information from the Mariner 9 ultraviolet
spectrometer by Schubert and Lingenfelter (27) have led them to conclude
that the center of figure is offset from the center of mass of Mars by
approximately 1 km. This conclusion results from the amplitude of the first
harmonic of the most accurate available Mars topographic information. The
phase of the first harmonic indicates that the direction of the displacement is
toward the Tharsis Ridge. Schubert and Lingenfelter believe that this result
supports the idea that Mars is differentiated and that internal convection is an
active process within the planet. The mere occurrence of large volcanic
constructs on the Martian surface also seems to support this idea.

SURFACE AND BULK COMPOSITION

There have not, as yet, been any direct chemical analyses of the composition of
the Martian surface. We can infer that some of the magma that has been
extruded onto the surface in the past has been of rather low viscosity,
comparing well with the viscosity of some terrestrial basalts. The morphologies
of other volcanic surface features suggest that somewhat more differentiated
rocks may occur also.

Surface water ice or hydrated minerals or both have been identified on the
surface of Mars by Pimentel et al. (28) from interpretation of Mariner 6 and 7
infrared spectrometer data. Our only clue to the bulk composition of the
planet is its density of approximately 3.9 gm/cc. This figure probably is a good
indicator of the ratio of iron to silicates, and suggests that Mars has a greater
proportion of iron than does the Moon, but less than the Earth.

ATMOSPHERE

The composition of the Martian atmosphere may offer important clues to the
planet's history and has been the subject of numerous investigations. Sir
William Herschel first suggested that Mars had a "considerable atmosphere"
in 1784. However, nothing was actually known of the composition or density

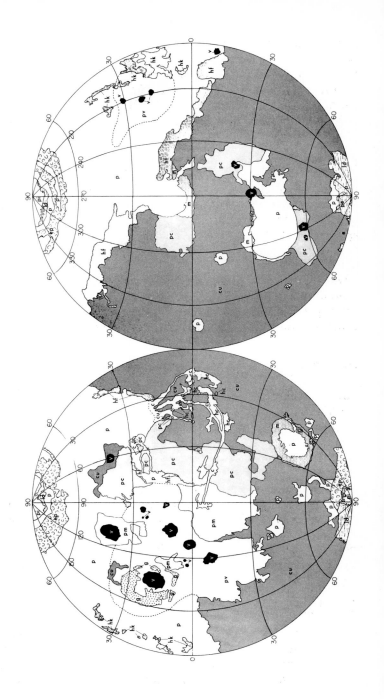

Figure 19 Geologic-physiographic map of Mars compiled by R. S. Saunders, R. Arvidson, K. Jones, and T. Mutch. Map units are as follows:

Polar Units

pi Permanent ice — predominantly H_2O ice.

ld Laminated deposits — thin (less than 100 m) continuous laminae at "B" frame resolution. Appear as shallow benches alternating with steeper slopes.

ep Etched plains — surface contains irregular pits, many of which coalesce.

Volcanic Units

v Volcanic constructs — shields, domes, or cones.

pv Volcanic plains — few craters, lobate scarps interpreted as flow fronts.

pm Moderately cratered plains — lacks volcanic features.

pc Cratered plains — most densely cratered plains unit, contains ridges resembling those of the lunar maria, also contains older eroded volcanic features.

Modified Units

hc Hummocky terrain, chaotic — chaotic terrain, disrupted, and tilted blocks of local surface material adjacent to channels or in closed basins.

hf Hummocky terrain, fretted — marginal unit to northern scarp, large blocks near contact with cratered terrain, smaller blocks out into plains.

hk Hummocky terrain, knobby — isolated regions of 10-km knobs, not clearly related to areas of chaotic or fretted hummocky terrain.

c Channel deposit — smooth floors of channels, probably composed of alluvial, windblown, or mass-wasted wall materials.

p Plains, undivided — sparsely to moderately cratered at "A" frame resolution, "B" frames show irregular ridges, scarps, and channels; generally scoured appearance.

q Grooved terrain — arcuate regions of linear mountains adjacent to Olympus Mons, 1 to 5 km wide and typically 100 km long; outer zone has finer texture.

Ancient Units

cu Cratered terrain, undivided — densely to moderately cratered uplands, most ancient of all surfaces.

m Mountainous terrain — rugged basin margin material, probably eroded basin ejecta.

Lambert equal area projection. (Courtesy of R. S. Saunders.)

of the Martian atmosphere until an experiment by G. P. Kuiper in 1947, who compared simultaneous infrared spectra of Mars and the Moon. Because the Moon essentially has no atmosphere, any difference in the two spectra must have been caused by the atmosphere of Mars. Kuiper detected absorptions at wavelengths characteristic of carbon dioxide, and he estimated that the mass of carbon dioxide above a unit area on Mars was approximately twice that of the Earth. However, it was later concluded from the intensities of radio signals from Mariner spacecraft as they were occulted by the planet that the atmosphere was much less substantial, with 8 to 10 mbar being the maximum surface pressure at most surface sites, and 6 mbar possibly closer to the average.

The composition of the atmosphere has been confirmed to be almost entirely CO_2. However, one very important result of the Mariner 9 atmospheric observation was the positive identification of water in the Martian atmosphere. Low clouds were observed over some of the prominent positive topographic features (volcanic mountains) of the Tharsis Ridge (29). These clouds occur at such low altitudes that they cannot be frozen CO_2. It was well known from years of telescopic observations that many of these features were of variable brightness in diurnal and seasonal cycles. Most telescopic observers agreed that the variations in brightness could only result from cloud formation, but the nature and composition of the clouds was unknown. From Mariner 9 imagery Leovy et al. (29) conclude that these clouds are most likely the product of orographic uplift, but convection and local degassing may also be involved. A maximum of 20 to 30 *precipitable* micrometers of water was detected over the northern polar cap by Conrath et al. (30) during the northern spring from their Mariner 9 infrared spectrometer data. Direct spectral evidence for the identification of H_2O ice clouds in the Martian atmosphere over the Tharsis Ridge has been obtained by Curran et al. (31). Little room is left for doubt that water vapor and water ice crystals are a small, but possibly exceedingly important, constituent of the atmosphere of Mars.

The gradual clearing of the dust storm during the Mariner 9 mission allowed atmospheric physicists to study the Martian atmosphere almost layer by layer. Wave clouds indicating strong winds have been observed in several parts of the planet, particularly in the south polar regions. Condensate clouds were identified in Hellas in the early morning at that site (29); however, most of the observations of the upper atmosphere are not relevant to geological interpretations or processes.

MOONS

The two moons of Mars were discovered by the American astronomer Asaph Hall at the Naval Observatory in Washington, D. C. in August of 1877.

However, a number of attempts to discover satellites of Mars in earlier years by several astronomers were not successful. Hall was confused by his first observations and thought that there must be several inner moons of Mars because their positions relative to each other and to the planet changed so rapidly. He named the inner satellite Phobos (fear) and the outer satellite Deimos (flight), after the horses that draw the chariot of Mars (also the attendants of Mars in other works).

Prior to Mariner 9, even our estimates of the sizes of these two very small moons were subject to considerable error. However, both satellites were imaged by Mariner 9 (Figs. 20 and 21), and we now have very good information on their sizes, volumes, and masses (Table II).

Mariner 9 data have confirmed that both satellites are in synchronous rotation. The orbits of both moons have very small eccentricities. Phobos has a period of revolution only a little more than 7½ hr, which is one third the length of a Martian day. Phobos is the only known satellite in the Solar System with a period of revolution that is less than the period of rotation of its parent planet. Deimos has a period of revolution of approximately 30 hr and 18 min. Both moons' orbits are very close to the surface of Mars; Phobos is nearly at the Roche limit. The orbital radii of Phobos and Deimos are approximately 9450 km and 23,500 km, respectively (1). Because the radius of Mars is approximately 3380 km, neither of these satellites would be visible to an observer at either of the Martian geographic poles.

The compositions of Mars' moons are not known; however, the Mariner 9 images do show detail of their surfaces. Both moons are densely cratered, and Hartmann (24) points out that these crater densities are much greater than on the surface of Mars itself. Thus, it appears that a substantial number of craters have been erased from the face of Mars by erosion and/or the satellites of Mars have seen a different flux history of meteoroids that would form craters on the order of 1 km diameter than has the surface of the planet. Photometric and other evidence for the existence of a regolith on the surface of both Phobos and Deimos, analogous to the lunar regolith, has been cited by Pollack et al. (32). They also suggest that the low geometric albedos of the two bodies may indicate that they are basaltic or composed of carbonaceous

Table II   Principal Axes of Phobos and Deimos[a]

Satellite	Largest Axis (radius, km)	Intermediate Axis (radius, km)	Smallest Axis (radius, km)	Volume (km³)	Mass[b] (10¹⁸ gm)
Phobos	13.5 ± 1	10.7 ± 1	9.5 ± 1	5748	17.2
Deimos	7.5 (+3, −1)	6.1 ± 1	5.5 ± 1	1054	3.16

[a]After Pollack et al. (32).
[b]A density of 3 gm/cc is assumed.

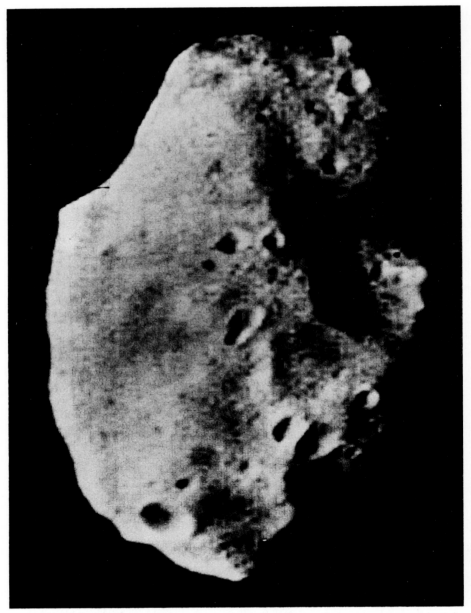

Figure 20 Phobos, Mars' innermost moon, as imaged by Mariner 9 at a range of approximately 5540 km. Numerous small craters, some with rims, are visible. The density of craters on Phobos is approximately the same as for craters the same size in the lunar highlands; however, the density of craters of this size on the surface of Mars itself apparently is less. The maximum diameter of Phobos is approximately 27 km. It probably is a captured asteroid. (Courtesy of NASA, Jet Propulsion Laboratory.)

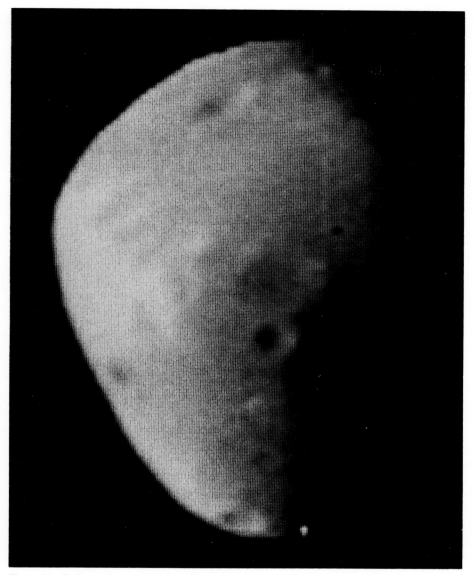

Figure 21 Deimos, the outer and smaller moon of Mars, as imaged by Mariner 9 at a range of 5490 km. The maximum diameter of Deimos is approximately 15 km. The prominent small crater near the terminator is named Swift, and the larger less distinct crater immediately above it is Voltaire. Observe the contrast in angularity, apparent crater density, and "sharpness" of surface morphology with Phobos (Fig. 20). (Courtesy of NASA, Jet Propulsion Laboratory.)

chondrite material. It does seem fairly certain that both moons are composed of well consolidated materials, otherwise they probably could not have sustained some of the large craters on their surfaces without disrupting (33).

The small sizes and irregular shapes of Phobos and Deimos, together with their unusually low and fast orbits, have led many observers to speculate that the two moons are captured asteroids.

A search was made for other possibly smaller moons of Mars during the Mariner 9 approach to the planet, but none were found. A complete set of the enhanced photographs of Phobos and Deimos, as well as associated relevant data, has been published by Veverka et al. (34).

FUTURE EXPLORATION

The presence of water vapor and water ice in the Martian atmosphere and the possibility of greater amounts of water having been present on Mars in the past make the possibility of life of Mars more likely than previously thought (35). The thin atmosphere and distance from the Sun offer some protection from radiation. The temperature ranges in some low latitude portions of the planet are within the tolerance ranges of some terrestrial organisms. The more we find out about Mars, the more likely it appears that life *could* have independently evolved and survived there. Whether or not there is life on Mars will be a prime object of investigation of exploration programs by the United States and probably also by the USSR.

Many other important questions also remain to be resolved. The presence of water in the Martian atmosphere means that we should expect different types of weathering and erosion on the surface of Mars than on the Moon. Are there hydrated minerals on the surface? What are the major rock types and minerals that compose the Martian surface and interior? What are the ages of surfaces and surface features on Mars? Is the basic chemistry of the planet similar to the Earth, Moon, or meteorites? These and *many* other questions remain to be answered. Unmanned spacecraft will provide even more information than they have to date. A long step toward many of the answers will be the eventual return of a small sample to terrestrial laboratories for sophisticated analyses, which will parallel our lunar information explosion.

REFERENCES AND NOTES

1. Glasstone, S. (1968) The book of Mars: National Aeronautics and Space Admin., SP-179, 315 p.
2. Flammarion, C. (1892, 1909) La Planete Mars et ses Conditions d'Habitabilite: Gauthier-Villars et Fils, vols. 1 and 2.

3. McDonald, T. L. (1971) The origins of Martian nomenclature: Icarus, vol. 15, p. 233-240.

4. Nicks, O. W. (1967) A review of the Mariner IV results: NASA, Spec. Pub., SP-130, 39 p.

5. Leighton, R. B., N. H. Horowitz, B. C. Murray, R. P. Sharp, A. H. Herriman, A. T. Young, B. A. Smith, M. E. Davies, and C. B. Leovy (1969) Mariner 6 television pictures: First report: Science, vol. 165, p. 684-690; also (1969) Mariner 7 television pictures: First report: Science, vol. 165, p. 787-795; and (1969) Mariner 6 and 7 television pictures: Preliminary analysis: Science, vol. 166, p. 49-67. However, the best reference on Mariner 6 and 7 is — Scientific Findings from Mariner 6 and 7 Pictures of Mars: Final Report, reprinted from Jour. Geophys. Res., vol. 76, no. 2, p. 293-472, January 10, 1971.

6. Leighton, R. B., and B. C. Murray (1966) Behavior of CO_2 and other volatiles on Mars: Science, vol. 153, p. 136-144.

7. Woronow, A., and E. A. King (1972) Size frequency distribution of Martian craters and relative age of light and dark terrains: Science, vol. 175, p. 755-757.

8. Masursky, H., R. M. Batson, J. F. McCauley, L. A. Soderblom, R. L. Wildey, M. H. Carr, D. J. Milton, D. E. Wilhelms, B. A. Smith, T. B. Kirby, J. C. Robinson, C. B. Leovy, G. A. Briggs, T. C. Duxbury, C. H. Acton, Jr., B. C. Murray, J. A. Cutts, R. P. Sharp, S. Smith, R. B. Leighton, C. Sagan, J. Veverka, M. Noland, J. Lederberg, E. Levinthal, J. B. Pollack, J. T. Moore, Jr., W. K. Hartmann, E. N. Shipley, G. De Vaucouleurs, and M. E. Davies (1972) Mariner 9 television reconnaissance of Mars and its satellites: Preliminary results: Science, vol. 175, p. 294-305.

9. Carr, M. H. (1973) Volcanism on Mars: Jour. Geophys. Res., vol. 78, p. 4049-4062.

10. Sagan, C., and J. B. Pollack (1968) A windblown dust model of Martian surface features and seasonal changes: Astron. Jour., vol. 73, p. S33-S34; also, Smithson. Astrophys. Obs., Spec. Rept. 255, 44 p.

11. Smith, H. T. U. (1972) Aeolian deposition in Martian craters: Nature Phys. Sci., vol. 238, p. 72-74.

12. Milton, D. J. (1973) Water and processes of degradation in the Martian landscape: Jour. Geophys. Res., vol. 78, p. 4037-4047.

13. Sharp, R. P. (1973) Mars: Fretted and chaotic terrains: Jour. Geophys. Res., vol. 78, p. 4073-4083.

14. Hodgkinson, R. J. (1972) personal communication. See especially Sengör, A. M. C. and I. C. Jones (1975) A new interpretation of Martian tectonics with special reference to the Tharsis region: Geol. Soc. Amer., Program with Abstracts, vol. 7, no. 7, Sept., 1975, p. 1264, *abstract*.

15. Sharp, R. P., L. A. Soderblom, B. C. Murray, and J. A. Cutts (1971) The surface of Mars, 2: Uncratered terrains: Jour. Geophys. Res., vol. 76, p. 331-342.

16. Soderblom, L. A., M. C. Malin, J. A. Cutts, and B. C. Murray (1973) Mariner 9 observations of the surface of Mars in the North Polar Regions: Jour. Geophys. Res., vol. 78, p. 4197-4210.

17. Cutts, J. A. (1973) Nature and origin of layered deposits of the Martian polar regions: Jour. Geophys. Res., vol. 78, p. 4231-4249; see also *Ibid.,* p. 4211-4221.

18. Sagan, C. (1973) Liquid carbon dioxide and the Martian polar laminas: Jour. Geophys. Res., vol. 78, p. 4250-4251.

19. Sharp, R. P. (1973) Mars: South polar pits and etched terrain: Jour. Geophys. Res., vol. 78, p. 4222-4230.

20. Soderblom, L. A., T. J. Kreidler, and H. Masursky (1973) Latitudinal distribution of a debris mantle on the Martian surface: Jour. Geophys. Res., vol. 78, p. 4117-4122.

21. Cutts, J. A. (1973) Wind erosion in the Martian polar regions: Jour. Geophys. Res., vol. 78, p. 4211-4221.

22. Öpik, E. J. (1950) Mars and the asteroids: Irish Astron. Jour., vol. 1, p. 221.

23. McGill, G. E., and D. U. Wise (1972) Regional variations in degradation and density of Martian craters: Jour. Geophys. Res., vol. 77, p. 2433-2441.

24. Hartmann, W. K. (1973) Martian cratering: 4, Mariner 9 initial analysis of cratering chronology: Jour. Geophys. Res., vol. 78, p. 4096-4116.

25. Wilhelms, D. E. (1973) Comparison of Martian and lunar multiringed circular basins: Jour. Geophys. Res., vol. 78, p. 4085-4095.

26. Oberbeck, V., and M. Aoyagi (1972) Martian doublet craters: Jour. Geophys. Res., vol. 77, p. 2419-2432.

27. Schubert, G., and R. E. Lingenfelter (1973) Martian center of mass — center of figure offset: Nature, vol. 242, p. 251-252.

28. Pimentel, G. C., P. B. Forney, and K. C. Herr (1974) Evidence about hydrate and solid water in the Martian surface from the 1969 Mariner infrared spectrometer: Jour. Geophys. Res., vol. 79, p. 1623-1634.

29. Leovy, C. B., G. A. Briggs, and B. A. Smith (1973) Mars atmosphere during the Mariner 9 extended mission: Television results: Jour. Geophys. Res., vol. 78, p. 4252-4266.

30. Conrath, B., R. Curran, R. Hanel, V. Kunde, W. Maguire, J. Pearl, J. Pirraglia, J. Welker, and T. Burke (1973) Atmospheric and surface properties of Mars obtained by infrared spectroscopy on Mariner 9: Jour. Geophys. Res., vol. 78, p. 4267-4278.

31. Curran, R. J., B. J. Conrath, R. A. Hanel, V. G. Kunde, and J. C. Pearl (1973) Mars: Mariner 9 spectroscopic evidence for H_2O ice clouds: *manuscript.*

32. Pollack, J. B., J. Veverka, M. Noland, C. Sagan, T. C. Duxbury, C. H. Acton, Jr., G. H. Born, W. K. Hartmann, and B. A. Smith (1973) Mariner 9 television observations of Phobos and Deimos, 2: Jour. Geophys. Res., vol. 78, p. 4313-4326.

33. Pollack, J. B., J. Veverka, M. Noland, C. Sagan, W. K. Hartmann, T. C. Duxbury, G. H. Born, D. J. Milton, and B. A. Smith (1972) Mariner 9 television observations of Phobos and Deimos: Icarus, vol. 17, p. 394-407.

34. Veverka, J., M. Noland, C. Sagan, J. Pollack, L. Quam, R. Tucker, B. Eross, T. Duxbury and W. Green (1974) A Mariner 9 Atlas of the Moons of Mars: Icarus, vol. 23, no. 2, p. 206-289.

35. The literature on the possibilities of extraterrestrial life is exceedingly large. For an introduction and source see Biology and the Exploration of Mars, C. S. Pittendrigh, W. Vishniac, and J. P. T. Pearman, eds., National Academy of Sciences, National Research Council, Pub. 1296, NAS/NRC, 1966, 516 p.

SUGGESTED READING AND GENERAL REFERENCES

Mutch, T. A., R. E. Arvidson, K. L. Jones, J. W. Head, III, and R. S. Saunders (1976) The Geology of Mars: Princeton Univ. Press, Princeton, New Jersey. A comprehensive treatment of the Geology of Mars through Mariner 9 and very well illustrated.

Journal of Geophysical Research, vol. 78, no. 20, p. 4007-4440, July 10, 1973. This entire volume is devoted to the interpretation of the Mariner 9 imagery and experiments and a good single source for our current ideas about the planet.

Glasstone, S. (1968) The book of Mars: National Aeronautics and Space Administration, SP-179, 315 p. This volume summarizes much of the early work on Mars and provides many references to the historical literature. As the date of the book suggests, only Mariner IV data were available at the time that this manuscript was prepared.

Journal of Geophysical Research, vol. 76, no. 2, p. 293-472, January 10, 1971. This volume contains the best collection of Mariner 6 and 7 results that appears in a single source. It is interesting to contrast some of the first impressions gained from these two missions with the results from the much more informative Mariner 9 mission.

Collins, S. A. (1971) The Mariner 6 and 7 pictures of Mars: NASA, Spec. Pub., SP-263, 159 p.

NASA (1974) Mars: As Viewed by Mariner 9: NASA, Spec. Pub., SP-329, 225 p. A good collection of Mariner 9 photographs with short interpretive texts.

The visit of Hermes was a near miss for earth inhabitants; if it ever strikes the earth — as it may in the future — the force of its impact will liberate the energy of one billion hydrogen bombs, and may destroy a substantial fraction of the population of the earth.

Robert Jastrow and Malcolm H. Thompson, 1974

8. Asteroids

INTRODUCTION AND HISTORY

There are many "minor planets" in the Solar System, some may be as much as 760 km diameter but most are much smaller. We have never closely observed any of these bodies except through the telescope. Two possible exceptions are the moons of Mars, Phobos, and Deimos (see Mars, Figs. 20 and 21), which may be asteroids that have been captured by that planet.

The first known asteroids were found in an attempt to locate a "missing planet" in the Solar System as predicted by Bode's Law. J. D. Titius, in 1772, discovered an interesting numerical relation that described the relative positions of the planets then known. He published this relation in an obscure footnote, but it gained recognition through the efforts of J. E. Bode and later became known as Bode's Law (1). Bode, who later became director of the Berlin Observatory, was intrigued by the large distance between Mars and Jupiter and by the predictions of the mathematical relation. The discovery of Uranus by Herschel[1] on March 13, 1781, solidified support for Bode's Law because it was found that Uranus was only about *2 percent* of the radius away from the orbit predicted by the law (19.2 A.U. compared with 19.6 A.U. predicted; 2). The discovery of Uranus stimulated an organized search for the "missing planet" at 2.8 A.U. by 24 European astronomers (3).

[1]This was the first new planet discovered in historic times. In recognition of this discovery King George III awarded Herschel an observatory and a stipend, which allowed him to concentrate on his research. Prior to this time, Herschel primarily had been a professional musician (2).

On New Year's Eve, 1800, a Sicilian astronomer, Giuseppe Piazzi spent several hours at his observatory in Palermo locating and mapping the stars in the constellation Taurus. While searching for a star that he was unable to locate,[2] he discovered a starlike object of the eighth magnitude (4). Piazzi tracked and observed the object for some time and observed that it "looped," as do the apparent motions of the planets. Piazzi named the new body Ceres, after the guardian goddess of Sicily and wrote of his discovery to Bode. Ceres was calculated to be at 2.77 A.U. by a young mathematician named Gauss (5). The discoveries of other small planetary bodies at approximately the same distance from the Sun followed rapidly. Approximately one year later, Wilhelm Olbers discovered Pallas. Juno and Vesta were discovered in 1804 and 1807, respectively, and by 1890, more than 300 asteroids had been observed and catalogued. Groeneveld and Kuiper (6) estimate that approximately 40,000 asteroids have been observed, although most of them have not been observed with sufficient precision to establish their orbits or brightnesses accurately (7). The total population of asteroids with absolute magnitude[3] >19 is thought by Kiang (8) to be more than 70,000.

ORBITS

BELT ASTEROIDS

The majority of the asteroids have orbits that lie between the orbits of Mars and Jupiter, that is, 1.5 to 5.2 A.U., but seven eighths of them lie between 2.3 and 3.3 A.U. However, the eccentricities of many asteroid orbits cause the asteroid belt to occupy a much wider zone. The orbits of all presently known asteroids are prograde, that is, counterclockwise if viewed from the north. The distribution of orbits of belt asteroids is far from random. In addition to the concentration of orbits with mean distances close to 2.8 A.U., as predicted by Bode's Law, there are absences due to "resonance" or "commensurability" with Jupiter. Because most of the asteroids have orbits inside that of Jupiter, they have shorter periods of revolution than Jupiter. Because of its very large mass compared with all other objects in the Solar System except the Sun, the gravitational field of Jupiter tends to dominate the middle portion of the Solar System. There are gaps in the distribution of asteroid orbits that correspond to some simple fractional periods of Jupiter's period, $1/2$, $1/3$, $2/5$, and $3/7$ (Fig. 1). These gaps in asteroid orbits are termed the Kirkwood gaps, after an American astronomer who called attention to these features in 1866. Schweizer (9) has studied the Kirkwood gaps extensively and has suggested that the gravitational attraction of Jupiter will

[2]Due to a printer's error in the Wollaston Star Catalog that he was using.
[3]Absolute magnitude is the intrinsic brightness of an object at a standard distance in a specified wavelength range assuming no absorption in the intervening space.

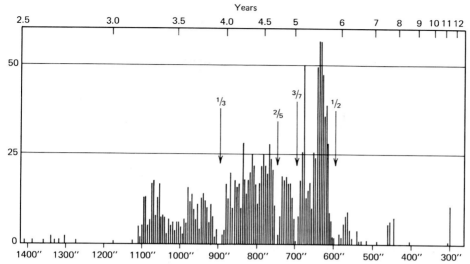

Figure 1 Histogram of the distribution of asteroids according to sideral period (upper scale) and daily mean motion in arc seconds (lower scale) showing the Kirkwood gaps and other commensurabilities. (After Hartmann, W. K., 1968, Astrophys. Jour., vol. 162, p. 337.)

tend to remove objects from these resonant orbits. However, some commensurate orbits tend to be stabilized by resonance effects (10, 11); thus, the asteroids with ³/₄ or ²/₃ the period of Jupiter are stable for long times, and belt asteroids cluster about these values. According to Alfvén (12), if the ratio of the mean motion of a minor planet is *not* a simple fractional relation to the motion of Jupiter, the perturbations tend to cancel out and the motion of the asteroid remains unchanged for relatively long periods of time. It should be emphasized that the "original" distribution of asteroids in this (or any other) portion of the Solar System is not known. Even the present distribution is known only for the largest objects. The inclinations of asteroid orbits are mostly low, but some range to as much as 30 deg, much greater than the inclination of any major planet's orbit.

TROJAN ASTEROIDS

Lagrange, a French mathematician and astronomer, was solving a difficult three-body problem in the year 1772. He pointed out that if a small body were revolving about the Sun in approximately the same orbit as a much larger planet, when the Sun, planet, and smaller body were at the corners of an equilateral triangle, the position of the asteroid with respect to the planet would remain unchanged. An asteroid that satisfied this condition was

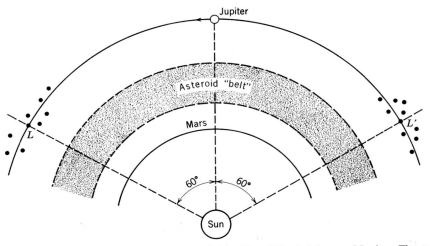

Figure 2 Relative positions of the Sun and the orbits of Earth, Mars, and Jupiter. The Lagrangian points (*L, L'*) are in the orbit of Jupiter 60 deg. ahead of and 60 deg. behind the planet. The Trojan asteroids ahead of Jupiter are termed the "Greek planets," and the following asteroids are termed the "Trojan planets."

discovered by Max Wolf in 1906 (Achilles), and numerous others subsequently have been identified (13, 14). Thus there are groups of asteroids 60 deg. of arc ahead and behind Jupiter in its orbit, at the Lagrangian points (Fig. 2). These are all named for the mythical heros of the Trojan wars (Trojan and Greek) as told by Homer. The eccentricities and inclinations of some of the Trojan asteroid orbits are high; thus, they perform complicated motions about the apex of the triangle and may at times be rather far from the ideal Lagrangian points (15). It is possible that new Trojan asteroids may appear, captured into the Lagrangian points by the immense gravity field of Jupiter over a long period of time. Also, it is possible that perturbations due to the proximity of Saturn may cause the escape of some Trojans (16, 17). That some of the Trojans may be escaped moons of Jupiter has been suggested by Dunlop and Gehrels (18) and Rabe (19).

MARS-CROSSING ASTEROIDS

More than 30 asteroids on the inner fringe of the asteroid belt presently are known that have orbits that cross the orbit of Mars, that is, these asteroids have perihelia that are less than the aphelion of Mars. Some of these asteroids may be the sources of fragments that can be further perturbed into Earth-crossing orbits. It has been suggested by Anders (16) that the

Mars-crossing asteroids could be the source of some of the meteorites recovered on the Earth.

APOLLO ASTEROIDS

Some asteroids have orbits that bring them into the central portion of the Solar System within the orbit of the Earth. At least 15 such asteroids are known at the present time (20). Öpik estimates that there must be approximately 40, based on discovery rates. More recent estimates suggest that the number is more like 100 and may be as much as 1000. All of these asteroids are small, with diameters on the order of half a kilometer.

That the half-life[4] of Apollo asteroids is only one percent of the age of the Solar System has been pointed out by Anders (16). Thus they must be the remnants of an immense early population, or they must be replenished. Many scientists in this field now believe that at least a portion, if not the majority, of the meteorites originate from these bodies. Because these asteroids are already in Earth-crossing orbits, a relatively small amount of energy is required to perturb or to knock a fragment of one of these bodies into an orbit such that it will collide with the Earth.

ASTEROIDS WITH CLOSELY SIMILAR OBITS

The Japanese astronomer Hirayama (21), in a series of papers from 1917 to 1933, pointed out that many asteroids can be grouped according to similar orbital elements. He used the semimajor axes, eccentricities, and inclinations of the orbits and showed that they clustered around certain special values. These groups are called "Hirayama families." A reinvestigation of the families in 1951, using orbital elements that were corrected for periodic perturbations caused by secular perturbations of the major planets and other corrections was accomplished by Brouwer (22). He redefined the 9 Hirayama families and identified 19 additional "groups." Brouwer's families contained from 9 to 62 asteroids and there were from 4 to 11 asteroids in each group. The whole problem was restudied by Arnold (23), using observational data from the literature and advanced numerical analysis techniques. This resulted in substantial change in the recognized families and in the addition of a number of new ones (Table I). Other groups of asteroids have been identified because of some other similarities of orbital properties. For example, the Hilda group (11), which clusters near the two thirds resonance. The Flora family contains groups of asteroids with similarity of all five orbital elements and Alfvén (24) developed the "jet stream" hypothesis to account for these groups. The literature of meteor and asteroid streams has been reviewed by Lindblad (25).

[4]The time required to reduce an initial orbit-related population to one half its original number through perturbations and capture.

Table I   Completely New Asteroid Families Recognized by Arnold[a]

Family	A-81	A-82	A-83	A-84	A-85	A-86	A-87	A-88	A-89	A-90	A-91
Total Number of Members	*9*	*7*	*8*	*9*	*17*	*17*	*14*	*26*	*10*	*12*	*7*
Members[b]	42	19	26	505	55	103	58	137	408	92	403
	67	21	37	563	59	110	125	199	552	318	519
	118	138	66	826	123	160	128	250	567	490	716
	474	435	77	869	127	203	210	314	874	844	824
	585	557	309	1178	197	206	267	448	942	1015	858
	647	1190	384	1277	200	272	301	493	979	1023	984
	889	1586	708	1525	237	308	340	545	1255	1086	1433
	902		1084	1555	359	363	380	665	1380	1161	
	1528			1681	371	460	395	690	1582	1163	
					481	847	868	762	1678	1351	
					527	1007	1039	768		1395	
					559	1020	1251	786		1558	
					614	1228	1502	788			
					741	1352	1541	791			
					872	1420		986			
					1176	1517		1030			
					1716	1726		1069			
								1113			
								1115			
								1165			
								1369			
								1452			
								1519			
								1572			
								1701			
								1735			

[a]From Arnold (23).
[b]Asteroids are given an identification number or catalog number once their orbital elements are well known. This number relates to the order in which the asteroids are discovered and their orbital elements determined.

OPTICAL STUDIES

There are numerous studies of the brightness variations, albedos, optical diameters, and the like of asteroids (6, 18, 20, 26, and many others). These studies have inferred that most asteroids rotate at rather high rates. This inference was first made by Gehrels (27) based on light intensity-time curves (Fig. 3). There are two obvious possibilities to explain a light-time curve of this type: (a) the asteroid has an approximately uniform surface albedo, but is

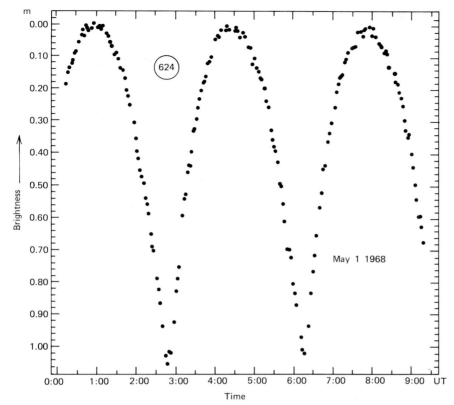

Figure 3 Brightness versus time curve for asteroid 624 Hector. Such observations have been interpreted to indicate that some asteroids are somewhat irregular in shape and are rotating in their orbits, assuming a uniform surface albedo. (After Hartmann, W. K., 1968, Astrophys. Jour., Vo. 162, p. 337.)

elongate or irregular in shape and is turning or tumbling, and (b) that the asteroid is roughly equant, but that it has surface properties that contribute to differences in albedo from place to place. Probably both of these possibilities are true for certain asteroids, but most workers seem to accept the tumbling or rotation of the asteroid as the more likely.

Only the diameters of the four largest asteroids can be measured directly on photographic plates. However, the diameters are more accurately measured by difficult optical measurements (28). More recent measurements are being made in the infrared. These measurements permit the calculation of the albedo and thereby the diameter, if the absolute magnitude is known and various assumptions are made. The sizes of smaller asteroids must be inferred from their brightness, especially by comparison with reference stars, other planets, or the moon, for which there are many observations. A mass-frequency distribution curve based on an asteroid survey was constructed by Kuiper (29), who found a "hump" or point of inflection at approximately 10^{20} grams (Fig. 4). Kuiper suggested that the point of inflection approximately separates those asteroids that have accreted (the larger ones) from the smaller asteroids that might be fragmentation products. Anders (16) further developed arguments relating to the size frequency

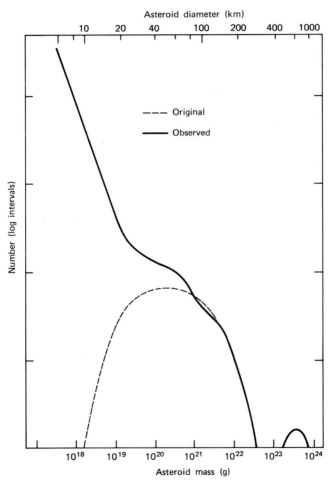

Figure 4   Mass-frequency distribution of asteroids after Kuiper (29). The dashed line is the inferred original distribution (Courtesy of the American Astronomical Society.)

distribution of asteroids by considering asteroids in different parts of the "belt." For example, Anders found that the absolute magnitude[5] distribution of asteroids between 2.15 and 3.16 A.U. approaches a Gaussian curve for the brighter (larger) asteroids and approximates a logrithmic curve for the fainter ones. Anders supported Kuiper's interpretation.

Reflectance spectral data of the asteroid Vesta have been taken by McCord et al. (30), who conclude that the data best match basaltic achondrites. This conclusion is based mostly on the 0.9 -μm band which arises from absorption of ferrous iron in the M_2 site of a magnesian pyroxene. A visible and near-infrared spectral reflectivity curve for (1685) Toro has been derived by Chapman et al. (31). They conclude that this asteroid is unique among those observed to date and that the observations most closely resemble equilibrated L-type (olivine-hypersthene) chondrites. Spectral data do not indicate a clear correlation of asteroid surface composition with heliocentric distance for the present distribution of large asteroids. Proposed matches of asteroid spectra with various meteorite types have been proposed by McCord and Chapman (32, Fig. 5, p. 286).

Although it now seems probable that, at least, some meteorites are asteroidal fragments, the exact bodies from which they originate are still in dispute. Resolution of these unknowns may have to await further advances in spectral work, remote analyses, or even sample returns.

REFERENCES AND NOTES

1. Flammarion, C. (1964) The Flammarion Book of Astronomy. Trans. A. B. Paget, Dir. by G. Flammarion, Simon and Schuster, New York, 670 p.

2. MacPherson, H. C. (1933) Makers of Astronomy. Oxford Press, 244 p. Also, Brandt, J. C., and S. P. Maran (1972) New Horizons in Astronomy. W. H. Freeman, San Francisco, 496 p.

3. Roth, G. D. (1962) The System of Minor Planets. D. Van Nostrand, New York, 128 p.

4. Leuschner, A. O. (1935) Research surveys of the orbits and perturbations of minor planets 1 to 1091: Lick Obs. Pubs., vol. 19, 519 p.

5. Payne-Gaposchkin, C. (1954) Introduction to Astronomy. Prentice-Hall, Englewood Cliffs, N.J., 507 p.

6. Groeneveld, I., and G. P. Kuiper (1954) Photometric studies of asteroids, I: Astrophys. Jour., vol. 120, p. 200-220.

7. See Kuiper, G. P., Y. Fujita, T. Gehrels, I. Groeneveld, J. Kent, G. van Briesbroeck, and C. VanHouten-Groeneveld (1958) Survey of asteroids: Astrophys. Jour., Suppl. 32, p. 289-427; also Gehrels, T. (1967) Minor planets — II,

[5]The brightness or magnitude of asteroids is taken to be proportional to their size or mean diameter. A density of about 3.3 grams/cm^3 is assumed to estimate the mass.

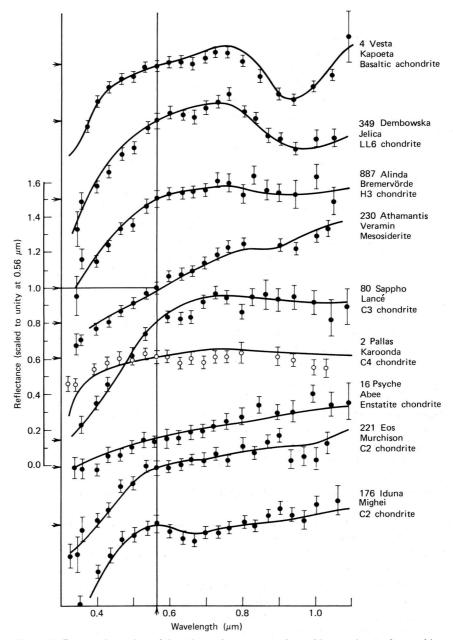

Figure 5 Proposed matches of the telescopic spectra (points with error bars) of asteroids with the laboratory spectra (solid lines) of meteorites. The matches suggest similar mineralogical compositions. The implication is that the proposed asteroid or a similar asteroid may have been the parent body of the specific meteorite or similar types of meteorites. Many meteorites have no asteroidal counterparts yet discovered. (From Chapman, C. R., 1974, The asteroids as meteorite parent bodies: Meteoritics, vol. 9, no. 4, p. 323; courtesy of C. Chapman and Meteoritics.)

286

Space missions to comets and to asteroids are the essential next steps toward understanding how the solar system came into being.   F. L. Whipple, 1974

9. Comets

INTRODUCTION AND HISTORY

There is a long history of observations of comets that dates back to the rather accurate records of ancient Chinese astronomers. The apparent motions of comets among the stars were carefully recorded. Old European records of comet observations are less ordered, but comets attracted great attention because the appearance of one was thought to herald impending disasters or great events.[1] Aristotle concluded that comets were phenomena of the upper atmosphere, but this view was corrected by Tycho Brahe in 1577. Tycho Brahe concluded that comets must be celestial objects, at least three times as distant as the Moon, because his observations indicated the absence of diurnal parallax (2). He also postulated that the comet probably revolved about the Sun. Kepler described Halley's Comet of 1607 in considerable detail, but he concluded that comets travel through the Solar System in straight lines.

Halley's Comet (Fig. 1) is the most famous of all comets. The recorded observations of this comet can be traced back to 467 B.C. Edmund Halley was the first astronomer to connect the comets of 1531, 1607, and 1682, and he predicted (successfully) that the comet would return in 1758, years after his death. It was found that Chinese records of observations of this comet have recorded each appearance for more than 12 centuries (1). The major

---

[1]Comets were thought to be responsible for plagues and other great catastrophes, and many wrote that the appearance of a comet indicated a change in the affairs of state or the futures of nations. The Bayeux tapestry depicts Halley's Comet during its apparition in 1066 on the eve of the Battle of Hastings. Another example is that the fall of Constantinople to the Turks was blamed on the comet of 1456, even though the city actually fell three years earlier (1)!

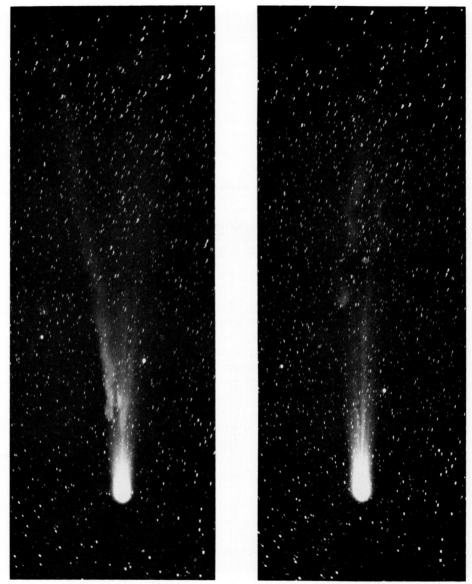

Figure 1   Halley's Comet as it appeared on June 6 (left) and June 7 (right), 1910. Notice the obvious rapid changes in the configuration of matter in the tail. (Courtesy of the Lick Observatory.)

conclusion that derived from observations of Halley's Comet was that comets moved through the Solar System in elliptical orbits with very great eccentricities.

In 1973 and 1974, Comet Kohoutek (Fig. 2) was discovered and attracted great public attention. However, it turned out to be far less spectacular than predicted and, in fact, was difficult to see with the unaided eye. The comet was discovered and tracked far in advance of its perihelion so that when it did approach it was the object of many scientific measurements and observations.

ORBITS

The most abundant type of cometary orbit is elliptical with very large eccentricity. Many cometary orbits have such eccentricities that the periods are extremely great, a few hundred to millions of years, and aphelion may be at 10^5 or more astronomical units from the Sun (3). These are the

Figure 2   Comet Kohoutek as photographed on November 28, 1973, through the 46 cm Schmidt camera at the Mt. Palomar Observatory by E. Helin. (Courtesy of the Hale Observatories.)

"long period" comets, which are by far the most numerous comet type. In addition, there are a number of "short period" comets, in orbits with periods of less than 10 to as much as 150 years, which apparently have resulted from perturbations of Jupiter on long period comets (4). More than 65 comets have been observed to have hyperbolic orbits; however, it is thought that these cometary orbits originally were ellipitical and have become hyperbolic by planetary perturbations (3). These comets have sufficient velocity to escape the Solar System. There are also a number of "annual" comets that have orbits with very small eccentricities. Some of these objects, such as P/Schwassmann-Wachmann and P/Oterma,[2] have orbits that are similar to some belt asteroids. The relations between the orbits of comets and asteroids have been investigated by Marsden (5), who concludes that there are some similarities. In fact, if the comets P/Arend-Rigaux and P/Neujmin 1 had not been observed until their second recorded apparitions, they would have been classified as asteroids. Several observers reported faint traces of comas at their first apparitions. Marsden suggests that comets that avoid encounters with Jupiter may be approaching the end of their cometary lives. Some of the Apollo asteroids may represent the remnants of such comets. The inclinations of comet orbits are scattered and many are at high inclinations, unlike the orbits of planets (3).

COMPOSITION

The compositions of comets are not well known. There are many spectral observations of the gaseous portions of comets from emission bands (molecular spectra) and a few atomic emission lines. Species reported include neutral molecules such as CN, HCN, CH_3CN, C_2H_6OH, C_2, C_3, NH, OH, NH_2, CH, Na, O_I, Fe, and Ni, which are found within the heads of comets, and ionized molecules such as CO^+, N_2^+, CO_2^+, H_2O^+, and CH^+, which are characteristic of the tails (6). It is assumed that these species are derived from frozen volatiles in the comet nucleus. Comets also contain substantial amounts of dust and larger solid material. Although this solid material probably is composed predominantly of silicates, it never has been knowingly observed or analyzed. Much of the dust-sized solid matter is left behind by the comet as it passes through the inner part of the Solar System, and these dust trails are responsible for the annual meteor showers. For example, the comet P/Tempel-Tuttle 1866 I is almost certainly the comet source of the Leonid meteors (3). It has not been demonstrated that any of the recovered meteorites are larger fragments of solid matter from comets. However, some investigators suspect that the volatile-rich carbonaceous chondrites may be

[2]Comets with elliptical orbits and periods of less than 200 years are termed short period comets and are designated by the prefix P/. Comets usually are named for their discoverers (Halley's Comet is a notable exception).

cometary material, and Wetherill (7) has correlated the orbits of bright fireballs and chondritic meteorites with those of short period comets and Apollo asteroids. It is in these solid materials of comets that the main geological interest lies.

STRUCTURE, SIZE, AND ORIGIN

In the inner part of the Solar System, comets generally have well defined heads and tails. The head can sometimes be subdivided into an apparently solid nucleus and a coma, or gas cloud surrounding the nucleus. Many comets are very faint objects and some develop only small, weak comas and may never develop tails (8). A few comets develop planes or "spikes" of material that precede the head in its orbit (Fig. 3). Some comets are immense objects whose comas may reach 10^5 km in diameter and whose tails may extend for 10^8 km; however, most comets are much smaller objects. Sekanina (9) has estimated that the average mass loss per perihelion for 23 short period comets is, at least, 10^{12} to 10^{14} grams, and that the nuclei of these comets should average approximately 10^{17} to 10^{18} grams. The total fraction of the mass of the Solar System represented in comets is not known. Cameron (10) has suggested that there must be a large amount of Solar System mass in small particles on the outskirts of the Solar System.

It is possible that most of this material may have been accreted by long period comets in the Oort Cloud.[3] Oro (11) has pointed out that the gaseous composition of many comets is comparable to that of interstellar gas clouds, and that this interstellar material may be represented in comets.

It seems clear that most comets spend the major portion of their existence beyond the range of terrestrial telescopes and beyond the outer planetary orbits of the Solar System. The orbital elements of many comets are well known, and the evolution of short period comets from long period comets seems well established, but their primary origin remains obscure. They may represent primary condensates from the outer reaches of the Solar System. Direct observation and analysis of known cometary materials would certainly help to resolve many of their mysteries.

REFERENCES AND NOTES

1. Brandt, J. C., and S. P. Maran (1972) New Horizons in Astronomy: W. H. Freeman, San Francisco, 496 p.
2. Abell, George (1969) Exploration of the Universe: Holt, Rinehart and Winston, New York, 722 p.

[3]The term Oort Cloud or Öpik-Oort Cloud is used to designate a "swarm" of comets that surrounds the Solar System whose semi-major axes extend out to approximately 50,000 A.U. The existence of this cloud was pointed out by the Dutch astronomer Jan Oort (4).

Figure 3   The comet Arend-Roland as it appeared on April 25, 1957, showing a prominent "spike" of material preceding the head. This spike may be a thin plane of concentrated particles that have a greater gravitational acceleration toward the Sun than the force of solar radiation, which tends to drive them away from the Sun. Thus these particles may be larger than those that compose most of the tail. (Courtesy of the Lick Observatory.)

3. Porter, J. G. (1963) The statistics of comet orbits: *in* The Solar System IV; The Moon, Meteorites and Comets, B. M. Middlehurst and G. P. Kuiper, eds., Univ. of Chicago Press, Chicago, p. 550-572.

4. Oort, J. H. (1950) The structure of the cloud of comets surrounding the Solar System, and a hypothesis concerning its origin: B.A.N. vol. 11, p. 91. See also, Whipple, F. L. (1964) Evidence for a comet belt beyond Neptune: Proc. National Acad. Sci., vol. 51, no. 5, p. 711-718.

5. Marsden, B. G. (1970) On the relationship between comets and minor planets: Astronomical Jour., vol. 75, no. 2, p. 206-217.

6. Wurm, K. (1963) The physics of comets: *in* The Solar System IV; The Moon, Meteorites and Comets, B. M. Middlehurst and G. P. Kuiper, eds., Univ. of Chicago Press, Chicago, p. 573-617. See also Whipple, F. L. (1975) Comets: Data, problems and objectives: *in* Proc. Soviet-American Conf. Cosmochem. Moon and Planets, Moscow, June 4-8, 1974, *in press*.

7. Wetherill, G. W. (1968) Stone meteorites: Time of fall and origin: Science, vol. 159, p. 79-82, also (1969) Origin of Prairie Network fireballs and meteorites: Trans. Am. Geophys. Union, vol. 50, p. 224, *abstract*.

8. Roemer, Elizabeth (1963) Comets: Discovery, Orbits, Astrometric Observations: in The Solar System IV; The Moon, Meteorites and Comets, B. M. Middlehurst and G. P. Kuiper, eds., Univ. Chicago Press, Chicago, p. 527-549.

9. Sekanina, Z. (1969) Total gas concentration in atmospheres of the short-period comets and impulsive forces upon their nuclei: Astron. Jour, vol. 74, no. 7, p. 944-950.

10. Cameron, A. G. W. (1962) The formation of the Sun and planets: Icarus, vol. 1, p. 13-69.

11. Oro, J. (1972) Extraterrestrial Organic analyses: Space Life Sci., vol. 3, p. 507-550.

SUGGESTED READING AND GENERAL REFERENCES

Marsden, B. G. (1974) Comets: *in* Annual Review of Astronomy and Astrophysics; G. R. Burbidge, ed., Annual Reviews Inc., Palo Alto, Calif., 1974, p. 1-21. This article has an excellent bibliography of recent articles on virtually all aspects of comet research.

Centuries hence, when current social and political problems may seem as remote as the problems of the Thirty Years' War are to us, our age may be remembered chiefly for one fact: It was the time when the inhabitants of the earth first made contact with the vast cosmos in which their small planet is embedded.   Carl Sagan, 1975

10. Other Planets and Moons

INTRODUCTION

The remainder of Solar System objects are either so distant that only the most fundamental facts are known concerning them, or they are inherently of lesser *geological* interest. We make no attempt to summarize the astronomical literature except for that of geological interest. Geological exploration of these objects is just beginning.

MERCURY

Our knowledge of the surface features of this closest planet to the Sun was greatly increased with the tremendously successful Mariner 10 mission, which made a total of three flyby encounters with the planet. Mariner 10 obtained more than 2000 images of Mercury, covering a substantial fraction of the planet's surface area. Some of the images have resolutions of less than 1 km, and many are useful for geologic description and interpretation of surface features (1; Figs. 1 through 6).

Mercury is small, with a mean diameter of approximately 4878 km, and has essentially no atmosphere, except for a small amount of helium detected by

Figure 1   Photomosaic of Mercury from approximately 20° W (terminator) to 110° W (bright limb) longitude, North is at the top. Bright rayed craters are clearly visible, and some craters are nearly 200 km in diameter in this heavily cratered terrain. Large linear features also can be seen. The images were taken at a range of 234,000 km by the Mariner 10 mission, approximately 6 hr before closest approach to the planet. (Courtesy of NASA, Jet Propulsion Laboratory.)

298

Figure 2 Photomosaic of Mercury from approximately 110° W (bright limb) to 200° W (terminator) longitude. North is at the top. This portion of the planet contains heavily cratered terrain as well as some lightly cratered areas. A large multiringed basin is partially visible near the terminator in the center of the left side of the image. These images were taken by the Mariner 10 mission approximately 5½ hr after closest approach to the planet at a range of 210,000 km. (Courtesy of NASA, Jet Propulsion Laboratory.)

the Mariner 10 UV experiment at a total pressure of less than 2×10^{-9} mbars. The bulk density of the planet is 5.45 gm/cm³, indicating a greater iron to silicate ratio for Mercury than for the Moon or the Earth (2). The rotational period of Mercury is 58.66 days, in close agreement with the value for the 3/2 commensurability between its axial rotation and orbital revolution (3). The overall impression from the Mariner 10 images is that this planet's surface is covered with large impact craters, and that it bears a striking resemblance to the Moon. However, in detail there are some significant differences. Although at least one large multiringed basin is present on the surface (Fig. 2), Mercury does not seem to have the marked highlands/mare dichotomy of the Moon. The pronounced albedo variations that are so common on the Moon appear much more subdued on Mercury. Rayed impact craters are common (Figs. 1 and 2), and a continuum from very young fresh craters to relatively old, greatly modified craters is plainly visible. The obliteration of older surface features around fresh impact craters is only approximately one sixth that

Figure 3 View of the northern limb of Mercury showing a prominent east-facing scarp on the northern limb, which extends for hundreds of kilometers. The notch in the limb at left is from missing data. The linear dimension along the base of the image is approximately 580 km. (Courtesy NASA, Jet Propulsion Laboratory.)

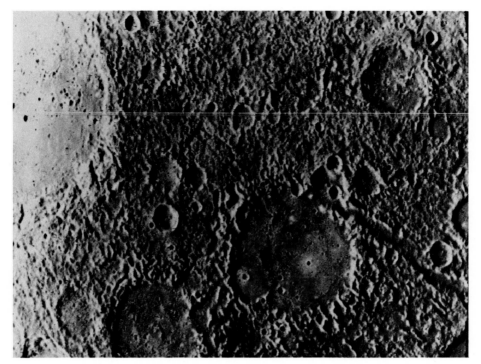

Figure 4 Close view of a portion of Mercury's surface 290 by 220 km, showing many old craters and low hills. The valley at right is 7 km wide and more than 100 km long. The large flat-floored crater near the center is approximately 80 km in diameter. (Courtesy of NASA, Jet Propulsion Laboratory.)

surrounding similar lunar craters, and the fields of secondary craters are closer to the primary crater on Mercury (4). These observations are consistent with the stronger gravity of Mercury than the Moon, but other factors may be important also. The bulk density and diameter of Mercury lead to the strong implication that, if differentiated, the metallic core of Mercury must be roughly ¾ of the planet's radius. Thus, the lithosphere can be no more than approximately 600 km thick. This thin lithosphere, as well as other possible variations in its composition and physical state, could respond differently to impacts than do the thick lithospheres of the Earth, Moon, and Mars. The tectonics of Mercury also must be greatly affected by the large proportion of iron and size of the core. The great scarps (Fig. 3) on the planet's surface may be related to Mercury's large core, but their origins are not yet understood.

Reflectance spectra (Figs. 7 and 8) indicate that the surface of Mercury is covered with regolith very similar to that of the Moon, according to McCord and Adams (5). They conclude that the optical properties of the surface are

Figure 5   Heavily cratered terrain on Mercury similar to portions of the lunar highlands. The flat-floored crater at right has interior peaks, and many of the small craters in the central portion of the image appear to be secondaries related to this crater. The rough terrain adjacent to the large crater probably is formed from crater ejecta. (Courtesy NASA, Jet Propulsion Laboratory.)

dominated by dark glass together with iron- and titanium-rich materials. Based on the cratered appearance of the surface alone it is reasonable to expect that some of the surface materials will be similar in texture to some of the clastic lunar rocks and unlithified regolith samples.

The exploration of the surface of Mercury by sophisticated spacecraft bearing analytical instruments or that are capable of returning samples is well within the current state of technology. Future missions should yield much more information about this geologically interesting terrestrial planet.

VENUS

Our best distant images of Venus also have been obtained by the Mariner 10 mission (6; Fig. 9), and atmospheric data, surface data and photography have been obtained from Russian atmospheric entry and surface landing probes. Venus is known to have a very dense atmosphere, approximately 90 Earth atmospheres of predominantly carbon dioxide, and the surface temperature appears to be approximately 450°C. The atmosphere contains

Figure 6   A high resolution image of a portion of Mercury's surface showing highly cratered terrain. A flow front extends across the floor of the 61 km diameter crater at lower left. The relatively fresh crater at center is approximatey 25 km in diameter. Craters as small as 1 km in diameter are visible in this image. The dark line represents a few missing lines of data. (Courtesy of NASA, Jet Propulsion Laboratory.)

numerous optically dense clouds; therefore, the surface is only visible in longer wavelengths. Radar imagery of the planet's surface has demonstrated the presence of large craters (7; Fig. 10). A Russian measurement of the natural gamma ray activity of Venus indicates a uranium to thorium ratio similar to the Earth. The bulk density of the planet is 5.16 gm/cm^3.

Because of the high temperature of the surface and the very dense atmosphere, it is unlikely that samples will be returned from Venus in the next decade; however, Venera 9 returned images from the surface showing angular to rounded rocks abundantly scattered about its landing site. We can expect further information gains from landers and improved radar imagery, including radar from spacecraft orbiting Venus, as well as atmospheric analyses.

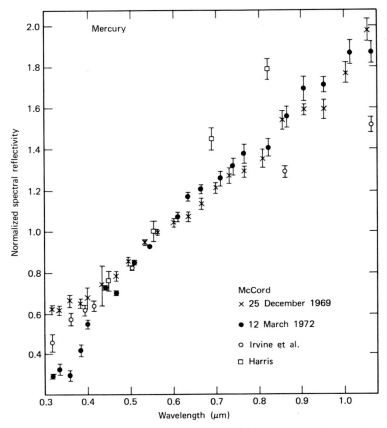

Figure 7   Spectral reflectivity of the integral disk of Mercury scaled to unity at 0.564 mm for each of two sets of observations (4). Earlier work also is shown. The wavelength positions of both new sets of data are the same, but one set has been deliberately displaced so that the error bars are visible. (Copyright 1972 by the American Association for the Advancement of Science.)

JUPITER

Jupiter and its moons have been the targets of several flyby missions, with excellent imagery of the huge planet coming from the Pioneer 10 mission (8; Figs. 11 and 12). This giant planet can be observed easily from the Earth, and considerable detail of the layered clouds and the Great Red Spot can be seen through even relatively small telescopes. However, the Pioneer 10 images are a very substantial improvement. Because of Jupiter's immense mass, approximately 1.9×10^{30} grams, soft landing in its atmosphere or on its "surface" is a task of heroic proportions, requiring substantial propulsion

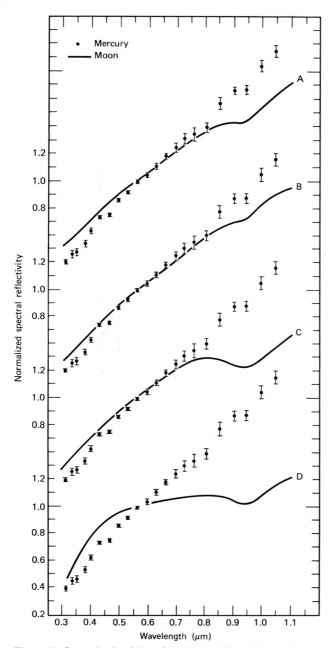

Figure 8 Spectal reflectivity of the integral disk of Mercury compared with four different lunar terrains: *A*, highlands; *B* maria; *C*, bright highland craters; *D*, bright mare craters. (From 4; Copyright 1972 by the American Association for the Advancement of Science.)

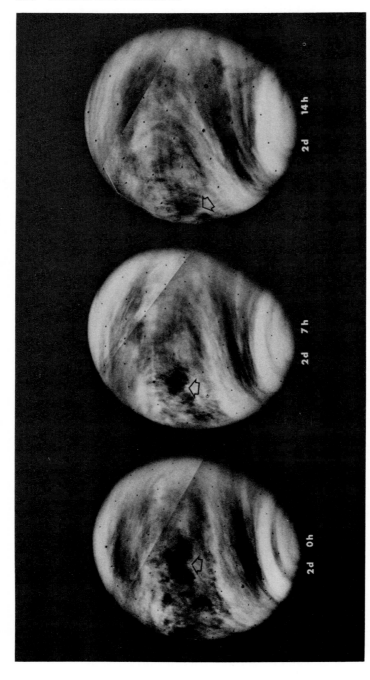

Figure 9 A series of photomosaics of Venus taken by Mariner 10 at 7-hr intervals two days after closest approch to the planet. The images are taken through ultraviolet filters and show the rapid rotation of Venus' thick cloud cover. The feature indicated by arrows is approximately 1000 km in diameter. (Courtesy of NASA, Jet Propulsion Laboratory.)

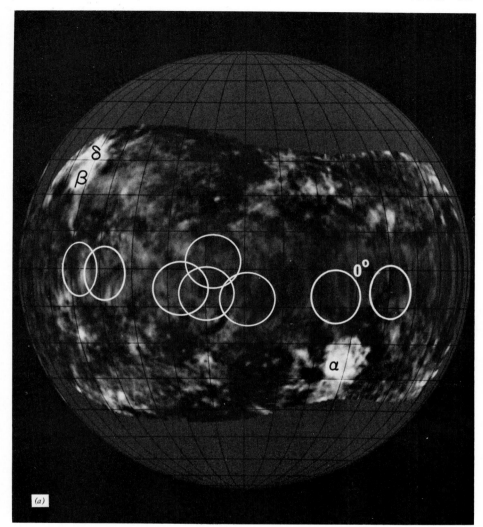

Figure 10 *(a)* A radar brightness image of a large portion of the surface of the planet Venus. The grid is spaced at 10° latitude and longitude intervals. The white circles indicate areas for which higher resolution radar brightness imagery is available, an example of which *(b)* clearly shows that craters are a dominant topographic form of the planet's surface. The largest crater visible is approximately 160 km in diameter, and the smallest craters that can be resolved are approximately 35 km diameter. (After Rumsey et al., 4; courtesy of Academic Press.)

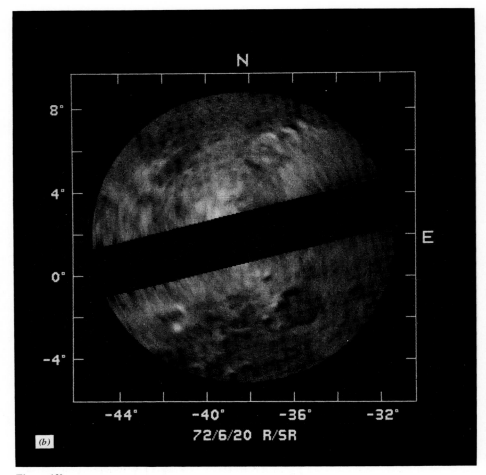

Figure 10b.

capabilities. The Jovian or outer planets are quite different from the terrestrial or inner planets in that they have retained large amounts of volatile elements such as hydrogen and helium. The Jovian planets generally are larger (except for Pluto), less dense, and have more satellites than the terrestrial ones.

The outer planets are not of prime *geological* interest, but some of their satellites may be extremely interesting bodies composed, at least in part, of silicate and metallic minerals. Jupiter is now known to have at least 14 moons.

Figure 11   A Pioneer 10 image of Jupiter showing details of the complex cloud structure, the shadow of the moon Io, and the Great Red Spot is just visible near the terminator. Image taken at a range of 2.5×10^6 km. (Courtesy of NASA, Ames Research Center.)

The inner four satellites, Io, Europa, Ganymede, and Callisto, were first observed by Galileo and are termed the "Galilean" satellites[2] These range from approximately 2920 to 5100 km in diameter. Spectral work with these satellites by Pilcher et al. (9) has led to the identification of water frost on Europa and Ganymede and detection of possible water frost on Io and Callisto as well. Furthermore, Pilcher et al. state that the spectral reflectivity

[1]There is an additional inner satellite, JV, which is actually the closest known moon to the planet, but as the number indicates it was not discovered until later.

Figure 12   Pioneer 10 image of Jupiter clearly showing the Great Red Spot and latitudinally layered cloud structure. (Courtesy of NASA, Ames Research Center.)

of the material interpreted to underlie the water frost on Europa, Ganymede, and Callisto resembles that of silicates. Other recent spectral work (10) has discovered the presence of a glowing sodium vapor cloud surrounding Io. In addition, Pioneer 10 results indicate a hydrogen cloud along the orbit of Io (11). Variations in the surface markings of Io have been observed for many years and preliminary charts of the surface have been published (12) together with extremely preliminary interpretations. Little other information is available (of a geologic nature) about the Galilean satellites.

The origin of the inner satellites of Jupiter is uncertain, but they are relatively large, close to the planet, and have circular orbits in the plane of rotation of Jupiter. The suggestion has been made by Kuiper (13) that the outer satellites are captured bodies. The capture problem for Jupiter has been studied by Bailey (14), who also concludes that satellites 6 throuth 12 (numbered in order of increasing distance from the planet) were captured. These satellites are much farther from Jupiter, their orbits have greater inclinations and eccentricities, and several of them are retrograde. These outer satellites probably are captured asteroids or possibly even comets.

SATURN

This huge planet is very far away from the Earth and is thus extremely difficult to study through Earth-based telescopes. It also is thought to be very rich in volatile elements such as hydrogen and helium. It has the unique rings that contain water frost (15), but little else of geological interest is known about Saturn except for the presence of its moons. Saturn is believed to have 10 moons, the largest of which is named Titan and is approximately the size of the planet Mercury. Titan has been observed to have absorption bands of methane in its own atmosphere. The most recently discovered satellite of Saturn was found in 1966 by the French astronomer Dollfus, who named the new moon Janus.

URANUS, NEPTUNE, AND PLUTO

The outermost planets of the Solar System are very far away indeed. Little is known concerning these planets except for their motions and approximate sizes. Uranus is known to have five moons and Neptune two, but even their approximate sizes are open to question. Triton, the innermost moon of Neptune is believed to be approximately 4000 km in diameter. Pluto is not known to have any moons, and it has been speculated that Pluto itself may be a captured comet or an escaped moon of Neptune that is now in a heliocentric orbit.

REFERENCES AND NOTES

1. Murray, B. C., M. J. S. Belton, G. E. Danielson, M. E. Davies, D. E. Gault, B. Hapke, B. O'Leary, R. G. Strom, V. Suomi, and N. Trask (1974) Mercury's surface: Preliminary description and interpretation from Mariner 10 pictures: Science, vol. 185, p. 169-179; see also other papers in this volume. Also see Jour. Geophys. Res., vol. 80, no. 17, p. 2341-2514, for interpretations of Mercury imagery through the first few months of 1975.

2. Kaula, W. M. (1968) An introduction to planetary physics: The terrestrial planets: Wiley, New York, 490 p. See also Reynolds, R. T., and A. L. Summers (1969) Calculations on the composition of the terrestrial planets: Jour. Geophys. Res., vol. 74, p. 2494-2511.

3. Klaasen, K. P. (1975) Mercury rotation period determined from Mariner 10 photography: Jour. Geophys. Res., vol. 80, p. 2415-2416.

4. Gault, D. E., J. E. Guest, J. B. Murray, D. Dzurisin, and M. C. Malin (1975) Some comparisons of impact craters on Mercury and the Moon: Jour. Geophys. Res., vol. 80, p. 2444-2460.

5. McCord, T. B., and J. B. Adams (1972) Mercury: Surface composition from the reflection spectrum: Science, vol. 178, p. 745-747.

6. Murray, B. C., M. J. S. Belton, G. E. Danielson, M. E. Davies, D. Gault, B. Hapke, B. O'Leary, R. G. Strom, V. Suomi, and N. Trask (1974) Venus: Atmospheric motion and structure from Mariner 10 pictures: Science, vol. 183, p. 1307-1315. See also other papers in this same issue, p. 1289-1321.

7. Rumsey, H. C., F. A. Morris, R. R. Green, and R. M. Goldstein (1974) A radar brightness and altitude image of a portion of Venus: Icarus, vol. 23, p. 1-7.

8. See several articles reporting Pioneer 10 results in Science, vol. 183, no. 4122, 25 January, 1974; also, a color image of Jupiter is on the cover of this issue.

9. Pilcher, C. B., S. T. Ridgway, and T. B. McCord (1972) Galilean satellites: Identification of water frost: Science, vol. 178, p. 1087-1089.

10. Brown, R. A., and F. H. Chaffee (1974) High resolution spectra of sodium emission from Io: Astrophys. Jour., no. 187, p. L125-L126.

11. Judge, D. L., and R. L. Carlson (1974) Pioneer 10 observations of the ultraviolet glow in the vicinity of Jupiter: Science, vol. 183, p. 317-318.

12. Katterfield, G. N., and E. I. Nesterovitch (1971) Nature and topography of the Galilean moons of Jupiter: Modern Geology, vol. 2, p. 41-48.

13. Kuiper, G. P. (1956) Satellites of the outer planets: Vistas Astron., vol. 2, p. 1631.

14. Bailey, J. M. (1971) Origin of the outer satellites of Jupiter: Jour. Geophys. Res., vol. 76, p. 7827-7832.

15. Pilcher, C. B., C. R. Chapman, L. A. Lebofsky, and H. Kieffer (1970) Saturn's rings: Identification of water frost: Science, vol. 167, p. 1372-1373. See also Kieffer, H. H. (1974) Ring particle sizes and composition derived from eclipse cooling curves and reflection spectra: in The Rings of Saturn, F. D. Palluconi and G. H. Pettengill, eds., NASA, Spec. Pub., SP-343, p. 51-63.

SUGGESTED READING AND GENERAL REFERENCES

Journal of Geophysical Research, vol. 80, no. 17, p. 2341-2514, June 10, 1975. This is the report of the Mariner 10 investigator teams for the planet Mercury. The preliminary report appeared in Science, vol. 185, no. 4146, p. 141-180, July 12, 1974.

Science, vol. 183, no. 4131, p. 1289-1321, March 29, 1974. The preliminary reports of the Mariner 10 experimenter teams for the Venus encounter.

Science, vol. 188, no. 4187, p. 445-477, May 2, 1975. Jupiter data and interpretations for both Pioneer 10 and 11 are included in this issue. Science, vol. 183, no. 4122, January 25, 1974 should be consulted for papers including only Pioneer 10 results.

Hartmann, W. K. (1972) Moons and planets: An introduction to planetary science: Bowden and Quigley, Belmont, Calif., 404 p.

Glasstone, S. (1965) Sourcebook on the Space Sciences: D. Van Nostrand, Princeton, N.J., 937 p. Somewhat out of date, but a good starting place for older observations.

Kaula, W. M. (1968) An Introduction to Planetary Physics: The Terrestrial Planets: Wiley, New York, 490 p.

Palluconi, F. D., and G. H. Pettengill, eds., (1974) The rings of Saturn: NASA, Special Publication, SP-343, 222 p.

Scientific American, vol. 233, no. 3, p. 22-173, September, 1975. An issue devoted mostly to summarizing current information and ideas about the Solar System and various objects in it, elementary and easy to read.

The present form of things, and the exact numerical determinations of their relations, has not hitherto been able to lead us to a knowledge of the past states, or a clear insight into the conditions under which they (planets) originated. These conditions must not, however, on that account be called accidental, *as men call everything whose genetic origin they are not able to explain.   Alexander von Humboldt, 1850*

11. Comparative Planetology

INTRODUCTION

We now have good imagery and a scattering of geophysical measurements from all of the terrestrial planets and one large moon. There is a strong tendency to develop general theories to account for many of the planetary properties and features observed. However, this developing science of planetology is still in its formative stages. The data are still too new and incomplete to be fully interpreted. There is a large number of possible variables, for example, heterogeneity in the Solar Nebula, position in the Solar System, bulk density and composition, amount of differentiation, size, moments of inertia, intensity and nature of magnetic field, radioactive element content, surface ages, impact history, tectonic styles, and so forth. Many of these (and other) variables are, or must be, interrelated in ways that range from very simple to exceedingly complex. However, our data are far from satisfactory or comprehensive, and the total sample population of planets within our present reach is small. The following qualitative discussions illustrate the *kinds* of problems that must be attacked.

BULK COMPOSITIONS

Our data for the Earth and Moon are considerably better than for the other planets, but are still only approximate. It should be remembered that even the composition of the deep interior of the Earth is not known from direct observation or sampling, but is inferred from seismic velocities, together with other seismic wave propagation characteristics, assumed Solar abundances, and possible analogies with certain meteorites. Much of the deep interior of the Earth is in pressure regimes where the phase changes, compositions, and identities are neither understood nor susceptible to experimental investigation with current technology. How, then, can we hope to compare the bulk compositions of planets if even our data about the Earth are poor? Fortunately, the situation is not hopeless for first-order comparisons. Orbital mechanics has given us the masses of planetary bodies, and direct telescopic observations and spacecraft imagery have yielded the diameters; thus, the bulk densities are known with fair precision (Table I).

As is frequently the case, a number of simplifying assumptions are necessary. Many authors have justified the use of bulk densities to estimate iron to silicate ratios in the terrestrial planets. This is based on the presumed abundance of iron, silicon, and oxygen in the Solar Nebula, as inferred from the presently assumed composition of the Sun and the analyzed compositions of chemically primitive meteorites, such as Type I carbonaceous chondrites. Thus, for planets of comparable size, one with a greater bulk density should have a larger proportion of iron to silicate than one with a lesser density, and so forth. For the terrestrial planets, the Moon is the smallest object and also has the least bulk density. Hence, it must contain a very small proportion of iron compared with the larger terrestrial planets, especially Mercury, Venus, and the Earth. It appears on the basis of bulk density alone that the Moon and Mars are one chemical class of object and that the Earth, Venus, and Mercury are rather different. The bulk chemistries of the Earth and Venus may be quite comparable owing to their very similar sizes and bulk densities, especially if the densities are corrected slightly for compression because of

Table I   Diameters, Masses, and Bulk Densities of the Terrestrial Planets[a]

planet	diameter (km)	mass (gm)	bulk density (gm/cm³)
Mercury	4,864	3.30×10^{26}	5.5
Venus	12,100	4.87×10^{27}	5.2
Earth	12,756	5.98×10^{27}	5.52
Moon	3,476	7.35×10^{25}	3.34
Mars	6,788	6.44×10^{26}	3.9

[a]After Hartmann (1).

the different masses. Mercury appears to be unique in that its bulk density is quite high and its size is rather small. Mercury should have the greatest proportion of iron to silicates and also perhaps the greatest relative volume of core to lithic crust (if the planet is well differentiated) of all of the terrestrial planets.

The bulk densities of the outer planets (ranging from about 0.7 to 1.6 gm/cm^3) are much less than those of the terrestrial planets, with the possible exception of Pluto whose remote location contributes to considerable uncertainties in the calculated density. Bulk density values of the outer planets may generally reflect proportions of iron plus silicates to hydrogen plus helium, ices, clathrates, and other volatiles. However, some of the moons of the outer planets appear to have bulk densities similar to terrestrial planets.

DEGREE OF DIFFERENTIATION

If we have reason to assume that a planetary body has a certain bulk composition, but we find that the surface rocks are *not* of that composition, the observation is extremely suggestive that the body is, at least, partially differentiated. The degree of differentiation is closely related to the interior thermal state and history, radioactive element content, rate and homogeneity of accretion, and time since accretion, as well as other possible factors. A good measure of the variation of density with depth within a planet, which should also be a function of the degree of differentiation, is the value of the moment of inertia ratio I/MR^2, where I is the mean moment of inertia, M is the mass, and R is the radius. Observations of the surface chemistry and seismic velocities within the Moon lead Kaula et al. (2) to conclude that the moment of inertia ratio I/MR^2 for the Moon is approximately 0.395. This value is not greatly different from the value that would be obtained for a sphere of uniform density distribution (0.4). It is therefore concluded that the Moon cannot have a large iron core, probably no more than 400 km in radius (<3% of the Moon's mass), and their preferred model would have no iron core at all (Kaula et al., 2). They further point out that the abundant chemical data that now exist for the Moon constrain the observational moment of inertia to a greater degree than the moment of inertia constrains the composition.

It also is apparent from numerous surface and shallow subsurface observations, as well as deep seismic data and the bulk density, that the Earth is a highly differentiated planet. Therefore, the moment of inertia (0.33) is not of prime importance to the argument, but is simply another indication of a substantial density increase toward the center of the Earth. However, such detailed observations do not yet exist for the other planets, and the moment of inertia is likely to be a critical data point in establishing the approximate degree of planetary differentiation.

INTERNAL STATE

A number of measurements contribute to the understanding of the internal state of a planet. Such measurements as heat flow are especially valuable when they can be combined with information derived from analyses of surface rocks, seismic data, and age dates. It may then be possible to attack problems such as thermal history of the planet, radioactive element content, times of massive magma generation, rate of accretion, and strength history of the body. Again, a fine example is the Moon. The Apollo 15 heat flow experiment measured the heat flow from the interior of the Moon as approximately 3.3×10^{-6} watts/cm^2 (see The Moon, page 221), and similar values have been obtained at the Apollo 17 site. These values are approximately one half the average terrestrial heat flow and are surprisingly large quantities for such a small planet as the Moon. If these data are representative of a large portion of the Moon, Langseth et al. (3) conclude that the radioactive element content of the Moon must be greater than that of the Earth or chondritic meteorites. These data combined with seismic observations (see The Moon, page 221) tend to support a hot deep lunar interior with possible partial melting of silicates at depths of approximately 1000 km and more. However, the observation that the mascons are not completely isostaticly compensated, even though the circular mare basin formations and floodings are old events, indicates that the outer shell of the Moon must be quite strong.

The heat flow of the Earth has been measured extensively and the seismicly active Earth has provided many energetic events that can be used to probe the deep interior of the planet. Detailed model cross sections of the structure of the Earth's interior and the approximate geothermal gradient can be found in most geology and geophysics textbooks. The observation of active volcanoes on the Earth's surface indicates that there are some locations at which the surface heat flow is very high indeed, in contrast with the surface of the Moon.

TECTONIC STYLES

Surface imagery and geophysical measurements show that the tectonic "styles" of the Earth, Moon, Mercury, and Mars are substantially different. Because of the dense atmosphere and cloud cover, it is difficult to make a supportable statement about Venus at the present time.

After a labored and tedious group of observations and arguments, and the passage of many years, it is now clear that the physical results of continental drift and sea floor spreading are well developed features of the crust and mantle of the Earth. These processes together with their accompanying fold-belt mountain ranges and island arcs do not appear to be duplicated to the same degree elsewhere among the terrestrial planets. However, the continent/ocean basin dichotomy may not be uniquely terrestrial. Mars may have a somewhat similar dichotomy as represented by the relatively low and smooth northern hemisphere of the planet as opposed to the relatively high and cratered southern "highlands." Plate tectonics may also have been active on Mars, but this matter remains to be resolved. The highlands and mare-flooded basin style of tectonism is best displayed on the Moon, but is also shared to a lesser extent by Mars. The great volcanic shields and giant canyons appear to be uniquely Martian. The immensely long and prominent scarps of Mercury differ from those of Mars and have been observed nowhere else. There are a few common threads, such as the ubiquitous impact craters that populate the surfaces of the terrestrial planets, although even these have been erased somewhat on the surface of the Earth. The terrestrial planets, at least, seem to have certain compositional similarities as do the outer planets. What are the fundamental physical/chemical reasons for the observed differences and similarities?

Obviously the histories of formation and general principles by which these planets have evolved must be tied together by the interrelations of the variables mentioned (and many other variables not mentioned) in the introduction to this brief chapter. The solutions to the interrelations between variables and the recognition and application of the now poorly understood general principles of planetology will require geologists and geophysicists of broad training and even broader interests.

REFERENCES AND NOTES

1. Hartmann, W. K. (1972) Moons and Planets: An Introduction to Planetary Science: Bowden and Quigley, Belmont, Calif., 404 p.
2. Kaula, W. M., G. Schubert, R. E. Lingenfelter, W. L. Sjogren, and W. R. Wollenhaupt (1974) Apollo laser altimetry and inferences as to lunar structure: Proc. Fifth Lunar Sci. Conf., Geochim. et Cosmochim. Acta, Suppl. 5, vol. 3, Pergamon Press, p. 3049-3058.
3. Langseth, M. G., Jr., S. P. Clark, Jr., John Chute, Jr., and Stephen Keihm (1972) The Apollo 15 lunar heat flow measurement: Lunar Sci. III, C. Watkins, ed., Lunar Sci. Inst. Contr. no. 88, p. 475-477. Also, Langseth, M. G., J. L. Chute, and Stephen Keihm (1973) Direct measurement of heat flow from the Moon: Lunar Sci. IV, J. W. Chamberlain and C. Watkins, eds., Lunar Sci. Inst., p. 455-456.

Glossary

The purpose of this abbreviated glossary is to provide help to the readers of this book who do not have all of the assumed technical background. Many of the definitions here are neither complete nor, in some instances, *strictly* technically accurate. For more exhaustive and complete definitions, the reader should consult the standard geological glossaries and other reference works. Additional terms are defined in the text where they are first used.

Accretion. The process whereby small particles and gases in the Solar Nebula came together to form larger bodies, eventually of planetary size.

Aeolian (syn. eolian). Referring to wind, usually wind deposited sediments or to surface features sculptured by wind blown particles.

Agglutinate (lunar). A term for certain particles in the lunar regolith that are held together by and largely composed of glass, probably spatter and melted ejecta from small hypervelocity impacts on the lunar surface, together with mineral grains and small rock fragments.

Albedo. The amount of light reflected by a surface; reflectance.

Aliphatic. Pertaining to organic compounds of relatively high molecular weight whose structures are open chains, such as paraffins.

Amorphous. No long-range atomic structure, lacking crystal structure; without form.

Anhedral. Without crystal form; missing crystal faces and polyhedral angles.

Anorthite. See plagioclase; a plagioclase feldspar group mineral with composition close to $CaAl_2Si_2O_8$.

Anorthosite. An igneous rock composed almost entirely of calcium-rich plagioclase (anorthite).

Anticline. A geological structure composed of rock strata that have been deformed such that originally horizontal rocks are convex upwards.

Apatite. A mineral that occurs in trace amounts in many terrestrial and lunar rocks, composed chiefly of calcium phosphate but with many minor constituents.

Aphelion. That point in the orbit of a planet, asteroid, comet, or other solar satellite farthest from the Sun.

Aromatic.   Pertaining to a group of organic compounds whose structures contain an unsaturated ring of carbon atoms, such as benzene and naphthalene.

Asterism (X-ray mineralogy).   An increase in the size, especially elongation, of X-ray diffraction spots caused by fine scale damage to crystal structure.

Astrobleme.   Impact structure; a geologic structure caused by the hypervelocity impact of a meteoroid, asteroid, or comet.

Astrogeology.   A term used to describe the more or less geological areas of lunar and planetary science; the space-related aspects of geology; syn., space geology, planetary geoscience.

A.U.   Abbreviation for astronomical unit, a measure of distance equal to the mean distance from the Earth to the Sun; approximately 1.496×10^8 km or 93 million miles.

Autometamorphism.   Metamorphism by its own heat; metamorphism caused by the heat contained in a rock unit as a result of its mode of origin, as in ash flow tuffs or impact base surge and fallback deposits.

Baddeleyite.   A mineral composed of monoclinic ZrO_2. Formed in impactites as one decomposition product of zircon.

Basalt.   A fine grained extrusive or shallow intrusive igneous rock of similar mineralogical composition to gabbro (chiefly plagioclase and clinopyroxene with or without olivine and minor amounts of other minerals).

Base surge.   Relatively low velocity ejecta from an impact or volcanic explosion crater that travels as a surface clastic debris flow radially away from the crater.

Breccia.   A clastic rock composed of angular fragments of rocks and/or minerals set in a finer grained matrix.

Bronzite.   An orthopyroxene mineral composed chiefly of $MgSiO_3$, but with a substantial amount of Fe substituting for Mg; the name derives from its "bronzy" color.

Caldera.   A volcanic collapse crater commonly developed in the central part of a volcanic construct, for example, the central composite collapse crater on Olympus Mons, Mars or Crater Lake, Oregon.

Chondrite.   A class of stony meteorite characterized by the presence of chondrules, except for the Type I carbonaceous chondrites which contain no chondrules but are grouped with the chondrites for chemical reasons.

Chondrules.   Spherical to subspherical silicate bodies found in certain stony meteorites and in a few lunar samples. Chondrules range in size from more than 10 to less than 1 mm and mostly are composed of glass, olivine, pyroxene, plagioclase, or some combination of these.

Chromite.   A chromium-rich spinel mineral; see spinel.

Clastic.   Containing clasts or fragments of preexisting rocks or metamorphosed fragments of preexisting rocks; as in clastic rock.

Clathrate.   A latticelike chemical structure in which one molecular group is surrounded or enclosed by others; for example, H_2O by CO_2.

Clinopyroxene.   A subgroup of the pyroxenes that have monoclinic symmetry; including diopside, hedenbergite, pigeonite, and augite.

Coesite.   A high pressure mineral polymorph of silicon dioxide formed usually from quartz during the brief very high pressures that occur during hypervelocity impacts, for example, at Meteor Crater, Arizona where the first natural occurrence was discovered.

Cohenite.   A minor mineral constituent of some meteorites with composition $(Fe,Ni,Co)_3C$; syn., cementite (for the artificial).

Crater density.   The number of craters, usually with crater sizes specified, per unit area; an indicator of the relative age of portions of planetary surfaces when properly used and only primary impact craters are counted.

Cristobalite.   A mineral polymorph of silicon dioxide that commonly occurs in certain terrestrial volcanic rocks and as a minor component of some meteorites and lunar samples.

Cryptovolcanic.   A term used to describe geologic structures that are thought by some workers to originate through an unknown explosive volcanic process; used to describe geologic structures, mostly containing disrupted rock strata and/or large breccia blocks, that have unknown or unproven modes of origin.

Crystal habit.   The general form or shape that a crystal tends to have, described by such terms as tabular, ascicular, bladed, equant, radiating, etc.

Cumulate.   A term for an igneous rock that is formed by the accumulation of crystals from a magma, usually by density difference — settling downward or floating upward, generally recognized by its texture.

Daubreelite.   A mineral that is a minor constituent of some meteorites, composition is $FeCr_2S_4$.

Diabase.   An igneous rock composed mostly of calcium-rich plagioclase (commonly labradorite) and clinopyroxene, characterized by ophitic texture.

Diamond.   A naturally occurring isometric polymorph of carbon which has great hardness and is stable at high pressure and temperature.

Diaplectic glass.   Glass formed in the solid state from a preexisting mineral by the action of a shock wave.

Differentiation (magmatic).   The general process whereby a homogeneous silicate magma is separated into fractions of different bulk composition; (planetary) the process whereby a relatively homogeneous planetary body becomes radially inhomogeneous, for example, develops a core and/or crust or other layers of different bulk composition.

Diktytaxitic.   A textural petrologic term used to describe igneous rocks that have a framework of crystals with little goundmass and numerous irregular voids; an igneous rock with irregular vugs with numerous crystals partially protruding into them.

Diopside.   A clinopyroxene mineral of the approximate composition $CaMg(SiO_3)_2$.

Dunite.   An igneous rock composed almost entirely of the mineral olivine with possible minor amounts of pyroxene, magnetite, chromite, and others.

Eccentricity (gen. astron.).   The departure of an orbit from a perfect circle; an orbit that is a long ellipse would be described as having "high" eccentricity, a nearly circular orbit would have a "low" eccentricity.

Ejecta blanket.   The layer of fallback, base surge, and other ejecta surrounding and within an impact crater and/or its associated structures, commonly thicker near the crater rim and thinning radially away from the impact site.

Electron microprobe.   An analytical instrument that can be used to analyze very small portions of minerals in polished section. The device impinges a finely focused beam of electrons on the mineral, which excites the characteristic X-radiation of the elements that compose it. The wavelengths of the fluorescent X-rays are determined by a spectrometer or detector and the intensities compared with standard samples, or the composition may be computed directly with various intensity corrections.

Enstatite.   An orthopyroxene mineral of the approximate composition $MgSiO_3$.

Escape velocity.   The velocity required for an object to escape the gravitational control of a planetary body; velocity required to change orbit from one primary to another.

Euhedral.   Good crystal form or presence of most crystal faces and polyhedral angles.

Extinction (min.).   The property of transpararent minerals *not* to transmit light, as viewed between crossed polarizers, in certain or all orientations, which depend on the crystal structure and composition of the mineral. Extinction is said to be "normal" if the intensity of the light is gradually and regularly diminished as the mineral is rotated between the polarizers, and "undulatory" if the transmitted intensity decreases irregularly and in patches.

Extraterrestrial.   Outside the Earth, not pertaining to the Earth or its materials.

Fall.   A meteorite that was observed, or whose accompanying phenomena were observed, during its passage through the atmosphere or during its impact with the Earth's surface.

Fallback.   Ejecta from impact explosion craters that is thrown into the atmosphere or near surface planetary space at less than escape velocity and falls back to the surface within or close to the crater; throwout is another term sometimes used for the ejecta that is deposited outside the crater.

Fault.   A physical break in rocks across which there has been movement of one side relative to the other.

Fayalite.   Iron-rich olivine; see olivine.

Feldspar.   A name for a very large and common group of silicate minerals to which aluminum and various monovalent and divalent cations also are essential components. The most common feldspars are the plagioclase group (see plagioclase) and the potassium feldspars, of which there are several minerals (orthoclase, microcline, sanidine). The most abundant group of minerals in the crust of the Earth.

Find.   A meteorite that is recognized by its physical properties, appearance, or peculiar location but whose exact time of fall is unknown or was not observed.

Fines.   A term used to designate the less than 1 mm fraction of the lunar regolith; generally, fine grained particles.

Fladen.   Impact-produced glassy ejecta that is sufficiently plastic to deform when it falls back to the surface.

Flux curve (meteoroid).   A curve that illustrates the size frequency distribution of meteoroids, either in sizes of meteoroids or sizes of craters produced, per unit area for a specified exposure time.

Fold.   A flexure or bending of planar rock strata or structures.

Forsterite.   Magnesium-rich olivine; see olivine.

Fulgurite.   A glassy fusion of terrestrial rocks resulting from lightning strikes, commonly hollow tubes or coatings on rock surfaces.

Fusion crust.   The remelted and cooled outer layer of a meteorite caused by atmospheric friction followed by cooling, mostly only a few millimeters or less in thickness, commonly black on fresh specimens and brown on weathered meteorites; ablation skin.

Gabbro.   An igneous rock composed chiefly of the minerals plagioclase (calcium-rich) and clinopyroxene; may also contain olivine, orthopyroxene, and lesser amounts of other minerals.

Gas chromatograph.   An analytical instrument that can be used to identify gases and volatile molecular species by their residence times in adsorption columns of various lengths, diameters, and materials.

Gibbous.   A term used to describe a phase of a planet or moon in which more than half, but not all, of the illuminated disk can be viewed by the observer.

Glass.   Solid material of wide range of possible chemical compositions and no long range atomic structural order.

Graben.   A down-dropped fault block, commonly elongate, between two upthrown (relative) blocks or stable regions.

Granular (petrologic).   A textural term for a rock with grainy texture, commonly composed of minerals of approximately equal size and rather equant dimensions.

Graphite.   A soft, black mineral polymorph of carbon.

Grossularite.   A mineral of the garnet group with composition close to $Ca_3Al_2 (SiO_4)_3$; grossular.

Groundmass.   Matrix, fine grained material in which occur larger crystals or clasts.

Harrisitic.   A textural petrographic term used to describe an igneous rock with olivine crystals, mostly larger than the other minerals present, whose long axes are oriented more or less perpendicular to the cumulate texture of the rock.

Hedenbergite.   A clinopyroxene mineral of the approximate composition $CaFe(SiO_3)_2$.

Hypersthene.   An orthopyroxene mineral of the approximate composition $(Mg, Fe)SiO_3$.

Hypervelocity.   Exceeding ordinary velocities; very high velocity; of sufficient velocity to make an explosion crater when the body intersects the surface of a planet. The minimum is assumed usually to be about 5 km/sec.

Igneous rock.   Rock that has formed from the cooling of magma.

Ilmenite.   A mineral composed of $FeTiO_3$, slightly magnetic, an abundant mineral on the Moon and common but less abundant at the Earth's surface.

Impactite.   Glassy fused or partially fused rock that has been formed by shock metamorphism accompanying meteorite impact.

Impact melt.   A portion of the target material that is melted by the impact of a hypervelocity projectile; rock melt produced on or within a planetary body by the hypervelocity impact of a meteorite, asteroid or comet.

Inclination (gen. astron.).   The angle that the orbital plane of a secondary body or satellite makes with the orbital plane or equatorial rotational plane of the primary.

Intersertal.   A petrologic textural term used to describe igneous rocks in which small amounts of glassy groundmass or mesostasis occur between larger crystals that compose the major portion of the rock, especially if the larger crystals are unoriented feldspars.

Intrafasciculate.   A textural petrologic term for plagioclase crystals that contain an elongate core of pyroxene.

Isentropic.   Maintaining the same entropy.

Isochron.   A line constructed by plotting isotopic ratios that connects points of equal age or time, or whose slope can be interpreted in terms of the last time of isotopic equilibration of the analyzed materials.

Isostatic compensation.   The process whereby rocks of different density and/or with different surface topography approach buoyant equilibrium; a balance of rock mass and density, analogous to ice floating in water.

Isotropic (min.).   A mineral or substance that does not transmit light when viewed in any position between crossed polarizers.

Joint.   A physical break in rocks across which there has been no relative movement of the two sides.

Kamacite.   Naturally occurring body centered cubic alpha iron (syn., ferrite); commonly occurs in meteorites.

Kilobar.   One thousand bars; a unit of pressure commonly used in shock metamorphism and crater mechanics.

Kink bands.   Angular changes across planar interfaces in minerals with good cleavage, notably micas, that offset the traces of cleavages in repeated chevronlike patterns.

Kurtosis.   A measure of the peakedness of a distribution curve; the terms leptokurtic (very peaked), mesokurtic (moderately peaked), and platykurtic (not very peaked) are commonly used.

Lechatelierite.   Natural silica glass, amorphous SiO_2.

Libration.   A minor motion (physical libration) of a planetary body; an apparent motion (geometric libration) caused, in the case of the Moon, by a slight eccentricity of the Moon's orbit about the Earth.

Liquidus.   The line in a temperature-binary composition phase diagram, usually at a given pressure, that indicates the temperature at which crystals of a specific mineral phase and composition begin to precipitate out of a melt under equilibrium conditions; a similar curved surface in a ternary composition phase diagram; a curve that separates a field of liquid stability from a field in which liquid plus crystals are stable.

Magma.   Molten or partially molten naturally occurring material from which igneous rocks crystallize as the temperature of the melt decreases; normally rich in silica and containing dissolved gases.

Magnetite.   An opaque, black, metallic mineral of the spinel group with composition close to Fe_3O_4, strongly ferromagnetic.

Magnetometer.   A geophysical instrument used to measure the intensity and/or direction of a planetary magnetic field or some portion or component thereof.

Mascon.   Mass concentration.

Maskelynite.   Diaplectic plagioclase glass, now known to occur in association with terrestrial and lunar impact craters and in highly shocked meteorites.

Mass spectrometer.   An analytical instrument that can be used to determine the amounts of various isotopes and molecular species present in a sample by separating them according to their mass to charge ratio as ions in motion in a strong magnetic field.

Mass wasting.   A term for several processes by which rock material moves downslope under the influence of gravitational force.

Megabar. One million bars; a unit of pressure commonly used in shock metamorphism and crater mechanics.

Melilite. A group of minerals composed of sodium, calcium, magnesium aluminum silicates; includes akermanite and gehlenite.

Mesostasis. The glassy and fine grained interstitial material that occupies the small spaces between larger crystals in some igneous rocks.

Metamorphism. A process, usually involving a change in temperature and/or pressure, whereby a rock or mineral is changed in texture or to a new mineral or assemblage of minerals that is stable under new physical/chemical conditions; change form.

Meteor. The bright streak briefly seen across a portion of the sky as a small meteoroid is heated to incandescence in its flight by atmospheric friction.

Meteorite. The iron and/or stony portion of a meteoroid that survives atmospheric passage and reaches the surface of the Earth; a fragment of a meteoroid that has survived atmospheric passage and impact on a planetary surface.

Meteoroid. A solid object in interplanetary space within the Solar System, probably a fragment of an asteroid or comet.

Microcline. A potassium feldspar with composition close to $KAlSi_3O_8$, but commonly containing minor amounts of sodium and even less calcium. Differs from orthoclase in the symmetry of its crystal structure.

Mineral. A naturally occurring inorganic substance with a definite range of chemical composition and a specific atomic structure with long-range order. Mineral phase is a synonym.

Monomict. Composed of clasts of one type; similar components; as in monomict breccia that is clastic rock composed of angular fragments of the same parent rock.

Nickel-iron. Native iron occurring in meteorites as kamacite and taenite; most commonly with a few percent nickel but rarely with as much or more than 50 percent.

Nicol. A specially constructed two-piece prism that transmits plane polarized light in a known orientation; polarizer.

Norite. An igneous rock composed mostly of calcium-rich plagioclase with orthopyroxene, most commonly hypersthene, as the chief mafic mineral.

Occult (astron.). To interpose a third body between the subject and the viewer; for example, during a solar eclipse the Moon is occulting the Sun; to hide from view.

Oligoclase. A plagioclase feldspar with composition in the range of An_{10} to An_{30}; see plagioclase.

Olivine. A mineral composed of $(Mg,Fe)_2SiO_4$, in which magnesium and ferrous iron can be present in any proportion; a mineral solid solution series between Mg_2SiO_4 (forsterite) and Fe_2SiO_4 (fayalite) in which the molecular percent of the two end members can vary in any proportion and commonly is designated by such notation as Fo_{85}, indicating forsterite 85 mole percent; or Fa_{15}, indicating fayalite 15 mole percent for the same composition.

Ophitic. An igneous rock texture in which crystals of one mineral totally enclose crystals of another, most typically plagioclase enclosed by clinopyroxene.

Orbit. The stable curved path through the Solar System followed by a solid body under the gravitational influence of the Sun and/or other large bodies.

Orographic. Pertaining to mountains and their affects, especially as in "orographic lift" in which relatively warm air with a greater water vapor content is raised to higher elevations because of flow over a mountain; the lower temperature at the higher elevation may cause precipitation of the water vapor as ice crystals or water droplet clouds.

Orthoclase. A potassium feldspar with composition close to $KAlSi_3O_8$, but commonly containing minor amounts of sodium and even less calcium. Differs from microcline in the symmetry of its crystal structure.

Orthopyroxene. A subgroup of the pyroxenes that have orthorhombic symmetry of their atomic structure; including enstatite, bronzite, and hypersthene.

Paragenesis. The chronologic order of formation, especially of minerals; the temporal, chemical, and mineralogical relations between minerals or rock types.

Parent body. The source body from which a smaller object originates, especially in relation to meteorites; the planet, asteroid, comet, or larger meteoroid from which a meteorite or other matter originally was derived.

Perihelion. The point in the orbit of a planet, asteroid, comet, or other solar satellite that is closet to the Sun.

Perovskite. A mineral with composition close to $CaTiO_3$.

Petrogenesis. The origins of rocks, the interrelations between different rock types and how they form.

Phase. A substance with a definite range of chemical composition and a definite atomic structure. Naturally occurring phases are called mineral phases or simply minerals.

Phase angle. Generally the angle defined by the path from a source to a reflecting surface to an observer, as in the Sun-Moon-Earth.

phi (ϕ) scale. A logarithmic particle size (diameter) scale on which $0\phi = 1$ mm, $1\phi = 500$ μm, $2\phi = 250$ μm, $4\phi = 62.5$ μm, etc.; in common use for grain size measurements of both terrestrial and lunar particles; phi size $= -\log_2$ diameter in mm.

Plagioclase. A group of feldspar minerals that ranges in composition from $CaAl_2Si_2O_8$ to $NaAlSi_3O_8$ forming a complete solid solution series at high temperature. Anorthite is the calcium-rich end member in this series and albite is the name for the sodium-rich end member. Other mineralogical names for members of this series of intermediate composition are labradorite, bytownite, oligoclase, and andesine. The composition of a plagioclase commonly is designated by the notation An_{35}, An_{60}, Ab_{21}, Ab_{90}, etc., indicating the molecular percent of one end member or the other. Thus a plagioclase feldspar whose composition is indicated by An_{65} contains 65 mole percent anorthite ($CaAl_2Si_2O_8$) and 35 mole percent albite ($NaAlSi_3O_8$). For the same example the designation Ab_{35} would be equally informative.

Planar features (syn., planar structures, planar elements). Sets of parallel, rationally oriented planes developed in such minerals as quartz and feldspar as a result of shock metamorphism; most probably are a glass phase derived from the host mineral or from a transitory phase during the passage of the shock wave.

Planetesimal. A presumed class of solid body of subplanetary dimensions which accreted with similar bodies to form planets.

Platykurtic.   A term used to describe the kurtosis of a size frequency distribution meaning that the distribution is not very peaked.

Plessite.   A fine-grained intergrowth of kamacite and taenite in meteorites, formerly thought to be a distinct mineral.

Poikilitic.   A textural petrologic term used to describe igneous rocks in which a crystal of one mineral encloses smaller crystals of another.

Polarization (astron.).   The specific polarization properties of light reflected from a surface; used to infer physical properties of the surface.

Polymict.   Composed of clasts of more than one type; dissimilar components; as in polymict breccia which is a clastic rock composed of angular fragments of different parent rocks.

Polymorphs.   Single chemical composition but more than one atomic structural form. Examples are quartz, stishovite, tridymite, cristobalite and coesite. All are essentially silicon dioxide but they have very different atomic structures thereby giving them different physical and optical properties.

Porphyritic.   A textural petrologic term used to describe rocks that have large crystals enclosed in a matrix or groundmass of distinctly smaller crystals or glass.

Pseudotachylite.   False basaltic glass; glassy or finely crystalline dark impact melt, commonly in veins injected into country rock along fractures and mixed with rock fragments and mineral grains which may or may not show evidence of shock damage.

P-wave.   A compressional seismic wave, the "primary" seismic wave.

Pyroxene.   A large group of minerals which primarily are silicates of magnesium, iron and calcium; enstatite, augite, hypersthene, diopside and pigeonite are common pyroxenes. Commonly divided into a monoclinic subgroup (clinopyroxenes) and an orthorhombic subgroup (orthopyroxenes).

Quartz.   The most commonly occurring mineral polymorph of silicon dioxide in the crust of the Earth. Also, a minor component of some meteorites.

Rarefaction wave.   The extensional wave immediately following a compressional or shock wave.

Ray (planetary surface).   Long deposits of ejecta that extend radially away from an impact crater, may appear as light or higher albedo streaks, beautifully developed and visible at high sun angles associated with the lunar crater Tycho.

Regolith.   A layer of fragmental and unlithified detritus that may overlie a planetary surface as in the case of the Moon.

Remanent magnetization.   A component of the magnetization of a rock that is not related to recently applied or extant planetary magnetic fields of moderate strength; the component of rock magnetization "frozen in" to an igneous or impact melt rock as its ferromagnetic mineral components were cooled through their Curie points.

Retrograde.   Motion of a body in orbit that is counter to the general direction of motion of the Solar System; if the Solar System is viewed from the north, the motions of the planets go counterclockwise. But some minor bodies move in orbits that are clockwise about their primaries when viewed in the same orientation. Such motion is termed retrograde; the normal motion is termed posigrade.

Revolution. The motion in orbit of a captured planetary body about its primary body, as in the Moon revolving about the Earth, or the Earth about the Sun.

Rille. A rectilinear or curvilinear trenchlike depression.

Roche limit. The distance from the surface of a larger primary planetary body to a smaller body at which the tidal stress across the smaller body exceeds its tensile strength. Named for Edouard Roche, a French mathematician, who defined this limit based on a body with no tensile strength; therefore, a body technically has to be inside the Roche limit before it will disrupt.

Rotation. The turning of a planetary body about an axis, as in roasting a chicken on a spit or as with the rotation of the Earth on its polar axis.

Scanning electron microscope. An instrument with a carefully focused and controlled electron beam that builds up an image by measuring and displaying on a cathode-ray tube the reflected and emitted electron intensity; useful for examination of fine surface morphology.

Schlieren. Streaks, especially in glass, that are visible because of a change in color and/or refractive index.

Schreibersite. A common mineral in meteorites, especially irons, that is white to light brassy in color and is composed of $(Fe,Ni)_3P$; also called rhabdite in the older literature.

Secondary crater. A crater formed by a block or mass of ejecta thrown out of another crater.

Seismic velocity. The velocity with which seismic waves travel through rocks, generally correlates with rock type, confining pressure, rock texture, and other geological factors.

Serpentine. A small group of minerals composed primarily of hydrous magnesium and iron silicate; also, a rock name for rock composed chiefly of serpentine minerals. Commonly formed by the hydration of olivine or olivine-rich rocks.

Shatter cones. Conical rock structures, ranging in size from centimeters to several meters, with braided striations streaming down the flanks of the cone from the apex, produced by shock waves in a wide variety of lithologies under the proper conditions.

Shield (crust). A large mass of continental crustal material mostly of granitic composition that has remained intact or relatively stable for long periods of geologic time; (volcanic) a broad volcanic construct with mostly low inclination slopes, typical of large basaltic volcanoes such as Mauna Loa and Mauna Kea.

Shield volcano. A broad volcanic construct with low slope angles (commonly a few degrees) in the general shape of an ancient warrior's shield, commonly built up from many low viscosity basalt flows; for example, Mauna Loa, Hawaii.

Shock wave. A high pressure transient compressional wave that lasts for a very short time period, resulting from the impact of a hypervelocity object or an explosion.

Skewness. A measure of the symmetry of the tails of a distribution curve, positively skewed means more fine than coarse material in the tails, negatively skewed means the opposite, low skewness means about the same amount (usually weight percent) of coarse and fine particles in the tails.

Soil (lunar). A term widely used to denote the finer rock and mineral material of the lunar regolith; no similarity to terrestrial soil, either in origin or content, is implied.

Solar Nebula.   The cloud of gases and dispersed solids from which the Sun, and other objects in the Solar System, condensed and/or accreted.

Solar System.   All of the matter and interstitial space within the gravitational retention of the Sun.

Spinel.   A group of minerals with the general formula type AB_2O_4; includes magnetite ($FeFe_2O_4$) and $MgAl_2O_4$ (sometimes called ''spinel'') together with many others.

Stishovite.   A very high pressure mineral polymorph of silicon dioxide formed usually from quartz during the brief but very high shock wave pressures that occur during hypervelocity impacts. First synthesized in the laboratory by Russian investigators and now known from several terrestrial impact craters, including Meteor Crater, Arizona and the Ries Crater, Germany.

Stratigraphy.   The portion of geology that deals with the sequence of strata and time sequence of events.

Subophitic.   Not quite ophitic; less than ophitic; plagioclase feldspar crystals partially enclosed by clinopyroxene, commonly because the crystals are approximately the same size.

Suevite.   Impact breccia, especially fallback breccia, that contains glassy inclusions (impactite) as well as shocked fragments of the preexisting country rock; first applied in the Ries Crater, but now in general use.

Supercool.   To cool at a rate such that equilibrium reactions or equilibrium crystallization cannot occur at their normal temperatures; undercool.

S-wave.   A seismic shear wave.

Synchronous rotation.   A planetary motion in which the period of rotation equals the period of revolution such that the same face is always toward the primary body, as in the case of the Moon about the Earth.

Taenite.   Naturally occurring face centered cubic gamma iron (syn., austenite); commonly occurs in meteorites.

Termination.   The end of an elongated crystal that is covered by planar surfaces or crystal faces, crystals with two such ends are said to be doubly terminated.

Terrestrial.   Pertaining to the Earth, Earthlike; the inner planets (Mercury, Venus, Earth, Moon, Mars) of the Solar System that appear to be like the Earth, that is, composed of iron plus silicate for the most part.

Thetomorph.   A mineral crystal that has lost its long-range atomic order (so that it is a glass) by the solid state action of a shock wave but has still retained its crystal form. Sometimes used synonymously with diaplectic glass; euhedral diaplectic glass.

Tridymite.   A mineral polymorph of silicon dioxide that commonly occurs in certain terrestrial volcanic rocks and as a minor component of some meteorites.

Troctolite.   An igneous rock composed mostly of calcic plagioclase and olivine with little or no pyroxene.

Troilite.   A mineral composed of FeS, similar to pyrrhotite but virtually stoichiometric. A common accessory mineral in iron and most stony meteorites.

Tuff (volcanic).   A pyroclastic volcanic rock which may contain rock fragments with a wide range of sizes of both the associated volcanic rocks and other country rocks.

Twinned crystal.   A crystal having parts in different atomic structural orientation, commonly related by a simple symmetry operation.

Ulvöspinel.   A mineral with composition Fe_2TiO_4; a common occurrence is as fine exsolution lamellae in magnetite.

Variolitic.   A textural petrologic term used to describe rocks that are composed of small radiating groups of crystals in a finer grained groundmass.

Vesicle.   A gas bubble or void space, ranging in shape from spherical to irregular.

Vitrophyre.   A general term for a rock composed largely of glass.

Widmanstätten figure (also Widmanstätten structure or pattern).   The pattern of intergrowth of nickel-rich and nickel-poor mineral phases observed on a fresh surface of most iron meteorites if polished and etched with dilute nitric acid.

X-ray diffraction (min.).   The scattering of X-rays by planes of high electron, hence, high atomic, density such that constructive interferences occur that can be used to deduce the mineral identity and structure. In some cases it can be used to identify the precise mineral composition of minerals that have a substantial compositional range.

Zircon.   A mineral composed of tetragonal $ZrSiO_4$. A common mineral of low abundance in granitic rocks, also occurs in some meteorites. It is chemically very stable at low temperatures and pressures and is a persistent minor component of many terrestrial clastic sedimentary rocks.

Name Index

Abell, G., 293
Abelson, P. H., 151
Abercrombie, T. J., 129
Abu-Eid, R. M., 234
Acton, C. H., Jr., 273, 274
Adams, J. B., 288, 301, 312
Adler, I., 79, 227, 242
Aggarwal, H. A., 115, 128
Ahrens, L. H., 66
Ahrens, W., 127
Albee, A. L., 240
Albritton, C. C., Jr., 96, 125
Aldermann, A. R., 119, 129
Aldrin, E. E., 179
Alfvén, H., 279, 281, 287
Alter, D., 169
Anders, E., 64-66, 217, 233, 239, 240, 280-282, 287
Anderson, A. T., Jr., 210, 233, 238, 239
Anderson, K., 241
Anderson, W. A., 232
Annell, C. S., 218, 235
Aoyagi, M., 264, 274
Arabian American Oil Co., 121
Arizona State University, 42, 99
Aristarchus, 153
Aristotle, 289
Armstrong, N. A., 179
Arndt, J., 151, 237
Arnold, J. R., 60, 66, 242, 281, 282, 287
Arrhenius, G., 287

Arvidson, R., 267, 275
Axon, H. J., 92

Bailey, J. M., 311, 312
Baker, G., 75
Baker, M., 98, 125
Baldwin, R. B., 82, 92, 126, 130, 156, 230, 243
Ball, R. S., 1, 61
Bansal, B. M., 236
Barlow, B. C., 129
Barnes, M. A., 80
Barnes, V. E., 69, 72, 79, 80, 137, 150
Barr, E., 169
Barrell, J., 156, 230
Barringer, D. M., 98, 125
Barringer, D. M., Jr., 102, 126
Bartrum, C. O., 129
Baschek, B., 243
Batson, R. M., 273
Beals, C. S., 105, 126
Bell, P. R., 235
Belton, M. J. S., 311, 312
Bence, A. E., 63, 182, 183, 233, 239
Bertsch, W., 128
Berzelius, J. J., 14
Biggar, G. M., 236
Birch, F., 150
Bjork, R. L., 126
Bjorkholm, P., 242
Black, L. P., 240

Blander, M., 63, 239
Blau, P. J., 92
Bloch, M. R., 114
Blodget, H., 242
Bode, J. E., 277, 278
Boeschenstein, H., 66
Bogard, D. D., 235
Boon, J. D., 96, 125
Boreman, J. A., 236, 237, 239
Borgström, L. H., 28, 62
Born, G. H., 274
Bottinga, Y., 236
Bowie, S. H. U., 234
Boyd, F. R., 233
Bradley, R. S., 92
Brahe, T., 153, 289
Brandt, J. C., 285, 293
Bray, J. G., 109, 127
Braziunas, T. F., 239
Brett, P. R., 118, 129, 236
Brezina, A., 12, 30, 51, 54, 61, 62
Briesbroeck, G. v., 285
Briggs, G. A., 273, 274
Bromwell, L. G., 238
Brouwer, D., 281, 287
Brown, A. R., 129
Brown, G. M., 174, 182, 233
Brown, H., 64, 65
Brown, R., 126
Brown, R. A., 312
Brown, R. W., 235
Bucher, W. H., 95, 124
Buchwald, V. F., 64
Bullard, F. M., 91
Bunch, T. E., 64, 151, 233
Burbidge, G. R., 295
Burke, T., 274
Burnett, D. S., 240
Burns, R. G., 234
Buseck, P. R., 52, 54, 65
Butler, J. C., 35, 63, 204, 205, 234, 237-239
Butler, L. W., 127
Butler, P., Jr., 234, 236

Cable, A. J., 92
Cameron, A. G. W., 62, 293, 295
Cameron, E. N., 233, 235
Cameron, K. L., 239
Campbell, M. J., 225, 242
Carlson, R. L., 312
Carman, M. F., 35, 63, 234, 237-239

Carr, M. H., 65, 273
Carrier, W. D., III, 206, 207, 238
Carron, M. K., 72, 218, 235
Carter, J. L., 237
Carter, N. L., 64, 65
Carter, W. J., 92
Cassen, P., 241
Cassidy, W. A., 115, 126, 128
Cavarretta, G., 89, 92
Ceplecha, Z., 66
Chaffee, F. H., 312
Chalmers, R. O., 128
Chamberlain, J. W., 205, 235, 241, 319
Chao, E. C. T., 72, 76, 79, 98, 126, 127, 132, 134, 149, 150, 151, 233, 236, 237, 239
Chapman, C. R., 285, 286, 288, 312
Chapman, D. R., 71, 79
Charters, A. C., 92
Chladni, E. F. F., 51, 65
Christian, R. P., 218, 235
Chodos, A., 65, 240
Church, S. E., 236
Chute, J., Jr., 241, 319
Clanton, U. S., 237, 238
Clark, J. F., 126
Clark, S. P., Jr., 241, 319
Clarke, R. S., Jr., 17, 61, 79
Coes, L., Jr., 132, 150
Colburn, D. S., 241
Collins, M., 179
Collins, S. A., 275
Collinson, D. W., 240
Columbo, G., 287
Comerford, M. F., 65
Conrath, B., 268, 274
Cook, P. J., 117, 128
Copernicus, N., 153
Coradini, A., 92
Costes, N. C., 238
Craig, H., 11, 12, 23, 25, 34, 40, 61
Crewe, A. V., 238
Criswell, D. R., 233, 237-243
Cuffey, J., 288
Curran, R. J., 268, 274
Currie, K. L., 95, 125
Cuttitta, F., 72, 218, 235
Cutts, J. A., 259, 260, 273, 274

Dachille, F., 150
Dahlem, D. H., 231
Dainty, A., 241

Daly, R. A., 95, 125
Danielson, G. E., 311, 312
Darwin, G. H., 227, 228, 242
Davies, M. E., 273, 311, 312
Da Vinci, L., 153
Davis, C. C., 240
Dawson, J. P., 240
DeCarli, P. S., 63, 150
Dence, M. R., 97, 106, 112, 123, 124, 127,
    129, 130, 141, 233
Dept. Energy, Mines and Resources, Ottawa,
    Canada, 104-111
Deprit, A., 287
Derham, C. J., 232
Desborough, G. A., 239
De Vaucouleurs, G., 273
Dickey, J. S., Jr., 62, 236, 238
Dietz, R. S., 1, 95, 110, 111, 125, 127, 146,
    151
Dixon, D., 236
Dollfus, A., 155, 288
Döring, W., 92
Dorman, J., 240, 241
Douglas, J. A. V., 62, 233
Drake, J. C., 233
Drake, M. J., 236
Drever, H. I., 234
Duennebier, F., 240, 241
Duke, M. B., 37, 40, 41, 55, 63, 65
Dunlop, L. J., 280, 287
Dunn, J. R., 240
Duvall, G. E., 89, 92
Duxbury, T. C., 273, 274
Dwornik, E. J., 79, 218, 235
Dyal, P., 221, 241
Dzurisin, D., 312

Eberhardt, P., 235
Eggleton, R. E., 230
Ehlers, E. G., 135
El Baz, F., 230
Eldridge, J. S., 61, 187, 235
El Goresy, A., 135, 136, 150, 180, 233, 234,
    236
Eller, E., 242
Elvius, A., 287
Emanuelli, P., 229
Emmert, R. A., 232
Engelhardt, W. v., 114, 127, 128, 139, 148,
    150, 237
Englund, E. J., 63

Epstein, S., 65
Eross, B., 274
Essene, E., 228, 242
Eugster, O. J., 240
Evans, G. L., 103, 126
Ewing, M., 240, 241

Fahey, J. J., 126, 150
Faul, H., 69, 79, 80
Fechtig, H., 238
Fielder, G., 156, 230, 243
Figgins, J. D., 126
Finger, L. W., 233
Fireman, E. L., 65
Fisher, R. M., 240
Flaherty, R. E., 65
Flammarion, C., 246, 272, 285
Floran, R. J., 239
Flory, D. A., 240
Földvari-Vogl, M., 62
Folinsbee, R. E., 62
Fontana, F., 245
Foote, A. E., 98, 125
Ford, C. E., 236
Forney, P. B., 274
Fowles, C. R., 89, 92
Franklin, F. A., 287
Franzgrote, E. J., 232
Fredriksson, K., 26, 239
Freeberg, J. H., 125, 243
French, B. M., 79, 92, 93, 95, 109, 125, 127,
    128, 131, 147-151
Freyer, R. J., 231
Fritz, J. N., 92
Frondel, C., 133, 233
Frondel, J. W., 243
Fuchs, L., 233
Fujita, Y., 285
Fulchignoni, M., 92, 237
Fuller, M. D., 240
Funiciello, R., 92, 237
Funkhouser, J. G., 235

Galileo, 153, 154, 245
Ganapathy, R., 217, 233, 239, 240
Garz, T., 243
Gast, P. W., 235, 236
Gault, D. E., 82-86, 92, 231, 237, 311, 312
Gauss, K. F., 278
Geake, J. E., 232
Gehrels, T., 280, 282, 285, 287, 288

Gehring, J. W., 92
Geiss, J., 11, 235
Gentner, W., 80, 114, 127, 128, 238
George III, King of England, 277
Gerard, J., 242
Gerdemann, P. E., 95, 125
Gerstenkorn, M., 228, 243
Ghose, S., 234, 236
Gibb, F. G. F., 234
Gifford, A. C., 155, 229
Gilbert, G. K., 98, 125, 155, 229
Gigly, P., 150
Glaser, P. E., 92, 230
Glass, C. M., 93
Glasstone, S., 246, 250, 272, 275, 313
Glikson, A. Y., 129
Goldberg, E., 64
Goldsmith, J. R., 238
Goldstein, J. I., 47, 49, 52, 54, 64, 65, 92
Goldstein, R. M., 312
Golopan, K., 66
Gomez, M., 61
Goodacre, W., 169
Gordon, S. G., 63
Gorenstein, P., 242
Görz, H., 238
Gose, W. A., 217, 240
Graf, H., 235
Grant, R. W., 30, 31-33, 62
Gray, C. M., 62
Green, D. H., 233, 236
Green, J., 156
Green, R. R., 231, 312
Green, W., 274
Greenacre, J. A., 169, 232
Greene, G. M., 238
Greenland, L. P., 218, 235
Grieve, R., 210, 239
Groeneveld, I., 278, 285
Grögler, N., 235
Groschopf, P., 128
Grove, T. L., 237
Guest, J. E., 312
Gümbel, C. W., 127
Guppy, D. J., 128
Gursky, H., 242

Hackman, R. J., 156, 164, 230
Hager, D., 124
Haggerty, S. E., 233
Hait, M. H., 231

Hale Observatories, 291
Hall, A., 268, 269
Halley, E., 289
Hanel, R. A., 274
Hapke, B., 311, 312
Harmon, R. S., 235
Harris, B., 242
Hartmann, W. K., 168, 231, 262, 269, 273,
  274, 279, 312, 316, 319
Hartung, J. B., 237, 238
Hawkins, G. S., 65
Hayes, J. F., 236
Hayes, J. M., 62
Head, J. W., III, 275
Heiken, G. H., 237, 238
Helin, E., 291
Heraclides, 153
Herr, K. C., 274
Herriman, A. H., 273
Herschel, W., 169, 265, 277
Hess, D. C., 11
Hess, W. N., 230-232
Hey, M. H., 4, 67, 129
Heymann, D., 64, 66, 218, 233, 235, 236, 240,
  243
Hibberson, W. O., 233, 236
Hinners, N. H., 236
Hintenberger, H., 63
Hipsher, H. F., 245
Hirayama, K., 281, 287
Hodge, P. W., 65, 119, 129
Hodgkinson, R. J., 273
Hohenberg, C. M., 66
Hollister, L. S., 234
Holweger, H., 243
Homer, 280
Hopkins, R. M., 128
Hornemann, U., 151
Horowitz, N. H., 273
Hörz, F., 138, 151, 237
Houston, W. N., 206, 238, 239
Hovland, H. J., 239
Howard, K. A., 164
Hubbard, N. J., 196, 235, 236
Huenecke, J. C., 240
Huey, J. M., 240
Huggins, F. E., 234
Humboldt, A. v., 315
Humphries, D. J., 236
Hurley, P. M., 80
Huss, G. I., 116, 128

Huygens, C., 245
Hyde, J. R., 61

Ihochi, H., 240
Innes, M. J. S., 126, 127

Jackson, E. D., 236, 237, 239
Jacoby, J., 239
Jaeger, R. R., 64, 150
Jaggar, T. A., 91
Jakeš, P., 235
James, O. B., 233, 236, 237
Jamieson, J. C., 150
Jarosewich, E., 61, 63, 235
Jastrow, R., 277
Jefferson, T., 1
Johnson, G. G., Jr., 238
Johnson, R., 234
Johnson, R. D., 240
Johnson, T. V., 288
Johnston, D. A., 235
Jones, I. C., 273
Jones, K. L., 267, 275
Judge, D. L., 312
Jung, K., 127

Katterfield, G. N., 312
Kaula, W. M., 226, 227, 242, 311, 313, 317, 319
Kaushal, S. K., 66
Kahle, H. -G., 127
Keihm, S., 241, 319
Keil, K., 26, 63, 64, 233, 239
Kemmerer, W. W., 240
Kennedy, G. C., 64, 65
Kent, J., 285
Kepler, J., 153, 289
Kerridge, J. F., 65
Kessler, D. J., 65
Keys, R. R., 239
Kiang, T., 278, 287
Kieffer, H., 312
King, E. A., 34, 35, 61, 63, 79, 80, 92, 204, 205, 210, 214-216, 230, 232-240, 243, 253, 262, 273
Kinslow, R., 92, 93
Kirby, D., 150
Kirby, T. B., 273
Kirkwood, D., 278, 279
Klassen, K. P., 312
Klein, C., Jr., 233

Knolle, K., 232
Kock, M., 243
Koestler, A., 229
Kohman, T. P., 240
König, H., 63
Kopal, Z., 155, 230, 232, 243
Kovach, R. L., 231, 241
Kozai, Y., 287
Kozyrev, N. A., 169, 232
Krähenbühl, U., 235, 240
Kranz, W., 127
Kreidler, T. J., 274
Kreiter, T. J., 231
Krinov, E. L., 126, 129
Kuiper, G. P., 92, 93, 125-127, 129, 155, 268, 278, 281, 282, 285, 288, 295, 310, 312
Kunde, V. G., 274
Kurat, G., 239
Kvasha, L. G., 26
Kvenvolden, K., 22

Ladle, G., 237, 238
Lagrange, J. L., 279, 280
Lammlein, D. R., 220, 223-225, 240-242
Lamothe, R., 242
Langseth, M. G., Jr., 241, 318, 319
LaPaz, L., 60, 66
Larson, E. E., 240
Larson, H. K., 79
Latham, G., 222, 240, 241
Latimer, W. M., 243
Laul, J. C., 233, 239
Lawless, J. G., 62
Lebofsky, L. A., 231, 312
Lederberg, J., 273
Leighton, R. B., 273
Leovy, C. B., 268, 273, 274
Leuschner, A. O., 285
Levin, E. M., 150
Levinson, A. A., 174, 232-243
Levinthal, E., 273
Lewis, C. F., 63
Liberty, B. A., 126
Lick Observatory, 172, 290, 294
Ligon, D. T., Jr., 218, 235
Lindblad, B. A., 281, 287
Lingenfelter, R. E., 242, 265, 274, 319
Lippolt, H. J., 80, 127
Lipschutz, M. E., 64, 65, 150
Littler, J., 79, 126
Lowman, P., 242

Longhi, J., 236, 237
Lonsdale, J. T., 63
Lovering, J. F., 52-54, 65, 233
Lowell, P., 246, 258
Lucchitta, B. K., 164

McCall, G. J. H., 116, 128
McCallum, I. S., 236
McCauley, J. F., 162, 165, 245, 273
McConnell, R. K., Jr., 236
McCord, T. B., 285, 288, 301, 304, 312
McCrosky, R. E., 3, 66
MacDonald, G. A., 91
MacDonald, G. J. F., 228, 243
MacDonald, T. L., 231, 246, 273
McGill, G. E., 262, 274
McHone, J., 1
McIntyre, D. B., 139, 151
McKay, D. S., 237, 238
McKay, G. A., 236, 239
McKee, E. D., 125
McKinney, S. R., 102
McLane, J. C., Jr., 240
McMurdie, H. F., 150
McNaughton, D. A., 128
MacPherson, H. C., 285
McQueen, R. G., 89, 92
Madigan, C. T., 128
Madsen, B. M., 126, 150
Malin, M. C., 273, 312
Manson, A. J., 240
Manwaring, E. A., 129
Maran, S. P., 285, 293
Marcus, A. H., 231
Markov, A. V., 243
Marsden, B. G., 292, 295
Marsh, S. P., 92
Martin, R. T., 238
Marvin, U. B., 17, 62, 236, 238
Maskelyne, N. S., 50
Mason, A. C., 230
Mason, B., 4, 6, 11, 12, 16, 22-25, 30, 36, 37,
    43, 44, 51, 52, 61, 62, 65, 67, 92, 186,
    188, 235
Masursky, H., 273, 274
Matheson, R. S., 128
Maxwell, J. A., 62
Maynes, A. D., 40
Mazor, E., 59, 66
Meen, V. B., 105, 126
Megrue, G. H., 63

Meissner, R., 241
Melson, W. G., 188, 235
Menzel, D. H., 230-232
Merek, E. L., 240
Merrihue, C., 65
Merrill, G. P., 98, 102, 126, 131, 150
Metzger, A. E., 242
Meyer, C., Jr., 236
Michaelis, A., 154
Michel, F. C., 129
Michel, H., 36, 63
Middlehurst, B. M., 92, 93, 125-127, 129, 295
Mikhailov, Z. K., 230
Millman, P. M., 66, 105, 126
Milton, D. J., 63, 117-120, 126, 129, 150, 254,
    255, 273, 274
Minkin, J. A., 233, 236, 237
Mitchell, J. K., 207, 238, 239
Mohorovičić, S., 155, 229
Monnig, O. E., 126
Moore, C. B., 43, 63, 65
Moore, H. J., 92, 237
Moore, J. T., Jr., 273
Moore, P. A., 169, 232
Moore, P. B., 233, 238
Moreland, G., 17
Morgan, J. W., 233, 239, 240
Morris, F. A., 312
Morrison, D., 288
Morrison, D. A., 237, 238
Moss, F. J., 129
Müller, O., 80
Muller, P. M., 224, 242
Müller, W. F., 151, 233, 237
Munford, C. M., 287
Munk, N. M., 66
Murray, B. C., 273, 311, 312
Murray, J. B., 312
Mutch, T. A., 243, 267, 275

Nagata, T., 240
Nakamura, Y., 240-242
Namiq, L. I., 239
Nava, D., 63
Nehru, C. E., 239
Nelen, J., 61, 239
Nelson, L. S., 63, 210, 239
Nesterovitch, E. I., 312
Newton, J. C., 238
Ng, S., 234, 236
Nichiporuk, W., 65

Nicks, O. W., 273
Nielson, B., 64
Nininger, H. H., 92, 102, 103, 116, 126, 128, 132, 150
Noland, M., 273, 274
Noonan, A., 239
Northcutt, K. J., 187, 235
Nur, A., 241

Oberbeck, V. R., 92, 93, 115, 128, 231, 264, 274
Offield, T. W., 167, 231
Ogilvie, R. E., 64
O'Hara, M. J., 236
O'Keefe, J. A., 60, 66, 72, 76, 79, 80, 228, 230-232, 242
O'Kelley, G. D., 187, 188, 190, 235
Olbers, W., 278
O'Leary, B., 242, 311, 312
Olsen, E., 63, 64, 238
Oort, J., 293, 295
Öpik, E. J., 155, 167, 228, 231, 243, 261, 274, 281
Oro, J., 23, 62, 293, 295
Orowan, E., 243
Ostic, R. G., 240
Oyama, V. I., 240

Padovani, E., 237
Paget, A. B., 285
Pallas, P. S., 51
Palluconi, F. D., 312, 313
Papanastassiou, D. A., 62, 235, 240
Papike, J. J., 182, 183, 233, 239
Parkin, C. W., 241
Patterson, C., 40, 63
Patterson, J. H., 232
Pavičevič, M., 234
Payne-Gaposchkin, C., 285
Pearce, G. W., 240
Pearl, J. C., 274
Pearlman, J. P. T., 275
Pearson, M. W., 238
Peckett, A., 233
Perry, S. H., 43, 64
Peterson, L. E., 242
Pettengill, G. H., 312, 313
Philby, H. St. J., 120, 122, 129
Phillips, R., 233
Piazzi, G., 278
Picard, L., 114, 128

Pickering, W. H., 169
Pieters, C., 288
Pilcher, C. B., 309, 312
Pimentel, G. C., 265, 274
Pinson, W. H., 80
Pirraglia, J., 274
Pittendrigh, C. S., 275
Plant, A. G., 233
Podosek, F. A., 240
Pohn, H. A., 167, 231
Pollack, J. B., 269, 273, 274
Pond, R. B., 93
Popova, S. V., 134, 150
Porter, J. G., 295
Posen, A., 66
Powell, B. N., 55, 65, 236, 238
Press, F., 240, 241
Prinz, M., 233, 238
Prior, G. T., 11, 12, 26, 28, 35, 36, 52, 61, 65
Ptolemy, 153

Quaide, W. L., 92, 93
Quam, L., 274
Quinlan, T., 128

Rabe, E., 280, 287
Raleigh, C. B., 236, 237
Rai, B. U. C., 1
Ramdohr, P., 233, 234, 236
Rankine, W. J. M., 88, 92
Ranneft, T. S. M., 128
Rayner, J. M., 119, 129
Reedy, R. C., 242
Reeves, F., 128
Reid, A. F., 233
Reid, A. M., 185, 186, 235, 236
Reiff, W., 128
Renard, M. L., 126
Reynolds, J. H., 66
Reynolds, R. T., 311
Rhodes, J. M., 236
Richardson, K. A., 61, 240
Richter, J., 243
Ridgway, S. T., 312
Ridley, W. I., 234, 235
Rinehart, J. S., 92, 126
Ringwood, A. E., 30, 62, 228, 229, 233, 236, 242
Roach, O., 102
Robbins, C. R., 150
Robinson, J. C., 273

Robertson, P. B., 127
Roedder, E., 193-195, 236
Roemer, E., 295
Ronca, L. B., 231
Roosa, S. A., 230
Rose, G., 12, 61
Rose, H. J., Jr., 186, 218, 235
Roth, G. D., 285, 288
Rottenberg, J. A., 126
Roy, S. K., 30, 62
Rumsey, H. C., 307, 312
Runcorn, S. K., 218, 231, 240
Russ, G. P., II, 240

Sagan, C., 242, 273, 274, 297
Salisbury, J. W., 92, 166, 206, 230
Salmon, D., 1
Sanz, H. G., 63, 240
Saunders, R. S., 265, 267, 275
Schaber, G. G., 231
Schaeffer, O. A., 127, 235
Schall, R., 92
Schardin, H., 92
Schiaparelli, G., 246
Schleicher, D. L., 231
Schmadebeck, R., 242
Schmidt, R. A., 92
Schneider, E., 238
Schneider, H., 151
Schnetzler, C. C., 76, 77, 80
Schonfeld, E., 61, 235
Schopf, J. W., 240
Schubart, J., 287
Schubert, G., 241, 242, 265, 274, 319
Schultz, L., 63
Schwaller, H., 235
Schwartz, G., 66
Schwartz, K., 241
Schwarzmüller, J., 235
Schweizer, F., 278, 287
Schwerer, F. C., 240
Scott, R. F., 238
Secchi, P. A., 246
Sedmik, E. C. E., 129
Seeman, R., 127
Sekanina, Z., 293, 295
Sekiguchi, N., 115, 128
Sellards, E. H., 102, 103, 126
Sengör, A. M. C., 230, 273
Shao, C. Y., 66
Sharp, R. P., 257, 259, 260, 273, 274

Shipley, E. N., 273
Shoemaker, E. M., 91-93, 98, 125-127, 150, 156, 167, 230, 231
Short, J. M., 64
Short, N. M., 79, 92, 93, 125, 127-129, 139, 141, 148-151, 238
Silver, L. T., 37, 40, 41, 55, 63, 65
Silverman, M. P., 240
Simons, P. Y., 150
Simpson, E. S., 116, 128
Simpson, P. R., 234
Sjogren, W. L., 224, 242, 319
Skaggs, S. R., 63, 239
Smalley, V. G., 166, 206, 230
Smith, B. A., 241, 273, 274
Smith, H., 239
Smith, H. T. U., 253, 273
Smith, J. V., 180, 233, 234, 238, 239
Smith, S., 273
Smithsonian Institution, 53
Snyder, F. G., 95, 125
Soderblom, L. A., 231, 259, 260, 273, 274
Sonett, C. P., 221, 241
Spencer, L. J., 69, 76, 79, 95, 120, 121, 125, 126, 129, 156, 230
Steinbacher, R. H., 245
Steinbrunn, F., 128
Stenzel, H. B., 79, 150
Stephenson, A., 240
Stettler, A., 235
Stishov, S. M., 134, 150
Stöffler, D., 88, 90, 92, 127, 133, 139, 144, 148, 150, 151, 237
Stokes, G. G., 88, 92
Stolper, E. M., 237
Storzer, D., 114, 128, 238
Strangway, D. W., 240
Strom, R. G., 311, 312
Stuart-Alexander, 164, 239
Suess, F. E., 69, 79
Suomi, V., 311, 312
Summers, A. L., 311
Sutton, G., 240, 241
Sutton, R. L., 231
Swann, G. A., 231
Sztrokay, K. L., 17, 62

Taddeucci, A., 92, 237
Takeda, H., 234, 236
Talwani, P., 241
Tani, B., 233

Taylor, H. P., 55, 65
Taylor, J. W., 92
Taylor, L. A., 234, 236
Taylor, S. R., 72, 80, 115, 120, 129, 189, 243
Taylor, R. M., 238
Tera, F., 191, 235, 240
Thompson, J. B., 133
Thompson, M. H., 277
Thornton, F. H., 169
Tilles, D., 65, 80
Titius, J. D., 277
Titulaer, C., 231
Toksöz, N., 240, 241
Tolnay, V., 62
Traill, R. J., 233
Trask, N., 311, 312
Trigila, R., 92, 237
Trombka, J. I., 242
Tschermak, G., 12, 34, 61, 67
Tucker, R., 274
Turkevich, A. L., 170, 232

Uchiyama, A., 64
Uhlig, H. H., 64
Ulbrich, M., 233
Urey, H. C., 11, 12, 23, 25, 34, 40, 60, 61, 66, 80, 153, 225, 229, 242, 243

VanHouten, C. J., 287
VanHouten-Groeneveld, C., 285
Van Schmus, W. R., 12, 13, 22, 26, 34, 35, 61
Van Son, J., 129
Veverka, J., 272-274
Vishniac, W., 275
Voshage, H., 66

Wackerlie, J., 150
Wahl, W., 36, 63
Walker, D., 196, 197, 236, 237
Walter, L. W., 76, 79, 234, 236
Wänke, H., 63
Ware, N. G., 233, 236
Wark, D. A., 233
Warner, J. L., 210, 211, 235, 239
Warnica, R. L., 92
Wasserburg, G. J., 62, 63, 235, 240
Wasson, J. T., 46, 48, 64, 67
Watkins, C., 205, 235, 236, 239, 241, 319
Watkins, J. S., 231, 241
Wegener, A., 156, 230

Weiblen, P. W., 193-194, 236
Weigand, P. W., 234
Weill, D. F., 236, 239
Welker, J., 274
Wells, A. T., 128
Wenk, H. -R., 233
Werner, E., 114, 127
West, M. N., 164
Wetherill, G. W., 66, 287, 293, 295
Whipple, F. E., 232
Whipple, F. L., 34, 62, 289, 295
White, E. W., 238
Widmanstätten, A., 46
Wiesman, H., 235, 236
Wiik, H. B., 14-16, 62
Wildey, R. L., 273
Wilhelms, D. E., 156, 161, 164, 165, 230, 273, 274
Wilkins, W. H., 169
Willard, G. E., 116
Williams, R., 236
Willmore, P. L., 126
Wilshire, H. G., 210, 239
Wilson, G., 236
Wise, D. U., 228, 242, 262, 274
Wolf, M., 280
Wollenhaupt, W. R., 242, 319
Wood, E. M., 67
Wood, J. A., 12, 13, 22, 26, 34, 35, 61, 62, 64, 67, 203, 236, 238
Wooley, B. C., 240
Woronow, A., 81, 253, 262, 273
Wosinski, J. F., 79
Wright, F. W., 65, 129
Wurm, K., 295
Wylie, P. J., 238

Yaniv, A., 235
Yavnel, A. A., 44
Yin, L., 242
Young, A. T., 273
Young, G. A., 129
Young, J., 231

Zähringer, J., 235
Zawartko, C., 118
Zechman, G. R., 233
Zellner, B., 288
Zook, H. A., 57, 65
Zygielbaum, J. L., 232

Subject Index

Names *in italics* are individual meteorites.

Abee, 6, 30, 33, 286
achondrites, 35-41
  classification, 35, 36
Achilles, 280
Adhi Kot, 33
Admire, 53, 54
agglutinates, 201-203
Agrell effect, 49
Alais, 14
Albin, 53
Al Hadidah, 122
Al Hadidah Craters, 120
Alinda, 286
Allende, 4-6, 8, 9, 14, 16-20
alpha backscatter analyses, 169, 170
Alphonsus Crater, 157-159, 169, 172
American Museum of Natural History, 41
amphoterites, 26
Anderson, 54
angrites, 36, 37
Aouelloul Crater, 96, 135, 136
Apennine Mountains, 168
Apollo asteroids, 281, 292, 293
Arend-Rigaux, 292
Arend-Roland, 294
Aristarchus Crater, 167, 169, 172
Arizona State University, 2, 42, 99
armalcolite, 179, 180
Ashanti Crater, 76

asterism, 138, 139
asteroid belt, 278, 279
asteroids, 277-288, 292
    Apollo asteroids, 281, 292, 293
    belt, 278, 280
    discovery of, 277, 278
    Hirayama families, 281
    jet streams, 281
    Mars crossing, 280, 281
    mass-frequency distribution, 284
    optical studies, 282
    orbits, 278-281
    reflectance spectra, 285, 286
    Trojan asteroids, 279, 280
ataxites, 43, 46
Athamantis, 286
Atlanta, 33
aubrites, 36
Australasian tektites, 70-76, 79
Australites, 70-72, 74, 75

baddeleyite, 76, 134-136
Barratta, 33
Barringer Crater, 98, 123, 124
Barringer Crater Company, 101
basaltic achondrites, 37-41, 58, 285, 286
Bath, 33
bediasites, 71, 72
Bendock, 53

Berlin Observatory, 277
Bethune, 33
Bishopville, 33
Bode's Law, 277, 278
bolide, 2
Bosumtwi Crater, 76, 78, 96
Boxhole Crater, 96, 116
breccia, lunar, 207-219
   autometamorphosed, 210-216
   geochemistry, 214, 217, 218
   magnetic properties, 217, 218
   soil, 208
Bremervörde, 286
Brenham, 53, 103
Brent Crater, 96, 105, 106, 112, 123, 124
British Museum, 2
buttons, 70, 74, 75

Cachari, 33
Callisto, 309, 310
Campo del Cielo Craters, 96, 103
Canyon Diablo, 40, 43, 49
Canyon Diablo Crater, 98
Cape York, 41
capture hypothesis, 228, 229
Carbo, 33
carbonaceous chondrites, 14-21, 187-189,
     217, 269, 272, 286, 292, 316
   chemical composition, 15, 16, 22, 187,
     188, 189
   inclusions, 17-19, 21
   mineralogy, 16-19
   organic chemistry, 19, 21-23
Carswell Structure, 96, 105, 106, 112, 123
Carthage, 42
Catherina Crater, 167
Cavendish Crater, 167
central peaks, 87
Ceres, 278
Charlevoix Structure, 96, 105, 112, 123, 124
chassignites, 36, 37
Chaves, 33
chondrites, 7-35, 58, 59, 187, 293
   amphoterites, 26
   carbonaceous, 14-21, 187-189, 217, 269,
     272, 286, 292, 316
   enstatite, 28, 30-33, 286
   olivine-bronzite, 21-29, 286
   olivine-hypersthene, 21-29, 285, 286
   ordinary, 11, 21-29, 286
chondrules, 7-10, 16, 17, 19-21, 26,

    27, 29-31, 33-36, 43
   lunar, 34, 35, 210, 214-216
   origin, 7, 30-35
Chubb Crater, 105
Clearwater Lake East, 96, 105, 106, 109, 112,
   124, 139
Clearwater Lake West, 96, 105, 106, 109, 112,
   124, 139
Clover Springs, 33
Coconino Sandstone, 98, 132, 134
coesite, 77, 132-135
Cold Bokkeveld, 15, 16
Colomera, 6
Colorado Museum of Natural History, 103
Comet Kohoutek, 291
Comets, 289-295
   composition, 293
   orbits, 291-292
   origin, 293
   size, 293
   structure, 293
comparative planetology, 315-319
Cone Crater, 202
Coon Butte Crater, 98
Coon Mountain Crater, 98
Coorara, 5, 6
Copernicus Crater, 167, 168, 172
Coya Norte, 43
Crab Orchard, 33
crater counts, 167, 168, 262-264
crater density, 167, 168, 262-264, 299, 302
crater formation, 82-91
   compression stage, 82-84
   energy partitioning, 89-91
   excavation stage, 85
   modification stage, 85, 87, 88
crater mechanics, 82-91
cratering, 82-91
Crater Mound, 98
craters, 81-130
Crooked Creek Structure, 96
cryptoexplosions, 95, 112
cryptovolcanic structures, 95, 114, 116
Cumberland Falls, 39

Dalgaranga Crater, 96, 116
Daniel's Kuil, 33
Davy Rille, 163
Dayton, 6
Decaturville Structure, 97
Deep Bay Crater, 97, 105-108, 112, 123, 124

Deimos, 268, 269, 271, 272, 277
Dellen Structure, 97, 123
Dembowska, 286
diaplectic glass, 139
diogenites, 36
double planet hypothesis, 229
doublet craters, 109, 264

enstatite chondrites, 28, 30-33
Eos, 286
equilibrated chondrites, 28, 29
Eratosthenes Crater, 168
Esterville, 33
eucrites, 36, 37, 187
   chemical composition, 40, 54, 55
Europa, 309, 310
europium anomaly, 187

Faucett, 27
Field Museum, 2
fireball, 2
fission hypothesis, 227, 228
Fladen, 114, 145
Flynn Creek Structure, 97, 146
Fra Mauro Formation, 34, 210
Frankfort, 33
fused rock glass, 143, 145, 146

Gagarin Crater, 226
Galilean satellites, 309, 310
Ganymede, 309, 310
Gassendi Crater, 167
glass spheres, lunar, 198-202
Goclenius Crater, 162
Glorieta Mountain, 53, 54
Gosses Bluff Structure, 97, 116-118, 123,
   124, 146
Gow Lake Structure, 105, 112
Grand Canyon, 256
Grant, 49
Great Red Spot, 309, 310

Hainholz, 33
Halley's Comet, 289-291
Haripura, 15, 16
Harvard University, 2
Haviland Crater, 97
heat flow, lunar, 221
Hector, 283
Hellas, 263, 268
Henbury, 42, 50, 116, 119

Henbury Craters, 69, 97, 119, 120, 123
herring bone structure, 168
hexahedrites, 43, 46
highlands, lunar, 172
   rocks, 173, 176-178, 207-219
Hilda, 281
Hirayama families, 281
Hoba, 41
Holleford Structure, 97, 105-107, 112, 123
Homestead, 33
howardites, 36, 37
   chemical composition, 40, 54, 55
Hugoniot equations, 88, 89
Hvittis, 28, 33
hydrocarbons, 21-23

Iduna, 286
Imbrian Basin, 34, 210, 214, 263
Imilac, 53
impact breccia, 143, 145, 146
impact craters, 81-130
   morphology, 87
impact criteria, 123, 124, 131-149
impact ejecta, 84, 85, 87, 89, 91, 100, 101,
   109, 110, 114, 115, 117, 143-145
impact flash, 84, 91
impact glass, 69, 101, 103, 109, 120, 121, 134-
   137, 139, 143-146, 148
impactite, 109, 143, 144, *see* impact glass
impact melt, 40, 87
impact metamorphism, 131-149
   mineralogic and petrologic criteria, 132-146
impact structures, 81-130
impact-triggered volcanism, 87
Indarch, 30, 32, 33
Indochinites, 72, 73
inverted stratigraphy, 99-101
Io, 309, 310
iron meteorites, 41-50, 58, 59
   chemical composition, 43, 44, 46-48
   classification, 43, 44, 46, 48
   inclusions, 43
   mineralogy, 43-46
   shock effects, 49, 50
iron-nickel phase diagram, 47
Ivory Coast tektites, 72, 76, 78
Ivuna, 4, 15, 16

Jajh deh Kot Lalu, 33
Jelica, 286
jetting, 84

Johnstown, 33
Juno, 278
Jupiter, 278, 279, 304, 308-311
  Great Red Spot, 309, 310
  moons, 308-311
Juvinas, 33, 40

Kaaba stone, 1
Kaalijarvi Crater, 97
Kaba, 17
Kaibab Limestone, 100
Kapoeta, 286
Karoonda, 5, 6, 286
Kendall County, 33
Kentland Structure, 97
Kepler Crater, 164, 172
Khairpur, 33
kink bands, 141, 142
Kirkwood gaps, 278, 279
Köfels Structure, 97
Kohoutek, 291
Kota-Kota, 33

Lac Couture Structure, 97, 105, 106, 111, 112
Lac La Moinerie Structure, 105, 112
Lagrangian points, 279, 280
Lake St. Martin Structure, 97, 105, 112, 123
Lance, 15, 16, 286
Lappajarvi Structure, 97
Laser altimetry experiment, 226, 227
lechatelierite, 76, 77, 136, 137
Leonid meteors, 292
Leoville, 19, 21
Libyan Desert Glass, 137
line broadening, 138, 139
Lodran, 51
lodranite, 51
Lost City, 3, 60
Lowell Observatory, 169, 246
lunar craters, 153-169
  origin of, 153-156, 166-168
lunar geophysics, 219-227
lunar ilmenite, 183, 184
lunar magmas, 196, 197
lunar minerals, 176, 179-184
lunar olivine, 184
lunar orbital gamma-ray exp., 227
lunar orbital X-ray fluorescence experiment,
  224
lunar plagioclase, 180, 181
lunar pyroxenes, 181-183

lunar regolith, 165, 166, 198-219
  grain size, 203-206
  particle types, 198-203
  physical properties, 206, 207
lunar samples, 169-219
  ages, 191, 192, 217
  breccias, 207-219
  igneous rocks, 171-197
  magnetic properties, 217, 218
  minerals, 176, 179-184
  organic content, 219
  remanent magnetization, 217, 218
  shock metamorphism, 196, 198-203
lunar passive seismic experiment, 221-225
lunar soil, 198-207
lunar surface ages, 167, 168
lunar surface magnetometer, 221

Manicouagan impact structure, 87, 97, 105,
  106, 112, 123, 124
Manson Structure, 97
mare basins, 158-160, 164, 172
Mare Imbrium, 155, 168, 172
Mare Nubium, 157
Mare Smythii, 226
Mare Tranquillitatis, 170, 172
maria, 156, 172, 185
  rocks, 173-176, 207, 208, 218, 219
Mars, 245-275
  aeolian features, 251-253, 255-257
  atmosphere, 252, 265, 268
  canals, 246
  channels, 254-258
  cliffs, 257, 258
  composition, 265
  craters, 246, 247, 261-264
  dust storms, 247, 250, 253, 264, 268
  geologic/physiographic map, 266, 267
  internal structure, 265
  moons, 268-272, 277
  permafrost, 255, 257
  physical data, 250
  plate tectonics, 256
  polar areas, 245, 248, 258-261, 268
  scarps, 257, 258
  surface features, 248-264
  telescopic observations, 246
  volcanic features, 248-254
  water, 256, 258, 259, 265, 268
mascons, 224, 225
Maurolycus Crater, 167

Mecca, 1
melt inclusions, lunar, 193-195
Mercury, 297-305
    reflectance spectra, 301, 304, 305
    regolith, 301
    surface morphology, 298-303
mesosiderites, 54, 55, 116, 286
meteor, 2, 3, 292
Meteor Crater, 82, 87, 89, 91, 95, 97-103,
    112, 119, 122-124, 131, 132, 134, 143,
    155
meteor showers, 61, 292
meteorites, 1-67, 186-189
    ablation surface, 7
    abundances, 2, 4
    achondrites, 35-41
    ages, 40, 41, 58-60
    amphoterites, 26
    carbonaceous chondrites, 14-21
    chemical compositions, 11, 12, 15, 16, 22,
        25, 40, 43, 44, 48, 53
    chondrites, 7-35
    classification, 2, 11-14, 16, 21-26
    collections, 2
    cooling rates, 46, 47
    enstatite chondrites, 28, 30-33
    fall, 2
    find, 2
    fusion crust, 4, 7
    general, 1-6
    irons, 41-50, 58, 59
    mineralogy, 4-6
    naming, 2
    olivine-bronzite chondrites, 11, 21-29
    olivine-hypersthene chondrites, 11, 21-29
    orbits, 60, 61
    ordinary chondrites, 11, 21-29
    organic chemistry, 19, 21-23
    origin, 1, 2, 58-61
    recognition, 2, 4
    stony, 7-41
    stony-irons, 51-54
meteoroid, 2
Mexico School of Mines, 41
Mezö Madaras, 5, 29, 36
microcraters, 57, 199, 202
micrometeorites, 55-58
Mien Lake Structure, 97
Middlesboro Structure, 97
Mighei, 15, 16, 286
Mincy, 33

minor planets, 277-288
Mississagi Quartzite, 109, 147
Mistastin Structure, 97, 105, 112
Moenkopi Formation, 100, 101
Mokoia, 15, 16
moldavites, 70, 72, 76, 114
Monturaqui Crater, 97
Moon, 153-243, 268
    astronomical observations, 153-169
    craters, 153-169
    crustal structure, 219-227
    general, 153-155
    geochemistry, 184-192
    geologic mapping, 156-162, 164, 165
    interior state, 219-227
    minerals, 176, 179-184
    origin, 227-229
    petrology, 173-187, 192-197
    physical parameters, 155
    regolith, 165, 166
    rock ages, 191, 192
    rock compositions, 184-191
    rock samples, 169-219
    stratigraphy, 156-162
    surface ages, 167, 168
    surface morphology, 153-168
    transient phenomena, 169
    see also lunar
moonquakes, 222-225
Moore County, 33
Morristown, 33
mosaicism, 139
Mount Padbury, 56
multiringed basins, 158-160, 164, 172, 263,
    299, 300
Murchison, 6, 22, 23, 286
Murray, 15, 16, 22

nakhlites, 36-38
Nawapali, 15, 16
Neptune, 311
Neujmin 1, 292
Neumann bands, 44, 49
Newport, 53
New Quebec Crater, 97, 105, 106, 112, 123
Nicholson Lake Structure, 97, 105, 106,
    112, 123
Nogoya, 16

octahedrites, 43-46, 52, 53, 56, 116, 122
Odessa Craters, 97, 102, 103, 116

Odessa, 4, 5, 44, 102, 103
olivine-bronzite chondrites, 21-29
olivine-hypersthene chondrites, 21-29, 285
Ollague, 4, 51
Olympus Mons, 248, 250, 251
Onaping Formation, 109, 110, 145
Oort Cloud, 293
Öpik-Oort Cloud, 293
ordinary chondrites, 11, 21-29
    chemical composition, 11, 21, 23, 25, 26
    mineralogy, 21-24, 26, 27
Ornans, 15, 16
Orgueil, 15, 16
Oterma, 292
outer planets, 304, 309-313, 315, 317
overturned flap, 99-101

Pallas, 278
pallasites, 51-54
Paris Museum of Natural History, 2
Pasamonte, 33, 37, 38, 40
Peña Blanca Springs, 37
penetration funnel, 103
penetration pit, 103, 104
Pesyanoe, 33
Petersburg, 40
Phobos, 268-272, 277
Pillistfer, 33
Pilot Lake Structure, 97, 105, 106, 112
planar features, 109, 118, 139, 141
planetology, 315-319
    bulk compositions, 316, 317
    degree of differentiation, 317
    internal state, 318
    tectonic styles, 318, 319
plate tectonics, 256, 319
plessite, 46
Pluto, 311
Pojoaque, 54
Prairie Network, 3, 67
Pribram, 60
Prinz Crater, 161
Prior's Rules, 11, 12, 22, 25
pseudotachylite, 146
Psyche, 286
Pueblito de Allende, *see* Allende
pyroxferroite, 174, 176, 179

Rankine-Hugoniot equations, 88
rays, 83, 120, 155, 263, 264, 298-300
regolith, lunar, 165, 166, 198-219

breccias, 207-219
    chemical composition, 214, 217-219
    grain size, 203-206
    physical properties, 206, 207
Revelstoke, 14
Ries Crater, 76, 97, 112-114, 118, 123, 124,
    132, 133, 135, 136, 139, 142, 144-146
rilles, 157-159, 161-163
Rochechouart Structure, 97
Rosebud, 7

Saint-Sauveur, 33
Salta, 52
Santa Cruz, 15, 16
Sappho, 286
Saturn, 311
Schwassmann-Wachmann, 292
secondary craters, 83, 161, 163, 168
Sedan Crater, 83
Serpent Mound Structure, 97
Sharps, 5
shatter cones, 101, 109, 114, 117, 118, 146,
    147
Shergotty, 40, 140
shock history, 147-149
shock metamorphism, 131-149, 196, 198-203
    mineralogic and petrologic criteria, 132-146
shock waves, 82-91
siderophyre, 51
Sierra Madera Structure, 97, 124, 146
Sikhote-Alin, 103, 104
Sikhote Alin Craters, 97, 104
silica phase diagram, 135
Siljan Structure, 97
sinuous rilles, 161
Sioux County, 33
Smithsonian Astrophysical Obs., 2
Smithsonian Institution, 2, 3, 39, 53
soil breccia, 208
Soko-Banja, 34
Soviet Academy of Sciences, 2
Springwater, 53
St. Mark's, 31-33
Stannern, 33
Steen River Structure, 97, 105, 112
Steinback, 33, 51
Steinheim Basin, 97, 112, 114, 123, 146
stishovite, 134, 135
stony meteorites, 7-41
stony-iron meteorites, 50-55
    classification, 50, 51

cooling rates, 54, 55
Stopfenheim Kuppel, 112, 114
Strangways Structure, 97
Sudbury Structure, 97, 105, 106, 108-112,
        123, 124, 145-147
suevite, 114, 118, 145

tektites, 69-80, 135, 143
    ages, 72
    chemical composition, 72, 78
    mineralogy, 76, 77
    origin, 69, 71, 76, 79
Tempel-Tuttle 1866 I, 292
Tenham, 5, 6
Tenoumer Crater, 97
terrestrial planets, 315-319
    physical data, 316
Texas Memorial Museum, 102
thetomorphic glass, 139, 140
Titan, 311
Tlacotepec, 43
Toluca, 5, 33
Tonk, 16
Toro, 285
Tourinnes la Grosse, 29
tranquillityite, 179, 181
Triton, 311
Tucson, 4
Turtle River, 45
Tycho Crater, 160, 167,
        172

undulatory extinction, 142
unequilibrated chondrites, 28, 29
Ungava Crater, 105
University of Texas, 7, 102
Uranus, 277, 311
urelites, 36

Vaca Muerta, 33
Van de Graff Crater, 218
Venus, 302, 303, 306-308
    atmosphere, 302, 306
    surface morphology, 307, 308
Veramin, 286
Vesta, 278, 285, 286
Vredefort Structure, 97, 124, 146

Wabar, 120
Wabar Craters, 69, 97, 120-122, 132, 143
Wanapitei Structure, 97, 105, 112, 123
Warrenton, 16
Weldona, 10
Wells Creek Structure, 97
West Hawk Lake Structure, 97, 105, 106, 110,
        112, 123, 124
Widmanstätten structure, 43-46, 52, 53
Wolf Creek Crater, 97, 115, 116

Xiquipilco, 45

Yale University, 2
Yurtuk, 40